Risk Assessment and Indoor Air Quality

Risk Assessment and Indoor Air Quality

Editor: Harlee Knight

MURPHY & MOORE
www.murphy-moorepublishing.com

www.murphy-moorepublishing.com

MURPHY & MOORE

Cataloging-in-Publication Data

Risk assessment and indoor air quality / edited by Harlee Knight.
     p. cm.
Includes bibliographical references and index.
ISBN 978-1-63987-759-1
1. Indoor air quality. 2. Indoor air quality--Risk assessment. 3. Air quality.
4. Air--Pollution. 5. Air quality management. I. Knight, Harlee.
TD883.17 .R57 2023
628.53--dc23

Murphy & Moore Publishing
1 Rockefeller Plaza,
New York City,
NY 10020, USA

ISBN 978-1-63987-759-1

Contents

Preface

Indoor air quality (IAQ) refers to the quality of air inside a school, house, office, or any other building environment. Poor IAQ has been associated with decreased productivity, sick building syndrome, and poor learning in schools. Nausea, sinus congestion, shortness of breath, headache and other symptoms are associated with poor IAQ. Indoor air pollution can cause a variety of health problems, including heart diseases, cancer, respiratory diseases and cognitive deficits. Risk assessment with respect to the IAQ involves the application of the tools and methodologies of risk assessment to the study of indoor air quality. The goal of an IAQ risk assessment is to detect problems with air quality in facility areas and develop long-term solutions to improve the safety and health of building occupants. It involves collecting samples from air and from building surfaces, studying the airflow, and determining the human exposure to pollutants. The main techniques for enhancing IAQ in the buildings include using air filters, eliminating or controlling the source of pollution, and using ventilation to dilute pollutants. This book contains some path-breaking studies on indoor air quality and its risk assessment. It will help the readers in keeping pace with the rapid changes in this field of study.

This book has been the outcome of endless efforts put in by authors and researchers on various issues and topics within the field. The book is a comprehensive collection of significant researches that are addressed in a variety of chapters. It will surely enhance the knowledge of the field among readers across the globe.

It gives us an immense pleasure to thank our researchers and authors for their efforts to submit their piece of writing before the deadlines. Finally in the end, I would like to thank my family and colleagues who have been a great source of inspiration and support.

Editor

Evaluation of Performance of Inexpensive Laser based PM$_{2.5}$ Sensor Monitors for Typical Indoor and Outdoor Hotspots

Sungroul Kim *, Sujung Park and Jeongeun Lee

Department of Environmental Health Sciences, Soonchunhyang University, Asan 31538, Korea; psj57732398@gmail.com (S.P.); l01025263029@gmail.com (J.L.)
* Correspondence: Sungroul.kim@gmail.com.

Featured Application: In consideration of relatively stable outcomes with the application of a correction factor for relative humidity, recently introduced inexpensive real-time monitors (IRMs), ESCORTAIR (ESCORT, Seoul, Korea) or PurpleAir (PA) (PurpleAir U.S.A.), our study supports their usage in PM$_{2.5}$ monitoring for various urban hotspots.

Abstract: Inexpensive (<$300) real-time particulate matter monitors (IRMs), using laser as a light source, have been introduced for use with a Wi-Fi function enabling networking with a smartphone. However, the information of measurement error of these inexpensive but convenient IRMs are still limited. Using ESCORTAIR (ESCORT, Seoul, Korea) and PurpleAir (PA) (PurpleAir U.S.A.), we evaluated the performance of these two devices compared with the U.S. Environmental Protection Agency (EPA) Federal Equivalent Monitoring (FEM) devices, that is, GRIMM180 (GRIMM Aerosol, Germany) for the indoor measurement of pork panfrying or secondhand tobacco smoking (SHS) and Beta-ray attenuation monitor (BAM) (MetOne, Grants Pass, OR) for outdoor measurement at the national particulate matter (PM$_{2.5}$) monitoring site near an urban traffic hotspot in Daejeon, South Korea, respectively. The PM$_{2.5}$ concentrations measured by ESCORTAIR and PA were strongly correlated to FEM (r = 0.97 and 0.97 from indoor pan frying; 0.92 and 0.86 from indoor SHS; 0.85 and 0.88 from outdoor urban traffic hotspot). The two IRMs showed that PM$_{2.5}$ mass concentrations were increased with increased outdoor relative humidity (RH) levels. However, after applying correction factors for RH, the Median (Interquartile range) of difference compared to FEM was (14.5 (6.1~23.5) %) for PA and 16.3 (8.5–28.0) % for ESCORTAIR, supporting their usage in the home or near urban hotspots.

Keywords: PM$_{2.5}$; sensor; correction; pan frying; secondhand smoke; urban traffic

---

1. Introduction

A large volume of previous epidemiological studies relied on the use of ground-based fixed national monitoring stations [1,2]. However, recently, inexpensive (<$300) particulate matter (PM) monitors (PM) have been introduced for home usage in South Korea. These devices can provide PM distribution patterns at high temporal and spatial resolution [3–5] which is a substantial improvement on establishing a pollution monitoring networking system as well as environmental epidemiologic study [6], as compared to traditional approaches that relied on relatively small number of ground-based fixed national air monitoring stations or mobile sampling techniques.

Most of these low-cost devices are classified into two groups, that is, optical particle counters (OPCs) or photometers. OPCs use the light scattered from individual particle to estimate the concentration of particles in different size ranges [7,8]. These data, along with assumptions of the particle shape and density, can be converted to estimate mass concentrations that compare favorably with reference instruments [7,8]. However, it has been reported that there may be bias when aerosol type or size is unknown [7,8].

Photometers use a light source to illuminate sensing zones that contain many particles at one time [9,10] and obtained that the mass concentration of aerosol scales linearly with the amount of light scattered by an assembly of particles captured at a discrete angle from the incident light [11]. The light scattered by the assembly of particles is measured by a photodetector at an angle specific to the photometer model, often $90°$ from the incident light [12]. The intensity of scattered light is directly proportional to gravimetrically measured mass concentration, although the relationship is dependent on the light scattering characteristics, density and size distribution of the particle [12].

The cost of research grade light scattering instruments (approximately, $10,000 or higher) limits their use to conduct studies at high temporal and/or spatial resolution. In Korea, a laser-based inexpensive OPC based IRM, for example, ESCORTAIR, was recently introduced. However, the reliability of ESCORTAIR have been unknown. Therefore, it may be necessary to test this new device before it is applied in a high spatial-temporal resolution exposure assessment study. As a proper calibration protocol providing a correction factor can have a dramatic impact on precision, accuracy and bias of a real-time monitor, researchers evaluated the OPC or photometers for use in the laboratory, outdoors or in the home in other countries [3,9,13–16]. A recent article reported that "one size fits all" approach to obtain $PM_{2.5}$ mass concentrations by OPC result in relatively high uncertainty in complex exposure situations. Although OPC Therefore, corresponding conversion curve approach may be most valuable when a relatively high contrast is expected in exposure levels for example, daytime home with indoor combustion sources, BBQ or secondhand smoke versus night time or day time outside with heavy traffic volume versus night time [16]. To our knowledge, no one has rigorously evaluated the performance of IRMs, operated with OPC or photometer, in Korea with comparison of the U.S. Federal Equivalent Method (US FEM) [17].

In this study, we evaluated the performance of inexpensive (less than $300) real-time PM monitors (IRMs), with high cost (about $2000–$10,000) and cross-comparisons between them and research grade PM monitors (RGMs). We used US FEMs as reference instruments (approximately $20,000 or higher) and provided a final error of mass concentration ($PM_{2.5}$) measurement after applying correction factors in this study.

2. Materials and Methods

2.1. PM$_{2.5}$ Real-Time Monitors

A laser-based light-scattering $PM_{2.5}$ sensor monitor (ESCORTAIR, ESCORT, Seoul, Korea) (weight <300 g, volume <510 cm^3) consisted of an optical particle counting (OPC) PM sensor (INNOSIPLE1), CO_2 sensor, temperature relative humidity sensor, data transfer networking module and light-emitting diode (LED) display screen (Figure 1). In the ESCORTAIR, the sensing volume is illuminated with a laser and airborne particles are counted and processed one at a time. There were various IRMs commercially available in South Korea. In this study, however, we chose ESCORTAIR as they allowed us to directly transfer data to our data server using its Wi-Fi function.

For comparison purposes, another inexpensive photometer type of PM monitor (PA, PurpleAir, Draper, UT, USA) (https://www.purpleair.com/), mounting two Plantower sensors in a monitor, was used. PA is recommended its usage by AQ-SPEC (Air quality sensor performance evaluation center, South Coast Air Quality Management District, CA, USA) or US EPA (Environmental Protection Agency, NC, USA) as an IRM. The inlet system of these IRMs did not have an impactor or a cyclone unlike that of the RGM.

The performance of the two IRMs (one OPC, that is, ESCORTAIR and one photometer, that is, PA) costed less than $300 were simultaneously compared with those of high-cost devices ($10,000 or so), that is, research-grade laser photometers including PDR-1500 (Thermo Scientific, Waltham, MA, USA) and SIDEPAK AM510 (TSI, Inc., Shoreview, MN, USA).

These research-grade monitors have a cyclone inlet or an impact inlet for the measurement of the respirable fraction of airborne particulate matters in different environments and can provide real-time data. The SIDEPAK is a portable battery-operated personal aerosol monitor with an impact inlet and light-scattering laser photometer that provides real-time aerosol mass concentration. The PDR-1500 (Personal DataRAM 1500) is a nephelometric monitor with a cyclone inlet for the measurement of the respirable fraction of the airborne particulate matters. The PDR-1500 can simultaneously collect particles on a 37 mm filter for the gravimetric analysis by passing through the sensing zone.

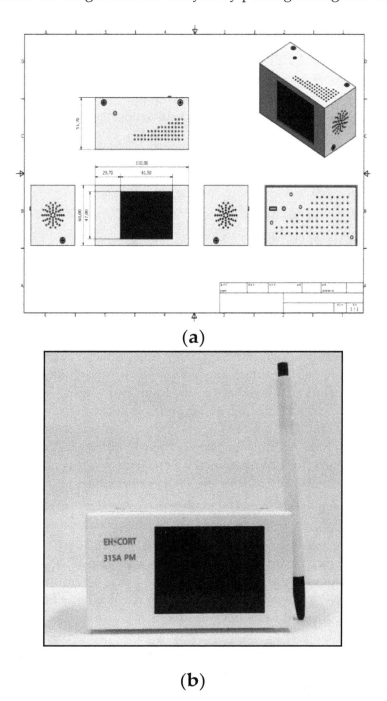

Figure 1. Layout of ESCORTAIR installed INNOSOPLE-1 PM sensor for measuring particulate matter (PM) concentrations.

2.2. Federal Equivalent Method

As mentioned above, to conduct a comparative measurement of IRMs, we used the U.S. Environmental Protection Agency (EPA) Federal Equivalent Method (FEM), that is, GRIMM 180 (GRIMM Aerosol, Technik Ainring GmbH & Co. KG, Ainring, Germany) for indoor testing and BAM-1022 (MetOne, Grants Pass, OR, USA) for outdoor field tests [17].

The Grimm Technologies, Inc. Model EDM 180 PM2.5 Monitor is a light scattering OPC monitor operated for 24 h at a volumetric flow rate of 1.2 L/min, configured with a Nafion®- type air sample dryer. BAM-1022, a beta-ray attenuation mass monitor has a $PM_{2.5}$ particle size separator. Using BAM, we obtained 24 1-h average measurements at the national $PM_{2.5}$ monitoring supersite operated by National Institute of Environmental Research, Daejeon, Korea.

2.3. Flow Rate Inspection

Before each indoor and outdoor experiment, RGMs, that is, the PDR-1500 and SIDEPAK were zeroed with an in-line high efficiency particulate air (HEPA) filter and the flow of each device (1.52 L/min, 1.7 L/min and 0.5 L/min) was checked by a mass flowmeter (TSI, Inc., Shoreview, MN, USA). The temperature and relative humidity data were also downloaded from the devices if the device is equipped with sensors for these data. Flow rate of GRIMM, one of FEMs, was also checked with a similar way. The values from BAM, the other FEM, were used as reference, since it is located at the KOREA national $PM_{2.5}$ monitoring site (Daejeon) and operated with high QAQC programs [18,19] to report hourly outdoor $PM_{2.5}$ data.

2.4. Experimental Setting

We collected $PM_{2.5}$ concentrations by performing both an indoor exposure test on March 2018 and outdoor $PM_{2.5}$ monitoring at the national supersite located in Daejeon, Korea from June to July 2018 including rainy days. We used the 2 sets of each IRM or RGM for indoor and outdoor testing (serial numbers of devices: Table 1).

2.4.1. Indoor Test

The indoor test included scenarios of frying pork in a pan and exposure to secondhand smoke (SHS). Indoor pan-frying tests were conducted at inside of an empty laboratory (4 m × 10 m × 3.5 m, W D H), according to the protocol described in detail in our previous article [20]. In brief, this experiment was carried out over a 2 h measurement period per trial including first 9 min simulating the barbequing of pork belly (100 g). Standard operating protocol: a portion of pork belly (100 g) was pan-fried for 9 min: 3 min on Side A, 3 min on Side B; then 1.5 min on Side A again and finally 1.5 min on Side B again. When we fried pork belly, after the first 9 min, we opened a window (0.5 m × 0.8 m) to allow the ventilating air to naturally reduce the $PM_{2.5}$ concentration.

Measurement of the $PM_{2.5}$ levels from the exposure to secondhand smoke was conducted with a lighted cigarette burned. We opened the same window after 30 min during our secondhand smoke exposure level test. Then, we collected concentration data over next two hours.

During our indoor test, we maintained a minimum distance of 20 cm among these devices, at least 50 cm from the emission sources and 1 m above the floor. To minimize the effect of additional source contribution to our $PM_{2.5}$ measurement results, we reported our $PM_{2.5}$ results after subtracting the field background $PM_{2.5}$ concentrations measured at the baseline. We collected with the frying pan test and the secondhand exposure test three times with GRIMM 180 on separate days. The indoor test data from each device (80-s interval for PA, 60 s for the remaining devices) were calculated to the 5 min average level to be compared with the outcomes from GRIMM. The final number of data for the 2 h panfrying test was approximately 50 (2 sets of each device × 12 data point/h × 2.0 h) and that for the 2.5 h secondhand smoke exposure test was about 60 (2 sets of each device × 12 data point/h × 2.5 h) for $PM_{2.5}$, as well as temperature and relative humidity.

Table 1. Summary of $PM_{2.5}$ measurement range of concentration, measurement interval, weight and Wi-Fi availability reported by manufacturers and unit cost in South Korea by 30 June 2018.

	Device Classification[a]	Sensor Type[b]	Measurement Range	Sampling Pump Flow Rate	Precision[c]	Log Interval[c]	Unit Price ($)	Weight (g)	Wi-Fi
GRIMM (EDM180)[1] (GRIMM Aerosol, Germany) S/N #: 11R15047	FEM	OPC	0~3000,000 particles/Liter	1.2 L/min,	97% over the whole measuring range	5 s to 1 h	19,000	20,000	No
BAM-1020[2]. (MetOne, OR) S/N #: N11181	FEM	Beta ray Attenuation	0~1000 mg/m³	16.7 L/min	Exceeds US-EPA Class III $PM_{2.5}$ FEM standards	1 min to 1 h	23,750	24,500	No
ESCORTAIR[3] (ESCORT, Seoul, South Korea) S/N #: 6a:c6:3a:c7:83:bf 6a:c6:3a:c7:88:b1	IRM	OPC	1000 µg/m³	NA	±10%@100~500 µg/m³	30 s	300	400	Yes
PA[4] (PurpleAir, CA, USA) S/N #: A0:20:A6:A:AD:1B. A0:20:A6:B:83:32	IRM	Photometer	0~500 µg/m³ as effective range	NA	±10%@100~500 µg/m³	80 s	300	450	Yes
PDR-1500[5] (Thermo Scientific, MA, USA) S/N #: CM17422007, CM17422017	RGM	Photometer	0.001~400 mg/m³	Adjustable 0 to 3.5 L/min	±2% of reading or ±0.005 mg/m³	1 s to 1 h	9000	1200	No
SIDEPAK[6] (TSI, MN, USA) S/N #: 11104037, 11008055	RGM	Photometer	0.001~100 mg/m³	Adjustable 0 to 1.8 L/min	±0.001 mg/m³ over 24 h as zero stability	1 s to 60 s	6000	460	No

[a] Federal equivalent method (FEM), Research Grade Monitor (RGM) and inexpensive real-time monitor (IRM); [b] Optical particle count (OPC); [c] Information from manufacture. 1. GRIMM: https://www.grimm-aerosol.com/fileadmin/files/grimm-aerosol/General_Downloads/The_Catalog_2018_web.pdf. 2. BAM: https://metone.com/wp-content/uploads/2017/08/bam-1020-9803_touch_screen_manual_rev_k.pdf. 3. ESCORTAIR: This study. 4. PA: https://www.purpleair.com/sensors. 5. PDR: https://www.newstarenvironmental.com/air-toxic-monitors/personal-dataram-pdr1500-aerosol-monitor.html?_vsrefdom=adwords&gclid=CjwKCAiA2fjjBRAjEiwAuewS_cYiCFBPpx0d0bTaRKtmxe-1Kf22JVs352pQKq9e63XyqT_pIbAZsRoCn9UQAvD_BwE. 6. SIDEPAK https://www.tsi.com/getmedia/84b5be22-c339-49bc-ab97-e3c4baee16c1/SidePak%20AM520_US_5001737_Web_1.

2.4.2. Outdoor Test

Using the same configuration of monitoring devices, we also measured the ambient $PM_{2.5}$ concentration at the outdoor Roof-top of one of the national $PM_{2.5}$ Supersites located in Daejeon, South Korea, operating BAM (MetOne, Grants Pass, OR, USA). Main body of BAM was installed at inside of an experiment laboratory of the Supersite while the inlet of BAM was located at the Roof-top. The outdoor temperature and RH during were measured by the sensors in ESCORTAIR, PA and PDR-1500 and the values were crosschecked. Measurements from IRMs were collected every minute, except for PA, which provided a response every 80 s. To compare the hourly concentration values provided by the Supersite, we calculated the hourly mean values using measured values acquired at hourly intervals from each IRM device. Final sample size for the outdoor data was 240 (2 sets of each device × 24 data points/day × 5 days) for $PM_{2.5}$, as well as temperature and relative humidity.

2.5. Statistical Analyses

The Spearman correlation tests were used to evaluate the associations among the outcomes of devices measured at indoor or outdoor environments, considering that variables were not normally distributed.

Using outdoor measurement data, we evaluate the association of device response with various relative humidity level. We also evaluated the associations of the daily mean concentrations from IRMs with those obtained with FEM methods using multivariate linear regression models. We used the values obtained with the FEM as dependent variables and those obtained with real-time devices as independent variables to obtain a correction factor. The hourly mean temperature and relative humidity data for each sampling date, which was previously compared with the nearest national meteorological monitoring sites, were used to adjust the effects of relative humidity on the association between IRMs and FEM outcomes. The corresponding slopes and coefficients of determinant (R2) for each RT monitor were also provided. The final measurement error (%) was calculated based on the U.S. EPA performance evaluation program for PM instruments;

$$Difference_i = \frac{Measurement_i - FEM_i}{FEM_i} \qquad (1)$$

EPA specifies that the percent bias goal for acceptable measurement uncertainty should be within ±20% [17]. Here, we reported median (IQR) value of the differences per device because the distribution of differences was not normal. We also provided mean of the differences just to compare with value in the guideline [21]. All analyses were conducted with SAS (Version 9.4) and R software (Version 2.15.3, R Development Core Team).

3. Results

3.1. $PM_{2.5}$ Concentration

In our indoor test, the median (IQR) $PM_{2.5.}$ concentration over the pan-frying test was 86.8 µg/m³ (17.8–254.4 µg/m³) by the real-time ESCORTAIR, 104.9 µg/m³ (43.9–228.2 µg/m³) and 236.2 µg/m³ (49.3–648.7 µg/m³) by the real-time PA or PDR-1500 devices and 153.2 µg/m³ (46.2–409.7 µg/m³) by using GRIMM (Table 2). The median (IQR) concentrations for secondhand smoke (SHS) test were 20.9 (17.4– 156.6) µg/m³, 31.2 (14.4–194.3) µg/m³ and 28.4 (12.8–314.0) µg/m³ for ESCORTAIR, PA and PDR, respectively, whereas GRIMM provided 23.5 (15.9–107.1) µg/m³.

Simultaneous outdoor $PM_{2.5}$ monitoring results are also provided in Table 2. The median (IQR) of the hourly average values of ESCORTAIR and PA were 13.7 (7.3~21.2) and 19.7 (9.3~35.8) µg/m³, which were an overestimation of the values obtained by BAM, one of U.S. EPA FEMs. During indoor testing, the median temperature and RH were approximately 20~22 °C and 37%. During outdoor testing, the median values (IQR) were 30.7 (25.6~40.7) °C and 56.4 (34.8~71.4) %, respectively (Table 2).

Table 2. $PM_{2.5}$ mass concentration ($\mu g/m^3$) (median, IQR) measured by real-time sensor devices and the federal equivalent method as well as the temperature and relative humidity throughout the sampling period.

	Indoor—Pan-Frying (n = 50)	Indoor—SHS (n = 60)	Outdoor—Urban Traffic Hotspot (n = 240)
GRIMM	153.2 (46.2–409.7)	23.5 (15.9–107.1)	NA
BAM	NA	NA	9.0 (4.0–22.0)
ESCORTAIR	86.8 (17.8–254.4)	20.9 (17.4–156.6)	13.7 (7.3–21.2)
PA	104.9 (43.9–228.2)	31.2 (14.4–194.3)	19.7 (9.3–35.8)
PDR-1500	236.2 (49.3–648.7)	28.4 (12.8–314.0)	13.8 (6.8–34.8)
SIDEPAK	261.3 (71.5–800.0	50.0 (21.0–652.0)	29.1 (15.6–59.9)
Temp. (°C)	21.7 (21.1–21.7)	20.1 (19.7–20.5)	30.7 (25.6–40.7)
RH (%)	37.0 (35.0–39.0)	37.0 (35.0–38.0)	56.4 (34.8–71.4)

3.2. Correlations among Devices and the Fem

High-level correlations were obtained between the IRMs (ESCORT, PA) and FEM. The Spearman correlation coefficients were 0.97 (P = 0.0001) or 0.92 (P = 0.0001) between ESCORTAIR and GRIMM for indoor pan-frying or the SHS test, respectively. The correlation coefficients between PA and GRIMM were similar (0.97 (P = 0.0001) or 0.86 (P = 0.0001)) for the indoor test (Table 3). Our outdoor test also showed high correlation between ESCORTAIR and BAM (0.84 (P = 0.0001)). The correlation coefficient for PA was 0.88 (P = 0.0001).

Similarly, the measurements by RGMs (PDR-1500, SIDEPAK) and FEM were highly correlated: 0.97–0.98 for the Pan-frying test, 0.88~0.96 for the SHS test and 0.84–0.91 for the outdoor test. Between IRMs, that is, ESCORTAIR and PA, the association of measurements were strong to each other (r = 0.93: Indoor pan frying; 0.85: Indoor-SHS; 0.93: Outdoor-urban traffic). In addition, they showed similar correlation patterns to PDR-1500 or SIDEPAK (Table 3).

Table 3. Scatter plots and Spearman correlation coefficients between real-time $PM_{2.5}$ monitoring devices and FEM.

	Indoor-Pan-Frying					Indoor-SHS					Outdoor Urban Traffic Hotspot				
	FEM	E	PA	P	S	FEM	E	PA	P	S	FEM	E	PA	P	S
FEM	1					1					1				
E	0.97	1				0.92	1				0.85	1			
PA	0.97	0.93	1			0.86	0.85	1			0.88	0.93	1		
PDR	0.98	0.95	0.99	1		0.96	0.93	0.94	1		0.84	0.93	0.99	1	
S	0.98	0.99	0.96	0.98	1	0.88	0.86	0.88	0.93	1	0.91	0.91	0.99	0.99	1

FEM: Federal Equivalent Method: In this study Indoor: GRIMM Optical particle courting, OPC), Outdoor: BAM (Beta ray attenuation monitor). E: ESCORTAIR, PA: PurpleAir, PDR: PDR-1500, S: SIDEPAK.

3.3. Effects of Ambient Humidity for Outdoor Measurement

In this study, as seen in Supplementary Figure S1, we observed that the $PM_{2.5}$ concentration of IRMs were significantly increased with the increase in relative humidity level. The slope obtained from a simple regression line of FEM on IMRs, that is, (BAM = Slope * ESCORTAIR + intercept) at a relative humidity above 80%, was smaller than the slope at < 20.0%, 20.1~40.0%, 40.1~60.0%, 60.1~80.0%. The degree of decreasing trend was larger with IRMs, compared to two RGMs (Figure 2).

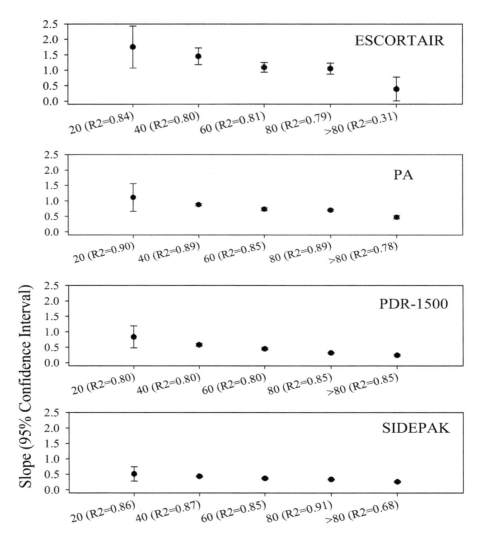

Relative Humidity [%] (Coef. of Determination)

Figure 2. Change of the slope with 95% confidence interval obtained from the regression line of BAM (Beta ray attenuation monitor) on IRMs (Inexpensive real-time particulate matter monitors) or RGMs (research grade monitors) by the degree of relative humidity.

3.4. Correction Factor

Since we found the effect of RH on the measurements of ESCORTAIR or PA, we developed our own correction models for those IRM devices by performing stepwise linear regression. The correction factors were obtained from the regression models (1.11 for ESCORTAIR and 1.92 for PA, $p < 0.01$), (1.00 for ESCORTAIR and 0.87 for PA, $p < 0.01$) and (1.15 for ESCORTAIR and 0.70 for PA, $p < 0.01$) for the measurement of $PM_{2.5}$ resulting from indoor pan-frying, SHS or the urban traffic hotspot of South Korea, respectively.

After adjusting for temperature and relative humidity, the results were unchanged for the indoor tests (1.10 for ESCORTAIR and 1.90 for PA, $p < 0.01$), (0.97 for ESCORTAIR and 0.81 for PA, $p < 0.01$) (Table 3). For the outdoor measurement, the change of slopes (0.72 for ESCORTAIR and 0.77 for PA, $p < 0.01$) was relatively small but the R^2 values were changed (Table 4).

Table 4. Slopes obtained from stepwise linear calibration models with adjusted R^2 (Dependent variable: US EPA FEM, Independent variable: IRM).

| | Indoor—Pan-Frying | | | | Indoor—SHS | | | | Outdoor—Urban Traffic Hotspot | | | |
| | Single | | Multivariate * | | Single | | Multivariate * | | Single | | Multivariate * | |
	β	R^2	β	R^2	β	R^2	β	R^2	β	R^2	β	R^2
ESCORTAIR	1.11	0.98	1.10	0.98	1.00	0.92	0.97	0.92	1.15	0.70	1.14	0.81
PA	1.92	0.94	1.90	0.94	0.87	0.89	0.81	0.90	0.70	0.83	0.71	0.87
PDR-1500	0.33	0.98	0.33	0.98	0.54	0.91	0.49	0.92	0.33	0.72	0.36	0.80
SIDEPAK 1	0.34	0.98	0.32	0.99	0.28	0.90	0.31	0.92	0.35	0.84	0.36	0.89

* results obtained after adjusting for temperature and relative humidity.

3.5. Bias after Application of Correction Factors

We then determined the extent to which the original PM$_{2.5}$ data measured by ESCORTAIR were improved after applying the correction factor obtained from the model by performing comparative analyses using the outcomes of multivariate regression models (Table 4). Using the corrected data, the final coefficient of determination (R2) between FEM (y) and ESCORTAIR (x) was 0.81. The coefficient for PA was 0.87. We found the difference (median (IQR)) with the calibrated data, compared to FEM, to be 16.3. (8.5~28.0)% for ESCORTAIR and 14.5 (6.1 to 23.5)% for PA for outdoor environments (Figure 3). The bias (mean of the difference) was 13.1% for ESCORTAIR and 7.8% for PA for outdoor. The bias for indoor data was at least similar or lower than the bias level obtained from the outdoor test.

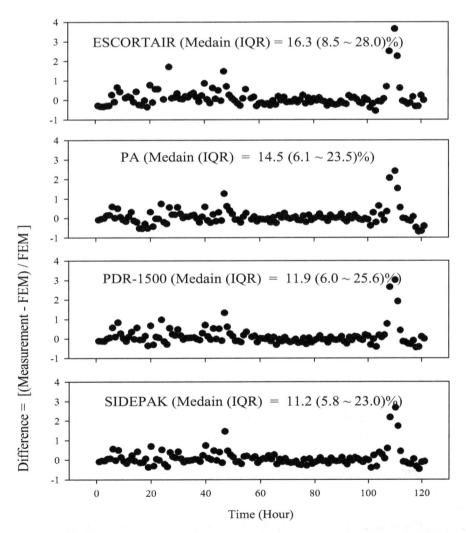

Figure 3. Distribution of difference of outdoor measurements against the FEM; BAM for outdoors.

4. Discussion

In this study, we compared $PM_{2.5}$ concentrations measured with IRMs and RGMs to those measured with U.S. EPA FEMs. A relationship between the real-time $PM_{2.5}$ concentration and FEM were acceptable (R = 0.97 to 0.98: Indoor pan frying; 0.86 to 0.96: Indoor-SHS; 0.84 to 0.91: Outdoor-urban traffic, respectively).

This study conducted indoor testing with common $PM_{2.5}$ sources of Korea, that is, frying pork in a pan or smoking. The indoor test was conducted in an indoor laboratory in which the room temperature was maintained at 20~22 °C and the relative humidity at 37% because we assumed that most low-cost PM sensing devices would be established inside susceptible populations' homes where the indoor temperature and humidity level would be relatively stable, compared to the outdoor environment.

In addition, in this study, we extended our comparison test by performing them outside with a federal equivalent method. We discovered that, on days with a high level of outdoor RH (80% or higher), our IRMs overestimated the $PM_{2.5}$ level. Thus, finally, we got the correction model providing a correction factor for ESCORTAIR to adjust for the effect of temperature and relative humidity as such a process has been conducted in a regulatory monitoring site (South Coast Air Quality Monitoring District, SCAQMD) for the field use of these kinds of inexpensive sensing devices including PA [22,23].

Several previous studies provided a correction factor for SIDEPAK monitors: 0.77 in Northern California, U.S.A. (ambient air), 0.43 or 0.52 in Italy (ambient air in urban or rural areas) and 0.42 in Italy (indoor-outdoor mixed environment) [24–26], which were comparable to our results (0.36). Our correction factor (0.36) for PDR-1500 was smaller than results reported by Wang et al. (2016) [27] (0.71 compared to PDR-1500 using its own filtering method) but very similar to the values obtained by Ramachandran et al. (2000) (0.33) and Wallace et al. (2011) (0.38), who conducted their studies on atmospheric environments [28,29].

A lack of quantitative information on the speciation of particles, traffic volume, type of vehicle or the particular sampling time or season limits further exploration of the basis for the differences in the correction factors between these studies and ours. Nevertheless, our correction model for ESCORTAIR, with consideration of the RH level, was derived in a similar way to that with which we obtained the factor for SIDEPAK or PDR-1500. Thus, we consider no significant systematic errors to have been involved in the calculation process. A good linear relationship has been obtained between the $PM_{2.5}$ mass concentration of FEM and the responses of low-cost PM sensors as reported previously in other country [8,9,25,30]. We demonstrated the urban hotspot specific correction factor for a light-scattering sensor in Korean urban environment of interest to enable our findings and methodology to be extended and replicated by researchers who are interested in the utility of low-cost sensing device, such as ESCORTAIR, in South Korea.

OPCs are reportedly good at estimating mass if they have numerous bins, such as GRIMM [31]. However, estimating the mass concentration from a limited number of bins may be subject to a measurement error during the conversion process with the factory-provided internal conversion algorithm. Therefore, we used an additional correction factor for ESCORTAIR after comparison to GRIMM for proper usage under Korean circumstances.

It is well established that the response of monitors based on light scattering varies with aerosol size distribution, composition and optical properties and need a proper calibration process [13,32,33]. No single calibration model (or correction model) can enable accurate performance for all particle sources in microenvironments. This challenge applies to both research and consumer monitors. Although gravimetric measurements may be used to determine a source- or environment-specific calibration for a research study, the approach is not practical for routine monitoring in homes. A key objective of continuous monitoring—to activate controls—can be achieved if the monitor reliably and clearly responds to sources that account for the majority of particles in the home even if responses are not quantitative.

We conducted this study by assuming that indoor $PM_{2.5}$ emission sources exist, that is, from frying pork in a pan and smoking, with consideration of Korean life style [20] and relatively high smoking prevalence [34]. We determined that our linear model obtained from the indoor test for each single-aerosol type showed excellent performance (R2 = 0.98 or 0.92 for ESCORTAIR; 0.94 or 0.90 for PA), compared to FEM responses. However, the responses of ESCORTAIR as well as PA were relatively less precise but good (R2 = 0.81, 0.87) for monitoring in urban traffic hotspots suggesting that IRMs need a site-specific calibration with a reference method before they are used [13,32,33].

The shape of the response curve can be related to the type of OEM sensors integrated with the monitor and relative humidity. Because we did not have information about the internal conversion factors (count to mass concentration for OPC type or light intensity to mass concentration for photometer), using GRIMM or BAM, known as US FEM, we tried to obtain our own correction factors for usage of ESCORTAIR or PA in urban indoor or outdoor settings.

The limitations of our study should be noted. First, the sample size in this study was relatively small. However, in our tests, we determined a $PM_{2.5}$ range of 10 to 3000 $\mu g/m^3$ including both indoor and outdoor measurement. This range ensures that the concentration distributions would not be systematically biased. Our $PM_{2.5}$ concentrations might not be representative of each sampling season or area as a result of spatial-temporal variations. Additional experiments are needed to understand the stability of our correction factor in different seasons and/or in other locations, that is, in industrial or rural areas of Korea. Extending the sampling periods for each season and location would ensure that our results are more representative. Furthermore, in future studies, measurements of the wind direction and speed are expected to provide improved correction factors between the IRM and FEM methods. As we mentioned in the method section, we checked flow rate for our RGMs or FEMs prior to our experiment. For IRMs, we have considered measuring flow rate but due to its open wide inlet and very low flow rate, we could not connect it to our mass flowmeter properly. Instead, especially, before our outdoor test, we operated 5 ESCORTAIRs and 5 PAs simultaneously and checked measurement errors between devices. Then, we selected 2 of them which provide best outcomes, compared to FEM. A preparation of QC/QA test program for massive products of IRMs are recommended. And for ESCORTAIR, like PA, application of weather proof design is suggested. In addition, future studies may be necessarily conducted to obtain site-specific correction factors including at coal power plants or in rural areas.

In this study, the performances of the IRM with RGM and that of the FEM operated with a high QCQA program were compared with one-hour monitoring intervals at national $PM_{2.5}$ monitoring site. This makes this study unique compared to previous studies, which were mostly conducted with one-day interval gravimetric methods at ordinary sampling sites.

Despite the growing public interest in reducing personal exposure levels to $PM_{2.5}$ in Korea, IRM monitoring still faces challenges in terms of providing real-time concentration information. Although the number of national $PM_{2.5}$ monitoring sites in Korea is increasing, additional IRMs in hotspots or communities are required because they can detect continuous spatial—temporal variations and identify nearby exposure sources on a real-time basis in a micro-environment of hotspots.

This study found that the measurement of $PM_{2.5}$ concentrations with recently developed laser-based IRM under- or over-estimates $PM_{2.5}$ concentrations obtained from FEM while its bias could be approximately 11 to 16% even at urban outdoor hotspots with traffic sources with high relative humidity levels. Therefore, the application of a correction factor is strongly suggested for inexpensive laser-based monitoring devices.

5. Conclusions

Our study determined that on days with a high level of outdoor RH (80% or higher), our IRMs overestimated the outdoor $PM_{2.5}$ level and showed the necessity of a correction factor for IRMs to adjust for the effect of temperature and relative humidity. $PM_{2.5}$ concentrations measured with IRMs need to be subjected to quality control and quality assurance evaluation before these monitors are used for the quantification of $PM_{2.5}$ levels in urban indoor or outdoor atmospheric environments in Korea. In consideration of relatively stable outcomes with the application of correction factors for recently developed new IRMs—ESCORTAIR or PA—our study supports their usage in networking monitoring for various urban hotspots.

Author Contributions: S.K. designed this study, wrote manuscript and conducted interpretation of the quantitative aspects of data analysis. S.P. performed modeling simulation and J.L. provided editorial efforts. S.K. supervised the whole study.

Acknowledgments: The authors deeply appreciate the technical comments for sensor evaluation from. Andrea Clements and Timothy Buckley at the national exposure research laboratory, U.S. EPA. The authors also appreciate the assistance of Sungmin Jung and the staff (Minhee Lee, Taekyung Hwang, Jieun Jeong) at the Korea PM2.5 supersite of Daejeon, South Korea for their support with data collection and the help of Juhee Kim in data screening. The content is solely the responsibility of the authors and does not necessarily represent the official views of the US EPA or Korea NIER (National Institute for Environmental Research).

References

1. Orellano, P.; Quaranta, N.; Reynoso, J.; Balbi, B.; Vasquez, J. Effect of outdoor air pollution on asthma exacerbations in children and adults: Systematic review and multilevel meta analysis. *PLoS ONE* **2017**, *12*, e0174050. [CrossRef]

2. Prieto Parra, L.; Yohannessen, K.; Brea, C.; Vidal, D.; Ubilla, C.A.; Ruiz Rudolph, P. Air pollution, PM$_{2.5}$ composition, source factors, and respiratory symptoms in asthmatic and nonasthmatic children in Santiago, Chile. *Environ. Int.* **2017**, *101*, 190–200. [CrossRef] [PubMed]

3. Gillooly, S.E.; Zhou, Y.; Vallarino, J.; Chu, M.T.; Michanowicz, D.R.; Levy, J.I.; Adamkiewicz, G. Development of an in~home, real-time air pollutant sensor platform and implications for community use. *Environ. Pollut.* **2019**, *244*, 440–450. [CrossRef] [PubMed]

4. Genikomsakis, K.N.; Galatoulas, N.; Dallas, P.I.; Ibarra, L.M.C.; Margaritis, D.; Ioakimidis, C.S. Development and On Field Testing of Low Cost Portable System for Monitoring PM$_{2.5}$ Concentrations. *Sensors* **2018**, *18*, 1033. [CrossRef] [PubMed]

5. Shao, W.; Zhang, H.; Zhou, H. Fine Particle Sensor Based on Multi Angle Light Scattering and Data Fusion. *Sensors* **2017**, *17*, 1033. [CrossRef] [PubMed]

6. Aneja, V.P.; Pillai, P.R.; Isherwood, A.; Morgan, P.; Aneja, S.P. Particulate matter pollution in the coal producing regions of the Appalachian Mountains: Integrated ground based measurements and satellite analysis. *J. Air Waste Manag. Assoc.* **2017**, *67*, 421–430. [CrossRef] [PubMed]

7. Burkart, J.; Steiner, G.; Reischl, G.; Moshammer, H.; Neuberger, M.; Hitzenberger, R. Characterizing the performance of two optical particle counters (Grimm OPC1.108 and OPC1.109) under urban aerosol conditions. *J. Aerosol. Sci.* **2010**, *41*, 953–962. [CrossRef] [PubMed]

8. Sousan, S.; Koehler, K.; Hallett, L.; Peters, T.M. Evaluation of the Alphasense Optical Particle Counter (OPC N2) and the Grimm Portable Aerosol Spectrometer (PAS 1.108). *Aerosol. Sci. Technol.* **2016**, *50*, 1352–1365. [CrossRef]

9. Dacunto, P.J.; Klepeis, N.E.; Cheng, K.C.; Acevedo Bolton, V.; Jiang, R.T.; Repace, J.L.; Ott, W.R.; Hildemann, L.M. Determining PM$_{2.5}$ calibration curves for a low cost particle monitor: Common indoor residential aerosols. *Environ. Sci. Process. Impacts* **2015**, *17*, 1959–1966. [CrossRef] [PubMed]

10. Lanki, T.; Alm, S.; Ruuskanen, J.; Janssen, N.A.; Jantunen, M.; Pekkanen, J. Photometrically measured continuous personal PM(2.5) exposure: Levels and correlation to a gravimetric method. *J. Expo. Anal. Environ. Epidemiol.* **2002**, *12*, 172–178. [CrossRef] [PubMed]

11. O'Shaughnessy, P.T.; Slagley, J.M. Photometer response determination based on aerosol physical characteristics. *AIHA J.* **2002**, *63*, 578–585. [CrossRef]

12. Baron, P.A. Aerosol Photometers for Respirable Dust Measurements. In *NIOSH Manual of Analytical Methods*; 1998. Available online: https://www.cdc.gov/niosh/docs/2003--154/pdfs/chapter~{}g.pdf (accessed on 25 March 2019).

13. Kelly, K.E.; Whitaker, J.; Petty, A.; Widmer, C.; Dybwad, A.; Sleeth, D.; Martin, R.; Butterfield, A. Ambient and laboratory evaluation of a low cost particulate matter sensor. *Environ. Pollut.* **2017**, *221*, 491–500. [CrossRef] [PubMed]

14. Semple, S.; Ibrahim, A.E.; Apsley, A.; Steiner, M.; Turner, S. Using a new, low cost air quality sensor to quantify second hand smoke (SHS) levels in homes. *TOB Control.* **2015**, *24*, 153–158. [CrossRef] [PubMed]

15. Zhou, X.; Aurell, J.; Mitchell, W.; Tabor, D.; Gullett, B. A small, lightweight multipollutant sensor system for ground mobile and aerial emission sampling from open area sources. *Atmos. Environ.* **2017**, *154*, 31–41. [CrossRef] [PubMed]

16. Franken, R.; Maggos, T.; Stamatelopoulou, A.; Loh, M.; Kuijpers, E.; Bartzis, J.; Steinle, S.; Cherrie, J.W.; Pronk, A. Comparison of methods for converting Dylos particle number concentrations to $PM_{2.5}$ mass concentrations. *Indoor Air* **2019**, *29*, 450–459. [CrossRef]

17. Williams, R.; Vasu Kilaru, E.; Snyder, A.; Kaufman, T.; Dye, A.; Rutter, A.; Russell, A.; Hafner, H. *Air Sensor Guidebook*; EPA/600/R-14/159 (NTIS PB2015-100610); U.S. Environmental Protection Agency: Washington, DC, USA, 2014.

18. National Institute of Environmental Research (NIER). *2015 Annual Report of Intensive air Quality Monitoring Station, NIER GP2016–160*; NIER: Incheon, Korea, 2016.

19. Yu, G.; Park, S.; Ghim, Y.; Shin, H.; Lim, C.; Ban, S.; Yu, J.; Kang, H.; Seo, Y.; Kang, K.; et al. Difference in Chemical Composition of $PM_{2.5}$ and Investigation of its Causing Factors between 2013 and 2015 in Air Pollution Intensive Monitoring Stations. *J. Korean Soc. Atmos. Environ.* **2018**, *34*, 16–37. [CrossRef]

20. Lee, S.; Yu, S.; Kim, S. Evaluation of Potential Average Daily Doses (ADDs) of $PM_{2.5}$ for Homemakers Conducting Pan Frying Inside Ordinary Homes under Four Ventilation Conditions. *Int J. Environ. Res. Public Health* **2017**, *14*, 78. [CrossRef]

21. Rosner, B. *Hypothesis Testing, Fundamentals of Biostatistics*; Duxbury: Parific Grove, CA, USA, 2000; Chapter 7.

22. Kim, B.M.; Teffera, S.; Zeldin, M.D. Characterization of $PM_{2.5}$ and PM_{10} in the South Coast Air Basin of southern California: Part 1 Spatial variations. *J. Air Waste Manag. Assoc.* **2000**, *50*, 2034–2044. [CrossRef]

23. Kim, B.M.; Teffera, S.; Zeldin, M.D. Characterization of $PM_{2.5}$ and PM_{10} in the South Coast Air Basin of southern California: Part. 2 Temporal variations. *J. Air Waste Manag. Assoc.* **2000**, *50*, 2045–2059. [CrossRef] [PubMed]

24. Borgini, A.; Tittarelli, A.; Ricci, C.; Bertoldi, M.; De Saeger, E.; Crosignani, P. Personal exposure to $PM_{2.5}$ among high school students in Milan and background measurements: The EuroLifeNet study. *Atmos. Environ.* **2011**, *45*, 4147–4151. [CrossRef]

25. Jiang, R.T.; Acevedo Bolton, V.; Cheng, K.C.; Klepeis, N.E.; Ott, W.R.; Hildemann, L.M. Determination of response of real-time SidePak AM510 monitor to secondhand smoke, other common indoor aerosols, and outdoor aerosol. *J. Environ. Monit.* **2011**, *13*, 1695–1702. [CrossRef]

26. Karagulian, F.; Belis, C.A.; Lagler, F.; Barbiere, M.; Gerboles, M. Evaluation of a portable nephelometer against the Tapered Element Oscilating Microbalance method for monitoring PM(2.5). *J. Environ. Monit* **2012**, *14*, 2145–2153. [CrossRef] [PubMed]

27. Wang, Z.; Calderon, L.; Patton, A.P.; Sorensen Allacci, M.; Senick, J.; Wener, R.; Andrews, C.J.; Mainelis, G. Comparison of real-time instruments and gravimetric method when measuring particulate matter in a residential building. *J. Air Waste Manag. Assoc.* **2016**, *66*, 1109–1120. [CrossRef]

28. Ramachandran, G.; Adgate, J.L.; Hill, N.; Sexton, K.; Pratt, G.C.; Bock, D. Comparison of short term variations (15 min averages) in outdoor and indoor $PM_{2.5}$ concentrations. *J. Air Waste Manag. Assoc.* **2000**, *50*, 1157–1166. [CrossRef]

29. Wallace, L.A.; Wheeler, A.J.; Kearney, J.; Van Ryswyk, K.; You, H.; Kulka, R.H.; Rasmussen, P.E.; Brook, J.R.; Xu, X. Validation of continuous particle monitors for personal, indoor, and outdoor exposures. *J. Expo. Sci Environ. Epidemiol.* **2011**, *21*, 49–64. [CrossRef] [PubMed]

30. Shi, J.; Chen, F.; Cai, Y.; Fan, S.; Cai, J.; Chen, R.; Kan, H.; Lu, Y.; Zhao, Z. Validation of a light scattering $PM_{2.5}$ sensor monitor based on the long term gravimetric measurements in field tests. *PLoS ONE* **2017**, *12*, e0185700. [CrossRef]

31. Peters, T.M.; Ott, D.; O'Shaughnessy, P.T. Comparison of the Grimm 1.108 and 1.109 portable aerosol spectrometer to the TSI 3321 aerodynamic particle sizer for dry particles. *Ann. Occup. Hyg.* **2006**, *50*, 843–850. [PubMed]

32. Njalsson, T.; Novosselov, I. Design and Optimization of a Compact Low Cost Optical Particle Sizer. *J. Aerosol. Sci.* **2018**, *119*, 1–12. [CrossRef] [PubMed]

33. Northcross, A.L.; Edwards, R.J.; Johnson, M.A.; Wang, Z.M.; Zhu, K.; Allen, T.; Smith, K.R. A low cost particle counter as a realtime fine particle mass monitor. *Environ. Sci. Process. Impacts* **2013**, *15*, 433–439. [CrossRef]

34. Kim, S.; Jung, A. Optimum cutoff value of urinary cotinine distinguishing South Korean adult smokers from nonsmokers using data from the KNHANES (2008–2010). *Nicotine Tob Res.* **2013**, *15*, 1608–1616. [CrossRef]

The Relevance of Indoor Air Quality in Hospital Settings: From an Exclusively Biological Issue to a Global Approach in the Italian Context

Gaetano Settimo [1], **Marco Gola** [2,*] **and Stefano Capolongo** [2]

[1]   Environment and Health Department, Istituto Superiore di Sanità, 00161 Rome, Italy; gaetano.settimo@iss.it
[2]   Architecture, Built Environment and Construction Engineering Department, Politecnico di Milano, 20133 Milan, Italy; stefano.capolongo@polimi.it
*   Correspondence: marco.gola@polimi.it.

Abstract: In the context of the architectures for health, it is an utmost priority to operate a regular and continuous updating of quality, efficacy, and efficiency's processes. In fact, health promotion and prevention take place through a proper management and design of healing spaces, in particular with regard to the most sensitive users. In recent decades, there has been increasing attention to indoor air quality in healthcare facilities. Nowadays, this issue must involve the implementation of a series of appropriate interventions, with a global approach of prevention and reduction of risk factors on users' health, which allows, in addition to a correct management of hospital settings, the realization of concrete actions. To date, in Italy, despite the indoor air being taken in consideration in numerous activities and studies aimed at understanding both building hygiene and environmental aspects, the greatest difficulty is strongly related to the absence of an integrated national policy. The scope of the paper is to underline the relevance of indoor air quality in hospital settings, highlighting the need of procedures, protocols, and tools for strengthening and improving interventions for health prevention, protection, and promotion of users.

Keywords: indoor air quality; healthcare settings; chemical and biological pollution; quality improvement; Italian context

1. The Relevance of Built Environment: The Case of Healing Spaces

In a strategic field such as care and assistance, diagnostics, prevention, research, training, and safeguarding of public health by architectures for health (hospitals, community health centers, clinics and outpatient centers, etc.), both public and private ones, it is utmost a priority to operate a regular and continuous updating of quality's processes, efficacy, and efficiency of healthcare practices. The approach should apply to its entirety with prevention techniques, training, health education, and promotion activities, in relation to the needs for the health protection of users (both patients, visitors, and staff), with particular attention to the most sensitive and vulnerable groups in hospital settings [1–4].

In this scenario the healthcare facilities, affected by the requirement of promoting greater innovation and improving the quality of services and processes, have given rise to a considerable amount of concrete actions and interventions, such as the improvement of staff's training, exceeding and updating the level of organizational, management, and structural standards of healthcare [1,5,6]. In several ways they contribute not only to the efficiency of territorial assistance and care [2] but also to the dissemination of the value of individual and public health prevention, with a broad perspective of citizens' health status, increasing as a consequence the years of life [5,7,8].

In particular, in an Italian context, in order to correctly respond to the healthcare needs of the population, in the Health Pact for the years 2014–2016, signed by the Permanent Conference for the

Relations between the State, the Regions, and the Autonomous Provinces, the state of health has been defined no longer as a source of cost, but as an economic and social investment, identifying a series of interventions to achieve and offer the best products for citizens' health and to promote the development of health and the competitiveness of the whole country [9]. This application has constituted a strategic and important opportunity to tackle some of the crucial and highly relevant issues of recent years with greater awareness, such as:

- improving the compulsory level of training and adequate preparation of healthcare and non-healthcare staff in prevention issues [10];
- the improvement of the investments for the technological and qualitative modernization of the healthcare infrastructures, so as to be able to operate effectively and efficiently (i.e., with a more careful attention to the correct selection of finishing and building materials, products, flexibility in use and ease use, and in the management of engineering plants, etc.);
- the enhancement of healthcare design and indoor air quality, which have a strong and direct impact on the quality of care [11,12];
- the humanization and hospitality of healing spaces [13,14].

The methodologies for assessing the healthcare costs incurred by the various countries were developed by the Organization for Economic Cooperation and Development (OECD), in which 20% of the total health expenditure, quantified in the report "Tackling Wasteful Spending on Health", does not contribute to a real improvement in populations' health status [15]. For this reason, several authors highlight the importance to promote health through design actions in the built environment (urban health strategies, healthy indoor spaces, etc.) [16–19].

Moreover, in relation to the Italian case, with the Decree no. 50/2015—Regulation for hospital assistance, structural, technological, qualitative, and quantitative standards relating to healthcare are aimed at promoting the expansion of the areas, increasing hospitable features of the environments, safety and security, and real and adequate quality of care, which must be adopted to create the conditions to produce benefits and high quality of the entire National Health System (NHS) network [2,20].

2. Design and Management Aspects that Affect Indoor Air in Hospital Settings

In this evolutionary context, there has been growing attention to indoor air quality's issue in healthcare facilities, which, in order to satisfy primarily the requests of patients, healthcare users and workers, administrative and non-administrative staff, etc., have been affected to a series of new adjustments and design approaches (i.e., configuration and rationalization of spaces and flows, the use of specific products and materials, etc.) [21,22], structural and functional actions (i.e., requalification, restructuring, energy efficiency improvement, etc.) [23], engineering plants' system (i.e., optimizing the performance of the centralized heating and cooling systems, energy performances, etc.), [24] and management strategies (i.e., the correct daily management of the ventilations systems, the reduction of costs, accounting for consumption, etc.) [25], with the aim of expanding the services supplied, the quality of healthcare services, obtaining greater organizational and working flexibility, and attempting to reduce the economic costs of healthcare facilities [26]. Gola et al. have highlighted the factors that mostly affect a healing space, as Figure 1 synthetizes [40].

In all these healthcare environments for different needs, the healthcare and technical and administrative staffs, and the users (caregivers, elderly people, children, volunteers, students, visitors, outsourcing services' staffs, maintenance workers and suppliers, etc.)—some of them with reduced mobility, too—interact, stay, live, and work [27,28]. For this reason, specific prevention measures are necessary, considering the exposure of key actors (from the users to hospital staff), whose roles, knowledge and background, motivations, and individual relationships have changed and evolved, becoming increasingly an informed, active, and willing participation to collaborate for improving the environments' quality, services, and treatments. Their exposure takes on particular significance and importance both for the vulnerabilities of the users (i.e., patients with various pathologies, with an acute health status, with different immune responses, people with disabilities elderly, etc.), and for the times of permanence in the hospital [29–33].

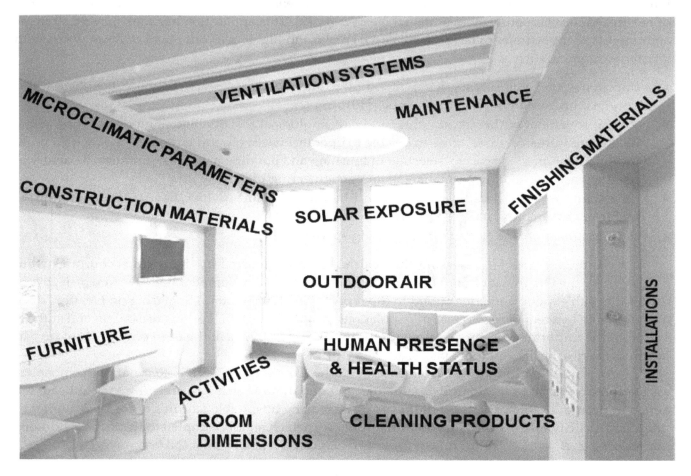

Figure 1. Factors that affect hospital environments.

In the specific case of the activities carried out in the healthcare facilities, it is essential to consider the close relationships between the behaviors and activities of medical e and technical-administrative staffs, and the different ones of patients, visitors, volunteers, students, professionals of external companies (i.e., cleaning, maintenance, suppliers, etc.), the quality of the spaces, and daily relationships with the organizational and management procedures of functional processes, that define the complex scenario of activities to be delivered [34,35]. The use of technological systems designed to perform and satisfy the various tasks in the best economic conditions, the technical furnishings, the level of use, the ordinary and extraordinary cleaning and sanitization activities (providing targeted actions according to the health status and the type of risk of patients, with different levels of contamination, and with microbiological monitoring), the maintenance, the procedures, and the organic management of the multiple routine prevention activities implemented and shared within the spaces, are all factors that contribute significantly to indoor air quality, and the health (this is even more concrete in view of the emergency period for SARS-CoV-2 virus that currently the population is experiencing) and satisfaction of all those users who attend the healing spaces [24,25,35].

In general, these interventions and initiatives have been adopted to address the significant change in healthcare needs, which affects the growth of requests for services and diagnostic treatments, as well as new fields of assistance and research, which require greater functionality of spaces, a reduction in the average length of hospitalization, the occupancy rate of beds, and inter-regional flows of healthcare mobility, overcoming social and territorial inequalities [36,37].

Specifically, on the operational level as regards the interventions carried out, it is necessary to highlight how often the choices of products and construction materials (i.e., paints, varnishes, etc.), finishing, (i.e., adhesives, silicones, etc.), furniture components (i.e., decors, curtains, etc.), products for cleaning and detergents for daily use, products for ordinary (methods and frequency that independently must always be adapted to the use of the area, to the flows of inpatients or medical staff, visitors, etc.) and extraordinary sanitization (i.e., use of more or less products concentrated, or not specific for cleaning surfaces, etc.), as well as engineering plant's management and maintenance activities (i.e., various air conditioning systems and centralized controlled mechanical ventilation systems), etc. were carried out in a disordered manner, without an adequate assessment of the emission behavior of pollutants from the materials and products used (i.e., VOCs—volatile organic compounds—and other substances emissions). In fact, the specificity and the protective value that the environments must respond to specific environmental conditions of use (i.e., temperature, relative humidity, air changes, etc.), the presence of patients, healthcare users, temporary visitors, volunteers, activities carried out by healthcare staff and not, and hygienic conditions of the environments depending on the health status, the type or risk of patients or, in general, of the daily flows (i.e., presence of microbial and fungal communities with a capacity for persistence, variability of concentration, and diversity in healthcare environments, which can generate an extension of the length of hospitalization stay, additional diagnostic and/or therapeutic interventions and additional costs, etc.) [38,39].

3. Chemical and Biological Concentrations in Indoor Air in Healthcare Environments

It should be underlined that, until a few years ago, in Italy, most of the activities and direct and indirect interventions of prevention and training were limited exclusively to select and identified healthcare environments with specific professional exposure to: chemical and biological agents (i.e., monitoring in the air of anesthetic gases in operating rooms, in laboratories dedicated to the preparation and administration of antiblastic drugs, in premises or areas of chemical sterilization, in histology and pathological anatomy departments for use of preservatives or disinfectants (i.e., formaldehyde, waste storage, and transport activities, etc.); ergonomic and physical factors (i.e., patient movement, sudden movements with efforts, critical or prolonged working posture, and in the administrative offices related to the workplace, etc.); video terminals (i.e., in administrative offices, call centers, back offices, departments, etc.); accidents (i.e., falls, etc.); psychosocial (i.e., excessive workload, stress and satisfaction levels, etc.); microclimatic factors such as temperature, relative humidity, air changes (both in health and administrative areas, etc.); implementation of programs of multidisciplinary hygiene surveillance and control such as those developed by the hospital infection committees for the control of infections, of the Supervisory Commissions, composed of a group of dedicated professional figures and with guidelines and protocols for the control of pollutants of biological origin (provided in compliance with ministerial acts), in order to prevent patient-related and non-healthcare staff and non-healthcare-related infections, which have always been a major concern for all hospitals [40–45].

For this reason, these aspects are increasingly integral components of the quality of services, therapies, healthcare services, activities and training, and information plans continuously provided, contributing to obtaining an effective and adequate indoor air quality, which responds to the main references elaborated for some time by the World Health Organization (WHO), and which currently constitute a valuable contribution worldwide.

In general, although the biological pollutants are constantly under analysis, they have already been studied and investigated by several research groups, and several countries have defined guidelines and very detailed protocols (that need to be improved more and more), such as Legionella, etc. [32,46].

Unlike the activities on biological compounds, investigations or monitoring activities of indoor air quality dedicated to the presence (or assessment) of the concentrations of chemical pollutants also to other environments have been carried out only recently and marginally, in some functional areas and environments of the hospital. Never before have such monitoring activities been brought to the attention of management by users, healthcare staff, etc., who complain of uncomfortable circumstances while living and working in the hospital settings or in carrying out their work activities that do not involve the use of chemical or biological agents [45,46]. Often at the operational level, these are requested that usually occur for complaints to situations related to an inadequate air exchange, the presence of new furnishings, the change of the room, during maintenance or renovation activities in specific areas and/or in punctual rooms, when the intended uses vary, when using cleaning and detergent products, or due to the inadequate or incorrect operation of the ventilation systems, etc. [45].

Therefore nowadays, this must entail the implementation of a series of appropriate and organized interventions (not limited to single and specific actions), with a global approach of prevention and reduction of risk factors on the health of all users, which allow, in addition to a correct management of the various environments of healthcare facilities, the realization of concrete actions on indoor air quality according to the priority principles and guidelines identified by WHO [47] and in part already listed as goals in various European and international programs of the prevention measures [12].

With particular attention to chemical pollutants, an examination of the current situation in the European Union (EU) shows that some Member States, such as France, Belgium, Finland, Portugal, Poland, and Lithuania have fully entered the quality of the indoor air in their national legislations with quantitative values (reference values, guidelines, etc.), and with practical guidelines which contain indications for the control, self-assessment sheets for identifying potential indoor sources (or close to the facilities), and the procedures for the development of indoor air monitoring, which are in many cases in line with the current WHO values published in 2009 and 2010 on the basis of the main scientific evidences [12].

In these countries, compliance with the legal requirements and the correct application of practical protocols remain one of the fundamental points for achieving good indoor air quality in the various healthcare environments [48]. In particular, France has foreseen a series of specific interventions including mandatory monitoring of indoor air quality in healthcare facilities as early as 2023 [49].

Until today, in Italy, despite being the quality of indoor air subjected to numerous activities and investigations aimed at understanding both the environmental and hygiene aspects, the greatest difficulty remains the absence of an integrated national policy about indoor air quality, with specific legislative references, which report the national references (i.e., guide values, references, etc.) and the rules for the data analysis of the results, and with documents that list the recommendations for an adequate management and evaluation of indoor air quality [50]. In the absence of national references, it is possible to use those present in the WHO documents related to indoor air quality or those in the legislation of other European countries or, by analogy, to other standards such as those relating to the ambient air for which specific legislative references have been issued on a limited number of pollutants, etc. [51].

There is no doubt that the current system of health prevention and protection laws has led to a confusion of language and knowledge that indeed has often confused and disoriented the practitioners, engaged in various capacities in the programs and evaluations in these environments and structures [37]. In this process of approach and strengthening of prevention actions, it is necessary to bring about a concrete harmonization, revision, innovation, updating and expansion on specific aspects, also to current standards [52]).

The aims and scope are to provide the procedures and tools necessary to strengthen, optimize, and improve interventions for the prevention, protection, and promotion of the health of users in healthcare environments that represent one of the priority objectives of the NHS's strategy in the prevention programs, with monitoring activities within the healing spaces [38,53].

Additionally, with regard to biological pollutants, although there are recommendations from international agencies and institutions, there are no legislative values or standards for the microbiological parameters of indoor air quality due to the difficulties encountered in correlating the data of the microbiological tests with those of the epidemiological investigations [46].

4. Future Perspectives

In conclusion, hospital facilities are complex constructions, with very different needs, users, and requirements compared to other building facilities, and they work 24/7, all year long. For this reason, every action should be assessed in relation to their performances and the aim to interrupt medical activities as little as possible.

It is clever that indoor air quality is a very broad topic in which any variable can affect the performances of air in indoor environments both in biological and chemical terms, as one of the goals of UN 2030—United Nations Sustainable Development. As several authors states, adequate design and management strategies, in relation to different procedures, can decrease or increase the quality performances of the healthcare environments.

The Scientific Community should continue to investigate the issue, define smart and efficient procedures, protocols for monitoring and tools, instrumentations for the investigations, etc. for strengthening and improving interventions, and guaranteeing protection and promotion of users. The new challenge should investigate the correlations between the chemical and biological pollutants and their effects in indoor air and the quality of the healthcare facility.

Author Contributions: Conceptualization, G.S. and M.G.; writing—original draft preparation, M.G. and G.S.; writing—review and editing, M.G.; supervision, S.C. All authors have read and agreed to the published version of the manuscript.

References

1. Brambilla, A.; Buffoli, M.; Capolongo, S. Measuring hospital qualities. A preliminary investigation on Health Impact Assessment possibilities for evaluating complex buildings. *Acta Bio-Med. Atenei Parm.* **2019**, *90*, 54–63.

2. Capolongo, S.; Mauri, M.; Peretti, G.; Pollo, R.; Tognolo, C. Facilities for Territorial Medicine: The experiences of Piedmont and Lombardy Regions. *Technè* **2015**, *9*, 230–236.

3. Settimo, G. Residential indoor air quality: Significant parameters in light of the new trends. *Ig. Sanità Pubblica* **2012**, *68*, 136–138.

4. Azara, A.; Dettori, M.; Castiglia, P.; Piana, A.; Durando, P.; Parodi, V.; Salis, G.; Saderi, L.; Sotgiu, G. Indoor Radon Exposure in Italian Schools. *Int. J. Environ. Res. Public Health* **2018**, *15*, 749. [CrossRef] [PubMed]

5. Brambilla, A.; Capolongo, S. Healthy and sustainable hospital evaluation-A review of POE tools for hospital assessment in an evidence-based design framework. *Buildings* **2019**, *9*, 76. [CrossRef]

6. Odone, A.; Bossi, E.; Gaeta, M.; Garancini, M.P.; Orlandi, C.; Cuppone, M.T.; Signorelli, C.; Nicastro, O.; Zotti, C.M. Risk Management in healthcare: Results from a national-level survey and scientometric analysis in Italy. *Acta Biomed.* **2019**, *90*, 76–86.

7. Capobussi, M.; Tettamanti, R.; Marcolin, L.; Piovesan, L.; Bronzin, S.; Gattoni, M.E.; Polloni, I.; Sabatino, G.; Tersalvi, C.A.; Auxilia, F.; et al. Air Pollution Impact on Pregnancy Outcomes in Como, Italy. *J. Occup. Environ. Med.* **2016**, *58*, 47–52. [CrossRef]

8. Vonci, N.; De Marco, M.F.; Grasso, A.; Spataro, G.; Cevenini, G.; Messina, G. Association between air changes and airborne microbial contamination in operating rooms. *J. Infect. Public Health* **2019**, *12*, 827–830. [CrossRef]

9. Signorelli, C.; Odone, A.; Ricciardi, W.; Lorenzin, B. The social responsibility of public health: Italy's lesson on vaccine hesitancy. *Eur. J. Public Health* **2019**, *29*, 1003–1004. [CrossRef]

10. Gianfredi, V.; Grisci, C.; Nucci, D.; Parisi, V.; Moretti, M. Communication in health. *Recenti Progress. Med.* **2018**, *109*, 374–383.

11. Moreno-Rangel, A.; Sharpe, T.; Musau, F.; McGill, G. Field evaluation of a low-cost indoor air quality monitor to quantify exposure to pollutants in residential environments. *J. Sens. Sens. Syst.* **2018**, *7*, 373–388. [CrossRef]

12. Settimo, G. Existing guidelines in indoor air quality: The case study of hospital environments. In *Indoor Air Quality in Healthcare Facilities*, 1st ed.; Capolongo, S., Settimo, G., Gola, M., Eds.; Springer Public Health: New York City, NY, USA, 2017; pp. 13–26.

13. Buffoli, M.; Bellini, E.; Bellagarda, A.; di Noia, M.; Nickolova, M.; Capolongo, S. Listening to people to cure people: The LpCp–tool, an instrument to evaluate hospital humanization. *Ann. Ig.* **2014**, *26*, 447–455. [PubMed]

14. Bosia, D.; Marino, D.; Peretti, G. Health facilities humanisation: Design guidelines supported by statistical evidence. *Ann. Dell'Istituto Super. Sanita* **2016**, *52*, 33–39.

15. Smith, D.; Alverdy, J.; An, G.; Coleman, M.; Garcia-Houchins, S.; Green, J.; Keegan, K.; Kelley, S.T.; Kirkup, B.C.; Kociolek, L.; et al. The hospital microbiome project: Meeting report for the 1st hospital microbiome project workshop on sampling design and building science measurements. *Stand. Genomic. Sci.* **2013**, *8*, 112–117. [CrossRef] [PubMed]

16. Rebecchi, A.; Buffoli, M.; Dettori, M.; Appolloni, L.; Azara, A.; Castiglia, P.; D'Alessandro, D.; Capolongo, S. Walkable environments and healthy urban moves: Urban context features assessment framework experienced in Milan. *Sustainability* **2019**, *11*, 2778. [CrossRef]

17. Congiu, T.; Sotgiu, G.; Castiglia, P.; Azara, A.; Piana, A.; Saderi, L.; Dettori, M. Built Environment Features and Pedestrian Accidents: An Italian Retrospective Study. *Sustainability* **2019**, *11*, 1064. [CrossRef]

18. Rebecchi, A.; Boati, L.; Oppio, A.; Buffoli, M.; Capolongo, S. Measuring the expected increase in cycling in the city of Milan and evaluating the positive effects on the population's health status: A Community-Based Urban Planning experience. *Ann. Ig.* **2016**, *28*, 381–391.

19. D'Alessandro, D.; Arletti, S.; Azara, A.; Buffoli, M.; Capasso, L.; Cappuccitti, A.; Casuccio, A.; Cecchini, A.; Costa, G.; De Martino, A.M.; et al. Strategies for Disease Prevention and Health Promotion in Urban Areas: The Erice 50 Charter. *Ann. Ig.* **2017**, *29*, 481–493. [CrossRef]

20. Campanella, P.; Azzolini, E.; Izzi, A.; Pelone, F.; De Meo, C.; La Milia, D.; Specchia, M.; Ricciardi, W. Hospital efficiency: How to spend less maintaining quality? *Ann. Dell'Istituto Super. Sanita* **2017**, *53*, 46–53.

21. Gray, W.A.; Vittori, G.; Guenther, R.; Vernon, W.; Dilwali, K. Leading the way: Innovative sustainable design guidelines for operating healthy healthcare buildings. In Proceedings of the ISIAQ–10th International Conference on Healthy Buildings, Curran Associates, Red Hook, NY, USA, 12 July 2012; pp. 1212–1217.

22. Setola, N.; Borgianni, S. *Designing Public Spaces in Hospitals*; Taylor and Francis Inc.: London, UK, 2016.

23. Pati, D.; Park, C.-S.; Augenbroe, G. Facility maintenance performance perspective to target strategic organizational objectives. *J. Perform. Constr. Facil.* **2010**, *24*, 180–187. [CrossRef]

24. Joppolo, C.M.; Romano, F. HVAC System Design in Health Care Facilities and Control of Aerosol Contaminants: Issues, Tools and Experiments. In *Indoor Air Quality in Healthcare Facilities*, 1st ed.; Capolongo, S., Settimo, G., Gola, M., Eds.; Springer Public Health: New York, NY, USA, 2017; pp. 83–94.

25. Moscato, U.; Borghini, A.; Teleman, A.A. HVAC Management in Health Facilities. In *Indoor Air Quality in Healthcare Facilities*, 1st ed.; Capolongo, S., Settimo, G., Gola, M., Eds.; Springer Public Health: New York, NY, USA, 2017; pp. 95–106.

26. Gola, M.; Settimo, G.; Capolongo, S. Indoor air in healing environments: Monitoring chemical pollution in inpatient rooms. *Facilities* **2019**, *37*, 600–623. [CrossRef]

27. Carducci, A.L.; Fiore, M.; Azara, A.; Bonaccorsi, G.; Bortoletto, M.; Caggiano, G.; Calamusa, A.; De Donno, A.; De Giglio, O.; Dettori, M.; et al. Environment and health: Risk perception and its determinants among Italian university students. *Sci. Total Environ.* **2019**, *691*, 1162–1172. [CrossRef] [PubMed]

28. Mosca, E.I.; Capolongo, S. Towards a universal design evaluation for assessing the performance of the built environment. In *Transforming Our World through Design, Diversity and Education*, 1st ed.; Craddock, G., Doran, C., McNutt, L., Rice, D., Eds.; Springer Studies in Health Technology and Informatics: Cham, Switzerland, 2018; Volume 256, pp. 771–779.

29. Gola, M.; Mele, A.; Tolino, B.; Capolongo, S. Applications of IAQ Monitoring in International Healthcare Systems. In *Indoor Air Quality in Healthcare Facilities*, 1st ed.; Capolongo, S., Settimo, G., Gola, M., Eds.; Springer Public Health: New York, NY, USA, 2017; pp. 27–39.

30. Borghini, A.; Poscia, A.; Bosello, S.; Teleman, A.A.; Bocci, M.; Iodice, L.; Ferraccioli, G.; La Milìa, D.; Moscato, U. Environmental pollution by benzene and PM10 and clinical manifestations of systemic sclerosis: A correlation study. *Int. J. Environ. Res. Public Health* **2017**, *14*, 1297. [CrossRef] [PubMed]

31. La Milia, D.; Vincenti, S.; Fiori, B.; Pattavina, F.; Torelli, R.; Barbara, A.; Wachocka, M.; Moscato, U.; Sica, S.; Amato, V.; et al. Monitoring of particle environmental pollution and fungal isolations during hospital building-work activities in a hematology ward. *Mediterr. J. Hematol. Infect. Dis.* **2019** *11*, e2019062. [PubMed]

32. Sotgiu, G.; Are, B.M.; Pesapane, L.; Palmieri, A.; Muresu, N.; Cossu, A.; Dettori, M.; Azara, A.; Mura, I.I.; Cocuzza, C.; et al. Nosocomial transmission of carbapenem-resistant Klebsiella pneumoniae in an Italian university hospital: A molecular epidemiological study. *J. Hosp. Infect.* **2018**, *99*, 413–418. [CrossRef]

33. Harvey, T.E., Jr.; Pati, D. Keeping watch. Design features to aid patient and staff visibility. *Health Facil. Manag.* **2012**, *25*, 27–31.

34. Leung, M.; Chan, A.H.S. Control and management of hospital indoor air quality. *Med. Sci. Monit.* **2006**, *12*, SR17–SR23.

35. Gola, M.; Settimo, G.; Capolongo, S. Chemical Pollution in Healing Spaces: The Decalogue of the Best Practices for Adequate Indoor Air Quality in Inpatient Rooms. *Int. J. Environ. Res. Public Health* **2019**, *16*, 4388. [CrossRef]

36. Mauri, M. The future of the hospital and the structures of the NHS. *Technè* **2015**, *9*, 27–34.

37. Signorelli, C.; Fara, G.M.; Odone, A.; Zangrandi, A. The reform of the Italian Constitution and its possible impact on public health and the National Health Service. *Health Policy* **2017**, *121*, 90–91. [CrossRef]

38. Gola, M.; Settimo, G.; Capolongo, S. Indoor Air Quality in Inpatient Environments: A Systematic Review on Factors that Influence Chemical Pollution in Inpatient Wards. *J. Healthc. Eng.* **2019**, 8358306. [CrossRef] [PubMed]

39. Śmiełowska, M.; Marć, M.; Zabiegała, B. Indoor air quality in public utility environments—A review. *Environ. Sci. Pollut. Res.* **2017**, *24*, 11166–11176. [CrossRef] [PubMed]

40. D'Amico, A.; Fara, G.M. The need to develop a multidisciplinary expertise for the microbiological safety of operating theatres. *Ann. Ig.* **2016**, *28*, 379–380. [PubMed]

41. Montagna, M.T.; Cristina, M.L.; De Giglio, O.; Spagnolo, A.M.; Napoli, C.; Cannova, L.; Deriu, M.G.; Delia, S.A.; Giuliano, A.; Guida, M.; et al. Serological and molecular identification of *Legionella* spp. isolated from water and surrounding air samples in Italian healthcare facilities. *Environ. Res.* **2016**, *146*, 47–50. [CrossRef] [PubMed]

42. ISIAQ (International Society of Indoor Air Quality and Climate). *Review on Indoor Air Quality in Hospitals and Other Health Care Facilities*; International Society of Indoor Air Quality and Climate: Philadelphia, PA, USA, 2003; Volume 43.

43. Ardoino, I.; Zangirolami, F.; Iemmi, D.; Lanzoni, M.; Cargnelutti, M.; Biganzoli, E.; Castaldi, S. Risk factors and epidemiology of Acinetobacter baumannii infections in a university hospital in Northern Italy: A case-control study. *Am. J. Infect. Control* **2016**, *44*, 1600–1605. [CrossRef]

44. Castaldi, S.; Bevilacqua, L.; Arcari, G.; Cantù, A.P.; Visconti, U.; Auxilia, F. How appropriate is the use of rehabilitation facilities? Assessment by an evaluation tool based on the AEP protocol. *J. Prev. Med. Hyg.* **2010**, *51*, 116–120.

45. Settimo, G.; Bonadonna, L.; Gherardi, M.; di Gregorio, F.; Cecinato, A. Qualità dell'aria negli ambienti sanitari: Strategie di monitoraggio degli inquinanti chimici e biologici. *Rapporti ISTISAN* **2019**, *19/17*, 1–55.

46. Bonadonna, L.; de Grazia, M.C.; Capolongo, S.; Casini, B.; Cristina, M.L.; Daniele, G.; D'Alessandro, D.; De Giglio, O.; Di Benedetto, A.; Di Vittorio, G.; et al. Water safety in healthcare facilities. The Vieste Charter. *Ann. Ig.* **2017**, *29*, 92–100.

47. WHO. *Guidelines for Indoor Air Quality: Selected Pollutants*, 1st ed.; World Health Organization: Copenhagen, Denmark, 2010.

48. Oddo, A. The intervention of the judicial system in order to maximise the prevention of chemical risk in operating theatres and to put personal health and safety first. *Med. Lav.* **2016**, *107*, 5–20.

49. ANSES. *Air Intérieur: Valeurs Guides*; Agence Nationale de Sècuritè Sanitaire: Paris, France, 2011.

50. Settimo, G.; D'Alessandro, D. European community guidelines and standards in indoor air quality: What proposals for Italy. *Epidemiol. Prev.* **2014**, *38*, 36–41.

51. Viviano, G.; Settimo, G. Air quality regulation and implementation of the European Council Directives. *Ann. Dell'Istituto Super. Sanita* **2003**, *39*, 343–350.

52. Oddo, A. The contribution of technology and the forefront role of technical measures in the prevention of infection risk in the healthcare sector. *Med. Lav.* **2016**, *107*, 21–30. [PubMed]

53. Settimo, G.; Gola, M.; Mannoni, V.; De Felice, M.; Padula, G.; Mele, A.; Tolino, B.; Capolongo, S. Assessment of Indoor Air Quality in Inpatient Wards. In *Indoor Air Quality in Healthcare Facilities*, 1st ed.; Capolongo, S., Settimo, G., Gola, M., Eds.; Springer Public Health: New York, NY, USA, 2017; pp. 107–118.

3

Application of Airborne Microorganism Indexes in Offices, Gyms and Libraries

Pietro Grisoli [1],*, Marco Albertoni [1] and Marinella Rodolfi [2]

[1]   Department of Drug Sciences, Laboratory of Microbiology, University of Pavia, 27100 Pavia, Italy; marco.albertoni@unipv.it

[2]   Department of Earth and Environmental Sciences, Mycology Section, University of Pavia, 27100 Pavia, Italy; marinella.rodolfi@unipv.it

*   Correspondence: pietro.grisoli@unipv.it.

Abstract: The determination of microbiological air quality in sporting and working environments requires the quantification of airborne microbial contamination. The number and types of microorganisms, detected in a specific site, offer a useful index for air quality valuation. An assessment of contamination levels was carried out using three evaluation indices for microbiological pollution: the global index of microbiological contamination per cubic meter ($GIMC/m^3$), the index of mesophilic bacterial contamination (IMC), and the amplification index (AI). These indices have the advantage of considering several concomitant factors in the formation of a microbial aerosol. They may also detect the malfunction of an air treatment system due to the increase of microbes in aeraulic ducts, or inside a building compared to the outdoor environment. In addition, they highlight the low efficiency of a ventilation system due to the excessive number of people inside a building or to insufficient air renewal. This study quantified the levels of microorganisms present in the air in different places such as offices, gyms, and libraries. The air contamination was always higher in gyms that in the other places. All examined environments are in Northern Italy.

Keywords: airborne microorganisms; bacteria; fungi; gyms; indoor air quality; libraries; offices

1. Introduction

The evaluation of microbiological air contamination is an important aspect of applied microbiology and industrial hygiene [1–3]. In Italy, several legislative decrees, in particular the Legislative Decree 81/2008, impose that employers are responsible for the assessment of risks of biological origin arising from the activities in the workplace; moreover, they must adopt suitable measures so that the air in the workplace be healthy [4]. Therefore, the realization of measuring methods of airborne contamination is a matter of prime importance in the most complex procedure of risk evaluation that must involve, necessarily, different scientific skills [5–8]. Air contamination, both in confined environments and outdoors, can be caused not only by chemical agents but also by biological sources such as pollen, dust, mites, insects, pet allergens, bacteria, fungi, and viruses [9]. The monitoring of bioaerosols represents an important tool both for assessing the risk in the working environment [10] and for the evaluation of the environmental impact of certain activities that take place outdoors, such as biopurification or waste disposal [11]. Several investigations have been directed to study the disease called Sick Building Syndrome (SBS), which is characterized by varied and non-specific symptoms observed in workers employed in confined environments [12–14]. The World Health Organization has defined SBS as "an increase in frequency in the occupants of non-industrial buildings, of not specific acute symptoms (irritation of eyes, nose, throat, headache, fatigue, nausea) that improve when building is left" [15]. The American Conference of Governmental Industrial Hygienists (ACGIH) does not suggest threshold limit values (TLV) for the environmental concentration of biological agents, because the existing information does not allow to establish a scientifically acceptable dose–response relationship [16].

However, the assessment of culturable microbial agents is currently the easiest and most practical measure to determine changes in the air quality of confined workplaces. The purpose of these controls may be to set guide values, technologically attainable by adopting containment measures of contamination established on the basis of repeated samplings in specific locations. With this aim, this study examined structures destined for different types of work and recreational activity such as offices, gyms, and libraries, measuring air-diffused contamination to identify and apply indices useful for the microbiological classification of the air quality [17].

The indoor air quality was evaluated by applying the following indices: the global index of microbiological contamination per cubic meter (GIMC/m^3), the index of mesophilic bacterial contamination (IMC), and the amplification index (AI). The GIMC considers different microbial types and emphasizes their ability to grow in a wide temperature range. GIMC may be a simple method to evaluate a potential biological risk in indoor and outdoor environments and to monitor sources of microbiological contamination. The IMC reveals the presence of obligated mesophilic bacteria, organisms of probable human origin in the indoor air. This index represents a practical instrument to underline bacterial growth caused by hypoventilation and overcrowding. The AI is important as it reveals microbial pollution in indoor ventilation systems. These indices showed their applicability in other types of environments [18,19].

2. Materials and Methods

2.1. Sampling Method

Environmental monitoring was performed in Northern Italy in 10 diverse offices, gyms, and libraries situated in buildings equipped with a ventilation system able to work in the following modes: heating, air conditioning, and simple ventilation. The offices are placed in bank edifices, the gyms in fitness centers, while the libraries belong to university buildings. The samplings were realized every month for one year, inside the various environments, during normal people activity, and outside the building. Quantitative data were determined in triplicate by means of an orthogonal impact Microflow Air Sampler (AQUARIA, Lacchiarella, Italy), kept 1.5 m above ground level, in the center of the room for offices, in the middle of the weight-training room for gyms, and in the reading rooms for libraries. Air samplers worked at a fixed speed of 1.5 Ls^{-1}, collecting a volume of 200 L. For the different types of environments, the number of persons (n. p.) and the air speed (m/s) were recorded, so that in three sampling periods, the following average values were obtained: offices (n. p. 4.7; 0.03 m/s), gyms (n. p. 21.3; 0.060 m/s), and libraries (n. p. 16.2; 0.05 m/s).

2.2. Microorganisms Assessed

Bacteria were collected using Tryptone Soya Agar (TSA, Oxoid, Basingstoke, UK), and the cultures were incubated at 37 °C for 48 h for mesophilic bacteria and at 20 °C for 6 days for psychrophilic bacteria. Fungi were collected on Sabouraud Dextrose Agar (SAB Oxoid, Basingstoke, UK), and the cultures were incubated at 20 °C for 6 days. The outdoor air quality used as control was analyzed following the same criteria. All total microbial counts are indicated as the number of colony-forming units per cubic meter of air (CFU/m^3), calculated as an average of three determinations from three samples collected serially. The assessment of microbial contamination was effected by using bacterial and fungal counts, and the following indexes were calculated: GIMC per cubic meter (GIMC/m^3), which is the sum of the values of the total microbial counts determined for mesophilic bacteria, psychrophilic bacteria, and fungi in all sampled areas; IMC, derived from the ratio between the values of CFU per cubic meter measured for mesophilic and psychrophilic bacteria at the same sampling point; AI, resulting from the ratio between the GIMC/m^3 values measured inside the building and those measured outdoor.

2.3. Data Analysis

The air microbial contamination values were expressed as colony-forming units (CFUs), and the limit of quantification was 1 CFU/m^3. The number of CFUs for contact plate after appropriate incubation was corrected using the positive-hole correction table provided by the supplier. Statistical analysis of the data was performed by comparing the results obtained in the different environments by means of analysis of variance one-way (post hoc test) values transformed into logarithms (natural log). The analyses were conducted using Prism 3.0. The significance level was $p < 0.05$.

3. Results

Table 1 shows the results of microbiological contamination of 10 different offices located in buildings with centralized systems for ventilation. The mean values were higher for mesophilic bacteria during the air conditioning phase compared to heating ($p = 0.034$) and during simple ventilation compared to heating ($p = 0.023$). The contamination of psychrophilic bacteria was greater during simple ventilation compared to air conditioning ($p = 0.016$) and to heating ($p = 0.032$). The mean values of CFU/m^3 for the fungi were not significantly different during the different periods of air treatment. Although the average contamination was low, there were high maximum values of CFU/m^3 for mesophilic bacteria in the periods of air conditioning and simple ventilation and for psychrophilic bacteria during simple ventilation.

Table 1. Total microbial concentrations measured in the offices during heating, air conditioning, and simple ventilation.

Functioning Modes	N	Mesophilic Bacteria CFU/m^3		Psychrophilic Bacteria CFU/m^3		Fungal Count CFU/m^3	
		M ± SD	Min–max	M ± SD	Min–max	M ± SD	Min–max
Heating	10	191.20 ± 144.53	8–450	176.70 ± 117.21	7–366	136.00 ± 32.80	22–654
Simple ventilation	10	667.60 ± 1602.60	6–5200	761.90 ± 1519.7	10–5000	131.50 ± 72.30	5–190
Air Conditioning	10	1089.70 ± 2392.20	19–7800	495.60 ± 307.20	76–860	119.60 ± 48.84	28–199

The number of samples (N), mean (M), standard deviation (SD), and range of values are indicated for each sampling period. CFU: colony-forming units.

The calculation of GIMC/m^3 confirmed the presence of higher values of contamination during conditioning ($p = 0.01$) and simple ventilation ($p = 0.01$) compared to heating. The average value of IMC during heating amounted to 7.30 and was higher than the values of the other modes of operation, corresponding to 0.80 during the simple ventilation ($p = 0.04$) and 1.60 during air conditioning (Table 2).

Table 2. Global index of microbial contamination per cubic meter (GIMC/m^3) and index of mesophilic bacterial contamination (IMC) measured in offices during heating, air conditionining and simple ventilation.

Functioning Modes	N	GIMC/m^3		IMC	
		M ± SD	Min–max	M ± SD	Min–max
Heating	10	503.90 ± 282.22	124–1111	7.30 ± 20.03	0.20–64.30
Simple ventilation	10	1561.00 ± 3154.80	21–10,450	0.80 ± 0.40	0.10–1.30
Air Conditioning	10	1704.90 ± 2537.57	229–8720	1.60 ± 2.67	0.02–9.10

The number of samples (N), mean (M), standard deviation (SD), the range of values are indicated for each sampling period.

The data regarding the microbiological contamination found in 10 different gyms with centralized ventilation systems are summarized in Table 3. The results show highest mean values for mycetic contamination during the air conditioning phase compared to simple ventilation ($p = 0.04$) and heating ($p = 0.01$). As regards bacteria, the average contamination values for psychrophilic bacteria were superior to those for mesophilic bacteria in the three air treatment modalities and were significantly different during conditioning ($p = 0.02$).

Table 3. Total microbial concentrations measured in gyms during heating, air conditioning, and simple ventilation.

Functioning Modes	N	Mesophilic Bacteria CFU/m^3		Psychrophilic Bacteria CFU/m^3		Fungal Count CFU/m^3	
		M ± SD	Min–max	M ± SD.	Min–max	M ± SD	Min–max
Heating	10	1096.60 ± 924.40	140–2850	1446.40 ± 1356.82	180–4650	187.50 ± 167.02	20–568
Simple ventilation	10	393.00 ± 257.00	120–980	819.00 ± 432.40	330–1800	1208.90 ± 1459.60	90–4848
Air Conditioning	10	666.50 ± 477.76	200–1800	1509.00 ± 1354.02	440–4400	2430.90 ± 3678.31	8–10,848

The number of samples (N), mean (M), standard deviation (SD), and range of values are indicated for each sampling period.

There was a notable increase in fungal count when the central heating was switched off. In fact, the highest maximum values of CFU/m^3 were found for fungi during simple ventilation and conditioning, in contrast to what observed for bacteria ($p = 0.016$).

Table 4 shows the results of microbial contamination based on the GIMC/m^3 and on the IMC measured in the gyms. Because of the elevated fungal count observed in this period, the mean of GIMC/m^3 was higher during air conditioning ($p = 0.019$), with a maximum value of 15,248. The mean of the IMC values was always <1, which implies that the counts of the psychrophilic bacteria were almost always higher than the counts of the mesophilic bacteria; however, no significant differences in the index were observed when central heating was on or off.

Table 4. GIMC/m^3 and IMC measured in gyms during heating, air conditioning, and simple ventilation.

Functioning Modes	N	GIMC/m^3		IMC	
		M ± SD	Min–max	M ± SD	Min–max
Heating	10	2703.50 ± 2144.84	1160–7090	0.90 ± 0.74	0.12–2.40
Simple ventilation	10	2420.90 ± 1645.70	910–6548	0.50 ± 0.30	0.17–1.20
Air Conditioning	10	4606.40 ± 4428.70	1480–15,248	0.60 ± 0.39	0.20–1.40

The number of samples (N), mean (M), standard deviation (SD), and range of values are indicated for each sampling period.

Similar sampling and monitoring methods were adopted for the evaluation of the air quality of libraries. The average levels of contamination detected during air sampling showed small oscillations with very restricted values of CFU/m^3 both for bacterial loads (mesophilic and psychrophilic) and for fungi. The results of airborne bacteria and fungi recovered from the libraries showed that the CFU/m^3 values were higher during heating than during simple ventilation or air conditioning. However, no significant differences in the concentrations of all microbial counts were evidenced in the three functioning modes of the ventilation system (Table 5).

Table 5. Total microbial concentrations in libraries during heating, air conditioning, and simple ventilation.

Functioning Modes	N	Mesophilic Bacteria CFU/m^3		Psychrophilic Bacteria CFU/m^3		Fungal Count CFU/m^3	
		M ± SD	Min–max	M ± SD	Min–max	M ± SD	Min–max
Heating	10	171.20 ± 72.43	112–320	301.10 ± 237.26	125–725	123.90 ± 54.52	62–220
Simple ventilation	10	65.10 ± 29.40	38–110	152.40 ± 90.70	81–380	66.00 ± 31.70	32–110
Air Conditioning	10	71.80 ± 30.50	30–120	200.60 ± 127.65	97–400	76.60 ± 36.55	35–120

The number of samples (N), mean (M), standard deviation (SD), and range of values are indicated for each sampling period.

In addition, the transformation of total microbial count values into the indices evidenced a general reduction of the GIMC/m^3 values and of the IMC values in air conditioning compared to heating, but the statistical analysis showed no significant differences. The IMC index assumed values lower than 1 and showed that air contamination was mainly due to environmental bacteria. All values were limited to a small range in every period of the monitoring campaign (Table 6).

Table 6. GIMC/m^3 and IMC measured in libraries during heating, air conditioning, and simple ventilation.

Functioning Modes	N	GIMC/m^3		IMC	
		M ± SD	Min–max	M ± SD	Min–max
Heating	10	596.20 ± 335.08	300–1250	0.70 ± 0.22	0.40–0.90
Simple ventilation	10	283.50 ± 131.70	153–578	0.50 ± 0.20	0.25–0.70
Air Conditioning	10	349.00 ± 176.43	182–652.50	0.40 ± 0.18	0.20–0.60

The number of samples (N), mean (M), standard deviation (SD), and range of values are indicated for each sampling period.

In indoor environments, microbiological air contamination can also be described by the Amplification Index of microbial contamination (AI). To calculate this index, it was necessary to use the GIMC values measured outside the buildings in the different sampling periods (Table 7). These values were significantly different in the three sampling points; in particular, the widest variations were detectable outside gyms in the air conditioning period ($p < 0.0001$).

Table 7. GIMC/m^3 measured outside gyms, libraries, and offices during heating, air conditioning, and simple ventilation periods.

Sampling Points	N	Heating GIMC/m^3		Simple Ventilation GIMC/m^3		Air Conditioning GIMC/m^3	
		M ± SD	Min–max	M ± SD	Min–max	M ± SD	Min–max
Gyms- outdoor	10	714.40 ± 294.00	328–1268	703.10 ± 226.40	484–1236	3586.10 ± 2326.42	1325–9240
Libraries outdoor	10	410.65 ± 187.81	266–880	596.55 ± 167.19	359–892	508.40 ± 141,01	364–830
Offices-outdoor	10	421.70 ± 179.62	248–890	619.90 ± 169.30	392–933	703.80 ± 226.20	485–1068

The number of samples (N), mean (M), standard deviation (SD), and range of values are indicated for each sampling period.

AI is determined by calculating the ratio between the GIMC/m^3 values measured inside a building and those measured outside. This index is greater than 1 when the microbial contamination inside a building is higher than the outdoor contamination. In most cases, the values of microbial contamination of indoor air are higher than the outdoor values. In the heating phase, AI showed a worsening of air contamination especially in gyms compared the other environments considered (3.7). On the contrary, during the conditioning period, the highest values were recorded in offices (3.0) (Figure 1).

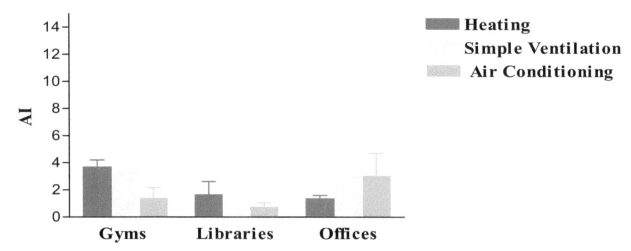

Figure 1. Amplification Index of microbial contamination (AI) measured in the three types of environments during heating, air conditioning, and simple ventilation.

4. Discussion

It can be reasonably assumed that the microflora existing in recreational and working indoor environments has a concentration lower than or equal to the external one detected in the same location and in the same climatic conditions [20,21]. In buildings equipped with centralized ventilation systems there is a reduction of microbial contamination compared to the outside when air filtration and treatment systems are maintained in appropriate conditions [22]. The air inside buildings may be contaminated by the growth of microorganisms on floors, walls, aeraulic plants or because of inadequate air changes [23]. The ventilation systems can have very different structural characteristics, ranging from the simple aspiration of exhaust air to integrated heating, cooling, and humidification. It is, therefore, clear that the air introduced from these systems can be considered as a matrix whose microbiological quality can be evaluated and classified, as for any other product that may have effects on human health. In addition, the activities carried out in a confined environment may be responsible for the spread of microorganisms.

Dissenting opinions have been expressed on the possibility to monitor microbiological environmental contamination in workplaces. Information on the dose–response relationship related to the exposure to microorganisms is still not available today. The ACGIH does not indicate TLV for biological agents. Furthermore, different analytical methods do not allow to recover and identify all the microorganisms present in the air. In fact, the percentage of viable, culturable bacteria recovered with normal sampling systems oscillates in a range of values that varies between 0.1% and 10% of the total bacteria present in the air [24–26]. The presence of microorganisms should also be sought in working environments because of their toxigenic potential and the possibility of spreading cell fragments and volatile organic compounds into the environment. The lack of viable, culturable cells does not necessarily indicate a healthy environment. Some studies, carried out in diversified working environments, report environmental contamination values referable to the number of viable cells belonging to a single class of microorganisms [27–29]. For example, a microbiological classification of the air quality in non-industrial environments and homes considers the contamination from bacteria that develop at 20–25 °C. In this classification, for non-industrial environments, a level <100 CFU/m^3 corresponds to the category of low contamination, while a value >2000 CFU/m^3 corresponds to the category of very high contamination [30]. It is clear that such assessment is incapable not only to describe all the factors that determine the accumulation and spread of microorganisms in the environment but also to identify possible risks for workers.

In this research, GIMC was used in order to include in the same data several categories of microorganisms. This index considers microorganisms proliferating at different temperatures, such as mesophilic and psychrophilic bacteria, and fungi capable of adapting to different types of environments. These characteristics make this quantitative measure of microbiological contamination significant and allow to evaluate the salubrity of a work environment; in fact, it considers microorganisms that can develop in a wide range of temperatures, including ambient temperatures, typical of saprophytic life, and 37 °C, which is the temperature of development of pathogens. This index is particularly useful because it is able to highlight even anomalous situations in indoor environmental microbiological contamination: in fact, in the offices we analyzed, the mesophilic bacteria during the conditioning phase were higher than the psychrophilic ones. This value represents an exception because it is usually reasonable to assume that psychrophilic bacteria, which might grow well in air conditioning systems, are more numerous than mesophilic bacteria during the cooling season, as shown in the results for gyms and libraries buildings and in previous researches [22]. In particular, GIMC attributes importance to bacteria that can proliferate in a wide range of temperatures. It is true that the two incubation temperatures (20 °C and 37 °C) do not differentiate the two categories of bacteria completely. Nevertheless, it is useful to determine the two total counts during environmental monitoring because they have different significances and allow a more complete evaluation of airborne bacterial contamination. In fact, the purpose of the index is to provide a measure of biological risk and to aggregate several "environmental indicators". IMC is mainly an index of anthropic contamination; it highlights the share of mesophilic bacteria in the microbial population examined. This index derives from the ratio between the value of CFU/m^3 at 37 °C and that at 20 °C. This value, determined outdoor, is always very close to 1, whereas in indoor environments, it may be higher, also depending on the number of people present. In fact, mesophilic bacteria derive from the normal bacterial flora of humans and can therefore constitute the predominant population in confined environments, as already verified in other works [9]. Finally, the amplification index is fundamental to detect the accumulation and proliferation of microorganisms in ventilation systems or in buildings. AI describes global indoor aerial modification. Generally, there are no relations between indoor and outdoor fungal and bacterial counts. However, it is important to judge Indoor Air Quality (IAQ) not only by the measurement of single parameters, but also using the global value of microbial contamination. High AI values may only be indicative of IAQ deterioration when caused by high total fungal and bacterial counts [10,22]. Moreover, AI is essential to determine the environmental impact of outdoor work activities with potential spread of pathogenic and non-pathogenic microorganisms; it must be calculated by referring to the contamination detected in a control point that must provide the background value [18]. AI and the other two indices could provide criteria to evaluate indoor environmental quality more uniformly [31]. In fact, these indices have different advantages: they consider several factors which contribute to the development of different types of microorganisms (determination of GIMC), they may reveal an inadequate functioning of the air conditioning system due to an excessive number of people in a building or insufficient ventilation (IMC determination), and they may suggest a malfunction of the air system due to the increase of microbial contamination in the ventilation ducts or in the building compared to the outside environment (AI measures). The utilization of these indices can enable the determination of threshold values as a function of the structures (buildings) analyzed. In previous research in other environments, such as university classrooms, a reference value of $GIMC/m^3 = 1000$ was connected to a correct maintenance of the aeraulic system [19]. In this work, the measured GIMC values were higher than 1000, with the exception of libraries that presented very low contamination. In particular, both offices and gyms had the highest values of pollution during the conditioning phase, corresponding to $GMIC/m^3$ of 1704.90 and 4606.40, respectively. GIMC exceeding 1000 does not necessarily indicate a health risk; however, it is appropriate to analyze the contamination levels through the calculation of IMC and AI. During air conditioning in the offices, IMC and AI showed that mesophilic bacteria underwent a real amplification due to few air changes or the accumulation of microorganisms in the air system. In fact, as reported by other authors, variations and fluctuations in indoor humidity and temperature have significant effects on microbial diffusion and growth [32]. For gyms, on the contrary, the GIMC values were higher in summer because the proportion of mesophilic bacteria decreased passing from the heating phase to the conditioning phase, but the portion of psychrophilic bacteria and mycetes increased consistently, as evidenced by the IMC.

It is important to note that the rise of the microbial load inside the gyms coincided with an increase of the microorganisms outdoors, as shown by the calculation of AI. Therefore, the presence of a greater microbial contamination during summer could depend on the incorrect functioning or poor controls of the ventilation systems, which resulted in the accumulation of microorganisms from the outside [33].

In conclusion, our research results show that the microbial contamination changes depending on the environment analyzed and highlight the easy applicability of the proposed indices. GIMC, IMC, and AI complete the information provided by the single classes of airborne microorganisms and, together, they allow the analysis of indoor air quality. On the basis of sufficient data, these indices classify environments in function of their microbiological contamination and identify guide values to be adopted for routine monitoring and for the implementation of containment and remediation measures.

Author Contributions: Conceptualization, P.G.; methodology, P.G.; formal analysis, P.G., M.R.; investigation, P.G., M.A., M.R.; funding acquisition, P.G., M.A.; writing—original draft preparation, P.G. and M.R.

References

1.   Anderson, A.M.; Weiss, N.; Raine, F.; Salkinoja-Salonen, M.S. Dust-borne bacteria in animal sheds, schools and children's day care centers. *J. Appl. Microbiol.* **1999**, *86*, 622–634. [CrossRef]

2. Bernasconi, C.; Rodolfi, M.; Picco, A.; Grisoli, P.; Dacarro, C.; Rembges, D. Pyrogenic activity of air to characterize bioaerosol exposure in public buildings: A pilot study. *Lett. Appl. Microbiol.* **2010**, *50*, 571–577. [CrossRef] [PubMed]

3. Maroni, M.; Seifert, B.; Lindvall, T. Indoor Air Quality: A comprensive reference book. In *Air Quality Monographs*; Elsevier: Amsterdam, The Netherlands, 1995; Volume 3, ISBN 9780444816429.

4. Testo unico in materia di tutela della salute e della sicurezza nei luoghi di lavoro. D.Lgs n. 81/2008. In *Gazzetta Ufficiale della Repubblica Italiana*; n. 101 (30 Aprile 2008); Ministero della Giustizia, Ufficio Pubblicazione Leggi e Decreti: Rome, Italy.

5. Brief, R.S.; Bernath, T. Indoor pollution: guidelines for prevention and control of microbiological respiratory hazards associated with air conditioning and ventilation systems. *Appl. Ind. Hyg.* **1988**, *3*, 5–10. [CrossRef]

6. Jensen, P.A.; Schafer, M.P. Sampling and characterization of bioaerosols. In *NIOSH Manual of Analytical Methods Niosh/DPSE*; NIOSH Publications: Washington, DC, USA, 1998; Volume 2, pp. 82–112.

7. Švajlenka, J.; Kozlovská, M.; Pošiváková, T. Assessment and biomonitoring indoor environment of buildings. *Int. J. Environ. Health Res.* **2017**, *27*, 427–439. [CrossRef] [PubMed]

8. Burge, P.S. Sick building syndrome. *Occup. Environ. Med.* **2004**, *61*, 185–190. [CrossRef] [PubMed]

9. Cabo Verde, S.; Almeida, S.M.; Matos, J.; Guerreiro, D.; Meneses, M.; Faria, T.; Botelho, D.; Santos, M.; Viegas, C. Microbiological assessment of indoor air quality at different hospital sites. *Res. Microb.* **2015**, *166*, 557–563. [CrossRef] [PubMed]

10. Dacarro, C.; Grisoli, P.; Del Frate, G.; Villani, S.; Grignani, E.; Cottica, D. Micro-organisms and dust exposure in an Italian grain mill. *J. Appl. Microbiol.* **2005**, *98*, 163–171. [CrossRef] [PubMed]

11. Prazmo, Z.; Krysinska-Traczyk, E.; Skorska, C.; Sitkowska, J.; Cholewa, G.; Dutkiewicz, J. Exposure to bioaerosols in a municipal sewage treatment plant. *Ann. Agric. Environ. Med.* **2003**, *10*, 241–248. [PubMed]

12. Mendell, M.J.; Lei-Gomez, Q.; Mirer, A.G.; Seppänen, O.; Brunner, G. Risk factors in heating, ventilating, and air-conditioning systems for occupant symptoms in US office buildings. *Ind. Air* **2008**, *18*, 301–316. [CrossRef] [PubMed]

13. Qian, J.; Hospodsky, D.; Yamamoto, N.; Nazaroff, W.W.; Peccia, J. Size-resolved emission rates of airborne bacteria and fungi in an occupied classroom. *Ind Air.* **2012**, *22*, 339–351. [CrossRef] [PubMed]

14. Wolff, C.H. Innate immunity and the pathogenicity of inhaled microbial particles. *Int. J. Biol. Sci.* **2011**, *7*, 261–268. [CrossRef] [PubMed]

15. World Health Organization. *Indoor Air Pollutants: Exposure and Health Effects*; Report on a WHO Meeting, EURO Reports and Studies no. 78; WHO Regional Office for Europe: Copenhagen, Denmark, 1983.

16. ACGIH. On-site investigation, pp. 1–8; Fungi, pp. 1–10; Bacteria, pp. 1–7. In *Guidelines for the Assessment of Bioaerosols in the Indoor Environment*; Committee on Bioaerosols, Ed.; American Conference of Governmental Industrial Hygienists: Cincinnati, OH, USA, 1989.

17. Traistaru, E.; Moldovan, R.C.; Menelaou, A.; Kakourou, P.; Georgescu, C. A comparative study on the quality of air in offices and homes. *J. Environ. Sci. Health A* **2013**, *48*, 1806–1814. [CrossRef] [PubMed]

18. Grisoli, P.; Rodolfi, M.; Villani, S.; Grignani, E.; Cottica, D.; Berri, A.; Picco, A.; Dacarro, C. Assessment of airborne microorganism contamination in an industrial area characterized by an open composting facility and a wastewater treatment plant. *Environ. Res.* **2009**, *109*, 135–142. [CrossRef] [PubMed]

19. Grisoli, P.; Rodolfi, M.; Chiara, T.; Zonta, L.A.; Dacarro, C. Evaluation of microbiological air quality and of microclimate in university classrooms. *Environ. Monit. Assess.* **2012**, *184*, 4171–4180.

20. Heildelberg, J.F.; Shahamat, M.; Levin, M.; Rahman, I.; Stelma, G.; Grim, C.; Colwell, R.R. Effect of aerosolization on culturability and viability of Gram-Negative bacteria. *Appl. Environ. Microbiol.* **1997**, *63*, 3585–3588.

21. Macher, J.M.; Chatigny, M.A.; Burge, H.A. Sampling airborne microorganisms and aeroallergens. In *Air Sampling Instruments for Evaluation of Atmospheric Contaminants*, 8th ed.; Cohen, B.S., Hering, S.V., Eds.; American Conference of Governmental Industrial Hygienists Inc.: Cincinnati, OH, USA, 1995; pp. 589–617.

22. Dacarro, C.; Grignani, E.; Lodola, L.; Grisoli, P.; Cottica, D. Proposta di indici microbiologici per la valutazione della qualità dell'aria degli edifici. *G. Ital. Med. Lav. Ergon.* **2000**, *XXII*, 229–235.

23. Gołofit-Szymczak, M.; Górny, R.L. Microbiological Air Quality in Office Buildings Equipped with Different Ventilation Systems. *Ind. Air* **2018**, *28*, 792–805. [CrossRef] [PubMed]

24. Jouzaitis, A.; Willeke, K.; Grinshpun, S.A.; Donnelly, J. Impaction onto a Glass Slide or Agar versus Impingement into a Liquid for the Collection and Recovery of Airborne Microorganisms. *Appl. Environ. Microbiol.* **1994** *60*, 861–870.

25. Stewart, S.L.; Grinshpun, S.A.; Willeke, K.; Terzieva, S.; Ulevicius, V.; Donnelly, J. Effect of Impact Stress on Microbial Recovery on an Agar Surface. *Appl. Environ. Microb.* **1995**, *61*, 1232–1239.

26. Walter, M.V.; Marthi, B.; Fieland, V.P.; Ganio, L.M. Effect of Aerosolization on Subsequent Bacterial Survival. *Appl. Environ. Microb.* **1990**, *56*, 3468–3472.

27. Flannigan, B. Indoor microbiological pollutants—Sources, species, characterization and evaluation. In *Chemical, Microbiological, Health and Comfort Aspects of Indoor Air Quality—State of the Art in SBS*; Knöppel, H., Wolkoff, P., Eds.; EDSC, EEC, EAEC: The Netherlands, 1992; pp. 73–98.

28. Miller, J.D. Fungi as contaminants in indoor air. *Atmos. Environ.* **1992**, *6A*, 2163–2172. [CrossRef]

29. Rao, C.Y.; Burge, H.A. Review of quantitative standards and guidelines for fungi in indoor air. *J. Air Waste Manag. Assoc.* **1996**, *46*, 899–908. [CrossRef] [PubMed]

30. Commission of European Communities. *Indoor Air Quality & Its Impact on Man*; Report n.12; Commission of European Communities: Luxembourg, 1993.

31. Sofuoglu, S.C.; Moschandreas, D.J. The link between symptoms of office building occupants and in-office air pollution: The Indoor Air Pollution Index. *Ind. Air* **2003**, *13*, 332–343. [CrossRef]

32. Flannigan, B.; Miller, J.D. Microbial growth in indoor environments. In *Microorganisms in Home and Indoor Work Environments: Diversity, Health Impacts, Investigation and Control*; Flannigan, B., Samson, R.A., Miller, J.D., Eds.; Taylor & Francis: Abingdon, UK, 2001; pp. 35–67.

33. Dacarro, C.; Picco, A.M.; Grisoli, P.; Rodolfi, M. Determination of aerial microbiological contamination in scholastic sports environments. *J. Appl. Microbiol.* **2003**, *95*, 904–912. [CrossRef] [PubMed]

4

BTEXS Concentrations and Exposure Assessment in a Fire Station

Wioletta Rogula-Kozłowska [1], Karolina Bralewska [1,*] and Izabela Jureczko [2]

[1] The Main School of Fire Service, Safety Engineering Institute, 01629 Warsaw, Poland; wrogula@sgsp.edu.pl
[2] Power Research & Testing Company ENERGOPOMIAR Ltd., 44100 Gliwice, Poland; bell_007@o2.pl
* Correspondence: kbralewska@sgsp.edu.pl.

Abstract: The aim of this study was to evaluate benzene, toluene, ethylbenzene, xylene, and styrene (BTEXS) concentrations in the changing room and garage in a fire station located in the Upper Silesian agglomeration (Poland), to compare them with the concentrations of the same compounds in the atmospheric air (outdoor background) and to assess the health exposure to BTEXS among firefighters and office workers in this unit. BTEXS samples were collected during the winter of 2018 in parallel in the garage, in the changing room, and outside, using sorption tubes filled with activated carbon. The average total BTEXS concentrations in the changing room and garage were over six times higher than those in the atmospheric air in the vicinity of the fire station. At each sampling site, toluene and benzene had the highest concentrations. According to the diagnostic indicators, the combustion of various materials and fuels was the source of BTEXS inside, while outside, the sources were the combustion of fuels and industrial activity. The carcinogenic risk related to benzene inhalation by the firefighters and office employees in the monitored unit exceeded the acceptable risk level value of 7.8×10^{-6} per 1 µg/m³ by more than 20 times.

Keywords: BTEXS; health exposure; occupational risk; markers of exposure; air quality

1. Introduction

Firefighters are often exposed to very high concentrations of various products of combustion and pyrolysis, including substances in a gaseous phase adsorbed on ambient particulate matter (PM-bound). The toxic substances found in fire smoke are most often polycyclic aromatic hydrocarbons (PAHs), volatile organic compounds (VOC) (including BTEXS), hydrogen cyanide (HCN), and several other organic and inorganic compounds [1–3]. Exposure to these compounds has been linked to a higher risk of specific cancers and cardiovascular diseases and thus acute and chronic effects that result in increased fire fighter mortality and morbidity [2,4–6]. The International Agency for Research on Cancer (IARC) assigns the profession of firefighter to group 2B, meaning "possibly carcinogenic to humans" [7]. Although firefighters use personal protective equipment during rescue and firefighting operations, such as gloves, coats, flash hoods, and breathing apparatus, this equipment can also serve as a secondary source of exposure.

Some non- or semi-volatile compounds released during fires may settle and/or condense on protective equipment and exposed skin, leaving a greasy residue or film; then (e.g., when removing the equipment), these compounds may penetrate the body directly through the skin and eyes or through inhalation. Volatile organic compounds, including BTEXS, usually remain in the vapor phase. However, some of them can partition into a solid phase and condense onto the skin where they become available for deposition into the human body [8]. In addition, gaseous substances, especially VOCs such as HCN and the most volatile PAH, can penetrate into the interior space of the turnout gear [9] and then undergo the phenomenon of off-gassing in fire truck cabins and storage areas, such as changing rooms and garages [2,10,11]. In this way, firefighters can be exposed to these substances not only during firefighting operations but also during their return from action or while resting in their fire stations. Leaving clothes and equipment in the changing room or garage without first decontaminating them may facilitate the accumulation of toxic substances and transfer them to other fire station rooms, such as bedrooms or offices [12–14]. Accordingly, not only firefighters extinguishing fires but also dispatchers, commanders, and office workers working in fire stations can be exposed to toxic combustion products.

Compounds of the BTEXS group are considered to be indicators of human exposure to volatile organic compounds [15]. Measuring the total concentration of BTEXS and the individual compounds from this group (as the main pollutants released during fires) is necessary to assess the threats to firefighter health, as well as the work environment. The most dangerous compounds from the BTEXS group are benzene and ethylbenzene, which are classified by the International Agency for Research on Cancer (IARC) as carcinogenic to humans (Group 1). Exposure to these substances is linked to an increased risk of leukemia and hematopoietic cancers [7,16–18]. Toluene and xylene are non-carcinogenic, but they may produce adverse reproductive effects, especially when exposure is chronic at low to high concentrations [18,19].

While research on the concentrations of pollutants released during fires and assessments of the health risks to firefighters have been the subject of many global studies since the 1980s [2], the presence of combustion products in fire stations is a less well-trodden topic. There is little research in this area. Studies that are available focus on PAH and VOC concentrations in structural firefighting ensembles [10] as well as in turnout gear [3,20] at fire stations in Brisbane (Australia), Philadelphia, and Illinois (United States); and the concentrations of polybrominated diphenyl ethers (PBDEs), polycyclic aromatic hydrocarbons (PAHs), and polychlorinated biphenyls (PCBs) in dust samples collected by a vacuum cleaner at twenty fire stations in California [13]. These studies were among the first to indicate the problem of high concentrations of toxic combustion products in fire stations.

The investigations described in this paper are the next stage of the multi-site study regarding the concentration of combustion products in fire station rooms. In the first stage of study, the concentrations of PAHs in garages and changing rooms at two selected fire stations belonging to the Polish National Fire Service were analyzed and compared to the concentrations of PAHs in the atmospheric air outside these units [14]. The goal of this work was to determine the ambient concentrations of BTEXS (benzene, toluene, ethylbenzene, xylene, and styrene) in a selected fire station in Poland, compare them with the outdoor background concentrations of BTEXS, and assess the health risks of the exposure to BTEXS (occupational carcinogenic and non-carcinogenic risks) between two groups of employees (firefighters and office workers) in the state fire service unit.

2. Materials and Methods

2.1. Sampling Site

The municipal headquarters of the state fire service, where the research was conducted, is located in the center of the Silesian voivodeship (50°16′1.401″ N, 18°51′40.607″ E). The building is located among old (100–150 years old) and low-rise (several-story) buildings. The apartments in the vicinity of the fire brigade building are mainly heated with hard coal and are largely inhabited by a low-income population. Therefore, a large number of people in the area burn poor quality fuel in their furnaces and produce waste. The property is located 50 m from voivodeship road no. 925 and about one kilometer from the A4 highway. The fire brigade building is a three-story building heated by a solid fuel stove that uses pellet fuel. The building was thoroughly rebuilt in the 1960s. The first floor of the building consists of a garage, changing rooms, a gym, and a workshop. The second floor contains commanders' offices, a common room, and social rooms for firefighters. On the third floor, there are a dispatch center and office rooms. The building also has two kitchens (with gas stoves) where the firefighters prepare meals. Each firefighter participating in fire-fighting operations has special clothing, gloves, and a balaclava, which offer external protection during all rescue and fire-fighting activities, as well as during exercises. Thus, this clothing is particularly exposed to the effects and absorption of all types of chemical compound. After a 24-h shift, the firefighter stores his or her clothes in lockers located in the changing room. Due to the relatively high cost of these clothes, firefighters usually have only one or two sets of clothes, and these clothes are washed no more than once a month.

In the fire brigade building, the highest BTEXS emissions likely occur when parking fire vehicles in the garages due to the lack of engine exhaust hoods. An additional internal source of BTEXS could also be the building and its finishing materials (paints, wallpaper, and floor coverings), as well as the varnishes, glues, and solvents used during maintenance work, which also release compounds from the VOC group into the environment.

The BTEXS concentrations were measured in November 2018. BTEXS samples were taken on five consecutive business days (Monday–Friday) simultaneously:

- In the garage: five available places for the parking of rescue and fire-fighting vehicles and special vehicles; the area is about 550 m^2 and has no air cleaning system or natural ventilation; the aspirator was placed on a shelf at a height of about 2 meters from the ground, in the middle of the room, in an area distant from parking cars.
- In the changing room: an area about 30 m^2, with no air purification system or natural ventilation; this area is used to store the firefighters' special clothes and personal equipment; the aspirator was placed on a shelf about 2 m above the ground in the center of the room.
- Outside the fire station: the aspirator was placed in a specially prepared casing on a platform (scaffolding) at a height of about 3.5 m from the ground and approximately 3 m from the building, with the building side shielded from the direct impact of the emissions from parking cars. The casing only covered the aspirator, which had no effect on sampling. It was intended to protect the device against rain. Tubes to collect BTEXS, attached to the device with a silicone tube, were placed outside the casing.

Three GilAir Plus aspirators from the Gilian company were used to conduct this research. These devices are designed to sample air, dust, and gaseous pollutants. The aspirator has the function of regulating the air flow in the range of 20 to 5000 mL/min. The flow stabilization system allows one to maintain a constant flow with an accuracy of 5% in a temperature range of 0–40 °C. During the tests, especially at night, the ambient air temperature dropped a few degrees below zero. Therefore, adequate protection (the casing with the heater) of the aspirator used for sampling outside was provided. Anasorb CSC Lot 2000 sorption tubes were used to collect compounds from the BTEXS group. These tubes were filled with sorbent-activated carbon from coconut shells and were intended to take up a wide spectrum of organic compounds [21,22].

After breaking the glass protection on both sides, the measuring tube was placed in a silicone tube connected to the aspirator. The manufacturer of the sorption tubes requires an air flow of 200 mL/min. Considering the capacity of the sorbent used and the expected BTEXS concentrations (previous tests), it was assumed that the BTEXS intake per tube can last a maximum of 4 h. After this time, the tube was changed. In total, 6 tubes were used in one measurement place for one day, which provided 24 h of measurement per day. In total, during a continuous period of five days of measurements, the samples were collected in 90 tubes. After a measurement, each tube was sealed with special plugs on both sides, wrapped with aluminum foil, and stored in a refrigerator at about 2 °C until analysis.

2.2. BTEXS Analysis

Details of the BTEXS analysis can be found in [22,23]. Briefly, to isolate the BTEXS adsorbed on the activated carbon, carbon disulphide extraction (Sigma Aldrich, St. Louis, MO, USA; CS$_2$ Chromasolv for HPLC; purity > 99.9%) was used. Sorbent from the 6 tubes used to collect BTEXS during one day was poured into a glass jar, and then 3 mL of solvent (CS$_2$) was added and extracted in an ultrasonic bath for 10 min. The extract thus prepared was analyzed using a gas ionization detector (GC-FID) Trace 1300 GC from Thermo Scientific (Waltham, MA, USA). The separation was carried out using a

Supelco Wax 10 capillary column (60 m $\times$ 0.53 $\times$ 10^{-3} m $\times$ 1 $\times$ 10^{-6} m). The carrier gas flow (helium) was 1 mL/min using the split function (split ratio 1:10), and the gas flow in the FID detector was as follows: air—350 mL/min; hydrogen—35 mL/min; and make-up (nitrogen)—40 mL/min. The detector temperature was 310 °C, and the dispenser temperature was 150 °C. In total, 1 μL of the sample was injected in each case for the GC-FID analysis. The determination of each sample was repeated twice. The difference in readings between the two measurements did not exceed 5%. The following temperature program was used in the chromatograph oven: 50 °C maintained for 10 min followed by a temperature increase at a rate of 30 °C/min up to 170 °C; the temperature was then maintained at 170 °C for one minute. The analysis time was approximately 15 min. The concentrations of individual BTEXS species (benzene, toluene, ethylbenzene, m-xylene, p-xylene, o-xylene, and styrene) were quantified based on an external calibration standard mixture (Sigma Aldrich; MISA Group 17 Non-Halogen Organic Mix). A five-point calibration using standards between 1 and 25 μg/mL was performed for quantifying the BTEXS species in collected samples. These standard solutions were used to produce calibration curves. The correlation coefficients for standards curves were 0.99. The retention times were 5.234, 10.124, 14.754, 14.854, 14.982, 16.350, and 17.463 min, for benzene, toluene, ethylbenzene, m-xylene, p-xylene, o-xylene, and styrene, respectively. The limit of quantification for the method was set to 0.05 μg/sample. The repeatability of the method was 5%; the expanded uncertainty was 25% (k = 2). The recovery values of the BTEXS constituents ranged from 90 ± 4% to 98 ± 3% (n = 10).

2.3. Health Risk Assessment

The health exposure assessment related to the inhalation of BTEXS compounds (excluding styrene) among the firefighters and office workers at a fire station in Poland was developed based on the methodology developed by the United State Environmental Protection Agency (US EPA) [24]. This assessment of the health exposure includes an assessment of the occupational carcinogenic risk (CR) associated with benzene inhalation, which was calculated according to Equations (1) and (2) [25], and an assessment of the occupational non-carcinogenic risk in terms of the threshold mechanisms of the toxic effects produced by compounds from the BTEXS group (expressed as hazard quotients, HQ), which were calculated according to Equations (1) and (3) [17].

$$EC = (CA \times ET \times EF \times ED)/AT \qquad (1)$$

$$CR = IUR \times EC \qquad (2)$$

$$HQ = EC/(RfC \times 1000 \ \mu g/mg) \qquad (3)$$

where CA is the chemical concentration (μg/m^3). The other variables used in Equations (1)–(3) are explained in Table 1.

In the calculations, we assumed that inhalation constitutes 50% of all intake [26]. The risk assessment was carried out for the period of professional activity of the two groups of firefighters: (1) active (i.e., those involved in firefighting) and (2) office workers. Exposure duration (ED) and exposure frequency (EF) were assessed on the basis of interviews and observations. It was assumed that firefighters participating in rescue and firefighting operations perform 24-h shifts three times a week, while office workers work eight hours a day, five days a week. In 2018, firefighters from the analyzed unit responded to 643 fires (33 times in November 2018) [27]. Moreover, we assumed that both firefighters and office employees spend 50% of the work shift in conditions such as in the garage, while the remaining 50% of time is spent in conditions such as in the changing room. During the day, firefighters use, among others, the workshop and gym, which are located next to the changing room and garage. Office rooms and a dispatch room are also located next to or directly above these rooms. Therefore, we have adopted a simplification that in these rooms, BTEXS concentrations are comparable to those in the changing room and garage. The values of the individual parameters used to calculate the carcinogenic and non-carcinogenic risks are presented in Table 1.

Table 1. Values of the factors used for the health risk assessment.

Variable	Definition	Firefighters	Office Workers	Reference
EC	Exposure concentration ($\mu g/m^3$)	Average from the changing room and garage concentrations		Sampling
ET	Exposure time (hours/day)	24	8	Interview
EF	Exposure frequency (days/year)	104	250	Interview
ED	Exposure duration (years)	20	20	Interview
AT	Averaging time = average time in hours per exposure period (h)	175,200	175,200	[24]
IUR	Inhalation Unit Cancer Risk	7.8×10^{-6} per 1 $\mu g/m^3$ *		[26]
RfC	Reference concentration (RfC) (mg/m^3)	3×10^{-2} for benzene; 5 for toluene; 1 for ethylbenzene; 1×10^{-1} for xylene		[17]

* The U.S. EPA provides two values of Inhalation Unit Cancer Risk (IURs) for benzene: 2.2×10^{-6} to 7.8×10^{-6} per 1 $\mu g/m^3$. The higher IUR was used to obtain the maximum estimate of cancer risk from benzene exposure.

3. Results and Discussion

3.1. BTEXS Concentrations

The mean concentrations and standard deviations for the BTEXS compounds are presented in Table 2. The highest total BTEXS concentration averaged over the whole measurement period ($\sum$BTEXS) was recorded in the changing room (925.2 $\mu g/m^3$). However, this value was only slightly higher than that in the garage (893.1 $\mu g/m^3$). The average $\sum$BTEXS in both the changing room and the garage was over six times higher than the $\sum$BTEXS outside the unit (139.6 $\mu g/m^3$). Most of the BTEXS compounds, except for xylenes, had higher concentrations in the changing room than in the garage. The concentration of individual BTEXSs in the changing room and garage on different measurement days varied notably. This is demonstrated by the high standard deviations of BTEXSs, especially toluene and ethylbenzene. However, every measurement day, the concentrations of benzene, toluene, ethylbenzene, xylenes, and styrene were many times higher both in the changing room and garage than in the atmospheric air (outdoors; the differences were statistically significant, $p < 0.05$; Table 2). Therefore, it can be assumed that the fluctuations in BTEXS concentrations may be related to the number and/or type of fires that the firefighters had to deal with on particular days. In November 2018, firefighters from the analyzed unit participated in extinguishing 33 fires and fighting 60 local threats [27].

In addition, the large differences between the BTEXS concentrations in the fire station rooms and in the surroundings suggest that the main sources of BTEXSs in the changing room and garage were the combustion products settled on uniforms, personal protective equipment, and the equipment used during firefighting. The standard deviations for the BTEXS concentrations in the garage were smaller than those in the changing room. Outside, the proportions of standard deviations relative to the average concentrations were very high—higher than those for the measurements inside. The BTEXS concentrations outside the fire station, although many times lower than those inside, were several times higher than, for example, the BTEXS concentrations observed in other urban areas, such as Tri-City, Tczew (Poland) [28], Hamburg (Germany) [29], or Pamplona (Spain) [30]. It is likely that this is related to the fact that the ambient air in the southern part of Poland (Upper Silesia) is more polluted than the air in other parts of the country. This pollution is caused by mining activities and intensified heating processes, especially since the measurements were conducted in winter [31].

Table 2. Mean concentrations of benzene, toluene, ethylbenzene, xylene, and styrene (BTEXS) ($\mu g/m^3$) in the changing room, in the garage, and outside the fire station in Poland.

	Changing Room		Garage		Outdoor	
	Mean	Standard Deviation	Mean	Standard Deviation	Mean	Standard Deviation
Benzene	201.3 *	48.9	196.6 *	92.0	37.8	16.1
Toluene	538.8 *	178.1	524.7 *	105.3	86.5	35.3
Ethylbenzene	71.6 *	45.8	58.9 *	16.2	3.1	1.5
m,p-Xylene	41.0 *	30.7	55.4 *	20.4	6.3	8.3
o-Xylene	17.2 *	11.1	20.3 *	10.5	4.4	2.9
Styrene	55.2 *	24.5	37.3 *	10.5	1.5	0.4
$\sum$BTEXS	925.1	339.1	893.1	254.9	139.6	64.5

* Indoor and outdoor concentrations of BTEXS in the case of changing room or garage are statistically significantly different (according to the Mann–Whitney U test; $p < 0.05$).

The BTEXS profiles (i.e., the percentages of individual BTEXS compounds in the total BTEXS concentration) in the changing room and garage were very similar. However, they differed from the BTEXS profile for the atmospheric air outside the fire station (Figure 1). In both rooms and outside, toluene had the highest share in $\sum$BTEXS (58% in the changing room, 59% in the garage, and 62% outdoors). In the whole measuring period, toluene concentrations were in the range 265–703 $\mu g/m^3$ in the changing room, 425–692 $\mu g/m^3$ in the garage, and 51–131 $\mu g/m^3$ in the atmospheric air. Benzene had the second largest share in $\sum$BTEXS (an average of 22% both in the changing room and the garage and 27% in the atmospheric air). It should be noted that toluene and benzene concentrations are reduced through their reactions with OH radicals, with the rate constant of toluene being approximately five times larger than that of benzene [32]. This explains the differences in the concentrations of these compounds. The average percentage of ethylbenzene and styrene among the BTEXSs inside the garage and the changing room was about four times higher than that outside. This suggests an internal source of these compounds in the analyzed rooms.

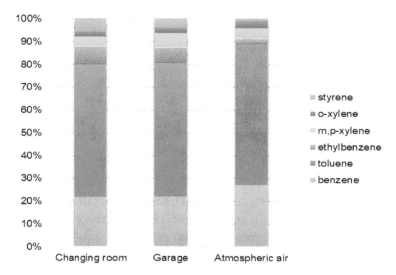

Figure 1. Mean profiles of the BTEXSs in the indoor (changing room and garage) and outdoor air at the selected fire station in Poland.

3.2. Origin of BTEXs Inside and Outside the Fire Station

Diagnostic indicators, which are the ratios of individual BTEXS concentrations, offer a preliminary assessment of the origin of individual compounds from the BTEXS group present in the indoor and outdoor air of the state fire service unit (Table 3). The ratio of toluene/benzene = 2.7 indicates local emissions of toluene and benzene in the changing room and garage, likely due to the combustion of liquid fuels [32–34]. However, the low values of the indicators m,p-xylene/benzene and o-xylene/benzene (<0.5) in comparison to [31,33,34] testify to other BTEXS sources in these rooms. Low m,p-xylene/benzene and o-xylene/benzene values indicate greater photochemical degradation and, therefore, suggest that a sampling site is being influenced by emissions that originated some distance away [32]. It can, therefore, be assumed that the main sources of BTEXS in the fire station rooms are the gases released during fires, which settled on the uniforms and equipment. For the outdoor measurements, the indicators toluene/benzene = 2.3 (outdoor) and m,p-xylene/ethylbenzene = 2 indicate the combustion of liquid fuels as a BTEXS source in the atmospheric air [32,34]. The obtained indicators are two times lower than other results in this field, such as those in Sarnia (Canada), where the impact of traffic emissions on BTEXS concentrations was clearly demonstrated [32]. High concentrations of benzene alongside relatively low indicators of m,p-xylene/benzene and o-xylene/benzene may also indicate additional industrial sources of BTEXS outside the fire station. An analysis of the environment of the sampling site indicates that these sources are likely related to the coal mining and storage processes taking place at a distance of about 5 km from the fire station, as well as the combustion of fuels in home boiler rooms [34,35]. It is difficult to clearly determine which of the sources listed, both inside and outside, has the greatest impact on shaping BTEXS concentrations. More detailed data could provide Pearson's correlations between the BTEXS concentrations, but this would require more measurements [32,36].

Table 3. Diagnostic indicators for the three sampling locations.

Indicator	Changing Room	Garage	Outdoors
Toluene/Benzene	2.7	2.7	2.3
Ethylbenzene/Benzene	0.4	0.3	0.1
m,p-Xylene/Benzene	0.2	0.3	0.2
o-Xylene/Benzene	0.1	0.1	0.1
m,p-Xylene/Ethylbenzene	0.6	0.9	2.0

Information on the origins of the BTEXS compounds is also provided by the I/O ratios calculated for the entire measurement period, presented in Figure 2. The average I/O ratios were in the range of 6.2–53.1. The values of the I/O ratios confirm that the BTEXS concentrations mainly originate from internal sources. The higher I/O ratios in the changing room than in the garage also suggest that the source of BTEXSs here could be residue on the uniforms, helmets, and gloves used during firefighting but also contaminated furniture (e.g., wardrobes for clothes and equipment shelves) [37]. The lower average I/O ratios obtained for benzene and toluene relative to the rest of the BTEXS compounds mean that, in addition to internal sources, the concentrations of these compounds are also likely affected by the infiltration of outdoor air, especially considering the fact that the changing room is located next to the garage, where the door is often opened. Ethylbenzene, styrene, and xylenes, whose I/O ratios are the highest and have the highest fluctuations relative to other BTEXSs, likely come from fires. The above observations are confirmed by the literature and other research conducted in this field around the world. Benzene is the second most frequently identified compound in over 80% of the fires tested, and the next most frequently occurring compounds during fires are toluene, xylenes, and ethylbenzene [3,10,38–40]. Furthermore, the presence of styrene may result from the thermal decomposition and combustion of polystyrene (plastics) [40]. The concentrations of individual compounds from the BTEXS group depend on the type of material burned, the phase of the fire [10], the type of fire (flame vs. flameless), the location of the fire, the meteorological conditions, and the distance from the fire [38,41]. The conducted research provides the basis for future research, which should also include an analysis of the ventilation solutions in fire stations and a study of other factors, such as PM concentration and the chemical composition of particulate matter, which would facilitate a more accurate assessment of the impact of the environment on BTEXS concentrations.

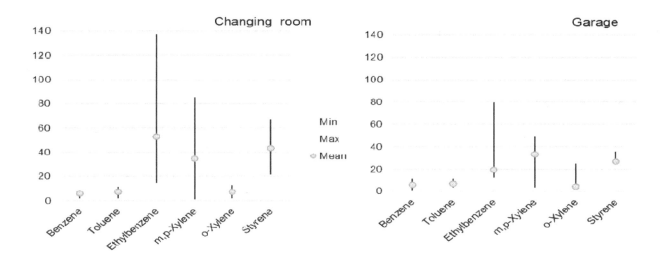

Figure 2. Ranges and average values of the I/O ratios in the two rooms of the fire station calculated on the basis of the average set of five 24-h BTEXS concentrations.

It is difficult to compare the obtained results with the results presented in the literature, where different methods were used, or sampling was conducted at different stages of the fire. Nevertheless, Table 4 summarizes several examples of such results from other researchers. The BTEXS concentrations in a fire station in Upper Silesia were several times higher than the BTEXS concentrations collected by Kirk and his team from the outer layer of the structural firefighting ensembles at various stages of a fire [10] and then from decontaminated and non-decontaminated turnout gear during the pre-fire and post-decon periods [3]. In addition, the toluene and ethylbenzene concentrations measured by these authors in both the garage and the changing room were higher than the concentrations of these compounds in the post-fire phase in Fent et al.'s study [3]. The BTEXS concentrations in the changing room and garage in the analyzed unit were also several dozen times higher than those in selected nursery schools in Poland [36] and Turkey [42], and the atmospheric air of urbanized areas in Poland, Germany, and Spain [28–30]. However, they were lower than the concentrations in the atmospheric air in the vicinity of waste dumps or at oil distribution stations [16,17].

Table 4. Concentration ranges of the BTEXS compounds ($\mu g/m^3$) measured at different locations.

References		Benzene	Toluene	Ethylbenzene	m,p-Xylene	o-Xylene	Styrene
Vehicle (engine) fire smoke—start-up of fire/overhaul [39]		5200/11,000	1400/3800	150/410	-	-	830/1600
Vehicle (cabin) fire smoke—start-up of fire/overhaul [39]		60,000/380	10,000/950	1400/120	-	-	14,000/450
Combustion of forest fuel—flaming phase [38]		93	45	18	21	-	
Outer layer of firefighting ensembles [10]	Pre-exposure	0.6–4.4	4.3–4.9	1.1–2.1	3.0–7.3		2.1–3.5
	Post exposure	13.0–88.0	38–80	1.7–15.0	7.7–20.0		41.0–88.0
	Post-laundering	0.4–0.7	0.3–18.0	0.9–2.4	3.6–7.9		1.3–3.9
Firefighter's personal protective equipment [3]	Decontaminated turnout gear during pre-fire/post-fire/ post-decon	0.5/250/9	0.5/150/5	0.5/20/0.5	0.5/15/1		0.5/400/9
	Non-decontaminated turnout gear during pre-fire/post-fire/ simultaneous with the post-decon periods	0.5/250/20	0.5/100/11	0.5/20/1	0.5/15/3		0.5/500/50

Table 4. *Cont.*

References	Benzene	Toluene	Ethylbenzene	m,p-Xylene	o-Xylene	Styrene
Stations of an oil distribution company (Iran) [17]	1847.00	3570.00	758.00	560.00		-
Landfill ambient air (Turkey) [16]	5.6–3137.8	23.4–10234.4	4.9–3717.1	7.9–7464.3		-
Nursery school (Turkey) [42]	1.60	26.20	0.70	1.10	0.81	-
Nursery school—outdoors (Turkey) [42]	1.23	6.11	-	-	-	-
Nursery school (Gliwice) [36]	1.37	1.19	2.11	0.72	3.31	0.44
Nursery school—outdoors (Gliwice) [36]	1.24	0.76	0.22	0.32	0.14	0.21
Gdańsk-Gdynia-Sopot (Poland)—tri-city urban area [28]	4–6	6–12	2–6	4–14	1–5	-
Hamburg (Germany)—urban area [29]	1.1–1.6	4.5–4.9	-	1.2–1.8	-	-
Pamplona (Spain)—urban area [30]	1.4–5.6	5.2–24.1	0.75–3.6	1.2–5.0	0.98–4.7	-

3.3. Assessment of Occupational Carcinogenic and Non-Carcinogenic Risks Associated with Exposure to BTEXS

The carcinogenic risk associated with the inhalation of benzene was calculated as 2.21×10^{-4} for firefighters participating in firefighting activities and 1.77×10^{-4} for office employees of the fire station (Figure 3). Both values are above acceptable cancer risk levels according to the Inhalation Unit Cancer Risk (IUR) ($>7.6 \times 10^{-6}$) [26,43,44]. The differences in risk values result from the different durations of working shifts between the individual exposure groups. Firefighters participating in rescue and firefighting operations perform 24-h shifts three times a week, while office workers work eight hours a day, five days a week. In the carcinogenic risk assessment, only benzene concentrations were used, while the concentrations of other pollutants, such as PAH or PM-bound substances, were not taken into account. In addition, the concentrations that prevail during fires were not taken into account. Firefighters use breathing apparatus and other personal protective equipment when extinguishing fires, but there are situations when they take that equipment off, such as during exterior operations (e.g., pump operations), immediately after extinguishing fires, when collecting equipment, or in fire truck cabins during their return from action. It can be assumed that then BTEXS concentrations are higher than in the garage or changing room [3]. Therefore, the risk may be even higher than calculated. Studies also show that the health exposure associated to hazardous combustion products does not only apply to firefighters extinguishing fires but also to dispatchers, commanders, and secretaries (i.e., people whose work rooms are often located near garages and changing rooms).

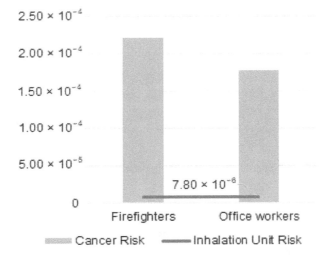

Figure 3. Cancer risk assessment results for exposure to benzene for various groups of fire service workers.

The occupational non-carcinogenic risk (adverse health effects)—expressed by the hazard quotient (HQ)—associated with exposure to compounds, for the BTEXS group of firefighters involved in firefighting and fire station office employees, was in the range of 0.01–0.76 (Table 5). This indicates an acceptable risk for non-carcinogenic effects in each scenario considered. However, as is the case for carcinogenic risk, risk modeling does not include the concentrations of other pollutants, such as PAH, which may also occur in fire station rooms [14].

Table 5. Non-cancer risk assessment results for exposure to BTEXS for various groups of fire service workers.

BTEXS	Firefighters	Office Workers
Benzene	0.94	0.76
Toluene	0.02	0.01
Ethylbenzene	0.01	0.01
Xylene	0.10	0.08

The carcinogenic risk calculated for firefighters and office workers was about 10 times higher than the risk calculated for fuel workers and cashiers at gasoline stations in Thailand [24] but about two times lower for firefighters and three times lower for office workers than the average lifetime cancer risk calculated for petroleum product distributors working at stations belonging to an Iranian company [17]. The non-carcinogenic risk for fuel workers (HQ = 0.80 for benzene) is lower than that for firefighters and higher than that for office workers, while that for cashiers at the gasoline station (HQ = 0.01 for benzene) is many times lower than that for firefighters and office workers [25]. Two-times higher non-carcinogenic risk values were recorded for petroleum product distributors in Iran [17] than for firefighters. The above comparisons are for reference only due to the different periods of exposure and averaging.

4. Conclusions

The concentrations of individual compounds from the BTEXS in the changing room and garage are several to several dozen times higher than the concentrations of these substances in the atmospheric air outside the fire station. Both firefighters and office workers staying under measured conditions are at risk of carcinogenic exposure that exceeds an acceptable level. Among the entire BTEXS group, toluene and benzene had the highest concentrations. According to the diagnostic indicators, the combustion of various materials and fuels was the source of BTEXS inside, while the combustion of fuels and industrial activity was the source of those outside. This research provides the following conclusions:

- Firefighters are exposed to combustion products not only during fires but also during rest between activities because they stay in rooms that are heavily contaminated with combustion products released during the off-gassing of stored clothes and equipment at the fire station.
- Although volatile compounds evaporate quickly, some of them are off-gassed during the storage of equipment. Therefore, the clothing and equipment used during fire extinguishing should be systematically decontaminated.
- The level of health risk (carcinogenic and non-carcinogenic) for office workers indicates that office rooms and dispatch rooms should be located as far as possible from the rooms in which the equipment is stored, i.e., far from changing rooms and garages. Moreover, in these rooms, efficient ventilation should be ensured.
- The obtained results demonstrate the need for more extensive research aimed at pollution control in various fire station rooms, such as offices and bedrooms, including measurements of the concentrations of gaseous pollutants, particulate matter, and its components, including toxic metals and polycyclic aromatic hydrocarbons.

Author Contributions: Conceptualization, W.R.-K. and K.B.; methodology, W.R.-K.; software, K.B.; validation, K.B., W.R.-K.; formal analysis, W.R.-K.; investigation, I.J.; resources, I.J.; data curation, K.B.; writing—original draft preparation, K.B.; writing—review and editing, W.R.-K.; visualization, K.B.; supervision, W.R.-K.; project administration, K.B. All authors have read and agreed to the published version of the manuscript.

Acknowledgments: The authors would like to acknowledge the Ministry of Science and Higher Education for their financial support as part of the statutory works.

References

1. Baris, D.; Garrity, T.J.; Telles, J.L.; Heineman, E.F.; Olshan, A.; Zahm, S.H. Cohort mortality study of Philadelphia firefighters. *Am. J. Ind. Med.* **2001**, *39*, 463–476. [CrossRef]

2. Tsai, R.J.; Luckhaupt, S.E.; Schumacher, P.; Cress, R.D.; Deapen, D.M.; Calvert, G.M. Risk of cancer among firefighters in California, 1988–2007. *Am. J. Ind. Med.* **2015**, *58*, 715–729. [CrossRef] [PubMed]

3. Fent, K.W.; Alexander, B.; Roberts, J.; Robertson, S.; Toennis, C.; Sammons, D.; Bertke, S.; Kerber, S.; Smith, D.; Horn, G. Contamination of firefighter personal protective equipment and skin and the effectiveness of decontamination procedures. *J. Occup. Environ. Hyg.* **2017**, *14*, 801–814. [CrossRef] [PubMed]

4. Brand-Rauf, P.W.; Fallon, L.F.; Tarantini, T.; Idema, C.; Zndrews, L. Health hazards of firefighters: Exposure assessment. *Br. J. Ind. Med.* **1988**, *45*, 606–612. [CrossRef]

5. Fent, K.W.; Evans, D.E. Assessing the risk to firefighters from chemical vapors and gases during vehicle fire suppression. *J. Environ. Monitor.* **2011**, *13*, 536–543. [CrossRef] [PubMed]

6. Faraji, A.; Nabibidhendi, G.; Pardakhti, A. Risk assessment of exposure to released BTEX in district 12 of Tehran municipality for employees or shopkeepers and gas station customers. *Pollution* **2017**, *3*, 407–415. [CrossRef]

7. Integrated Risk Information System (IRIS). Chemical Assessment Summary: Benzene. Available online: cfpub.epa.gov/ncea/iris/iris_documents/documents/subst/0276_summary.pdf (accessed on 26 February 2020).

8. Thrall, K.D.; Poet, T.S.; Corley, R.; Tanojo, H.; Edwards, J.A.; Weitz, K.K.; Hui, X.; Maibach, H.I.; Wester, R.C. A real-time in-vivo method for studying the percutaneous absorption of volatile chemicals. *Int. J. Occup. Environ. Health* **2000**, *6*, 96–103. [CrossRef]

9. Wingfors, H.; Nyholm, J.R.; Magnusson, R.; Wijkmark, C.H. Impact of fire suit ensembles on firefighter PAH exposures as assessed by skin deposition and urinary biomarkers. *Ann. Work Expo. Health* **2018**, *62*, 221–231. [CrossRef]

10. Kirk, K.M.; Logan, M.B. Structural firefighting ensembles—Accumulation and off-gassing of combustion products. *J. Occup. Environ. Hyg.* **2015**, *12*, 376–383. [CrossRef]

11. Fent, K.W.; Toennis, C.; Sammons, D.; Robertson, S.; Bertke, S.; Calafat, A.M.; Pleil, J.D.; Wallace, M.A.G.; Kerber, S.; Smith, D.; et al. Firefighters' absorption of PAHs and VOCs during controlled residential fires by job assignment and fire attack tactic. *J. Expo. Sci. Environ. Epidemiol.* **2020**, *30*, 338–349. [CrossRef]

12. Brown, F.R.; Whitehead, T.P.; Park, J.S.; Metayer, C.; Petreas, M.X. Levels of non-polybrominated diphenyl ether brominated flame retardants in residential house dust samples and fire station dust samples in California. *Environ. Res.* **2014**, *135*, 9–14. [CrossRef] [PubMed]

13. Shen, B.; Whitehead, T.P.; McNeel, S.; Brown, F.R.; Dhaliwal, J.; Das, R.; Israel, L.; Park, J.S.; Petreas, M. High levels of polybrominated diphenyl ethers in vacuum cleaner dust from California fire stations. *Environ. Sci. Technol.* **2015**, *49*, 4988–4994. [CrossRef] [PubMed]

14. Rogula-Kozłowska, W.; Majder, M.; Jureczko, I.; Ciuka-Witrylak, M.; Łukaszek-Chmielewska, A. Polycyclic aromatic hydrocarbons in the firefighter workplace: The results from the first in Poland short-term measuring campaign. *E3S Web Conf.* **2018**, *45*, 00075. [CrossRef]

15. Marčiulaitienė, E.; Šerevičienė, V.; Baltrėnas, P.; Baltrėnaitė, E. The characteristics of BTEX concentration in various types of environment in the Baltic Sea Region, Lithuania. *Environ. Sci. Pollut. Res.* **2017**, *24*, 4162–4173. [CrossRef]

16. Durmusoglu, E.; Taspinar, F.; Karademir, A. Health risk assessment of BTEX emissions in the landfill environment. *J. Hazard. Mater.* **2010**, *6*, 870–877. [CrossRef]

17. Heibati, B.; Pollitt, K.J.; Karimi, A.; Yazdani Charati, J.; Ducatman, A.; Shokrzadeh, M.; Mohammadyan, M. BTEX exposure assessment and quantitative risk assessment among petroleum product distributors. *Ecotoxicol. Environ. Safety* **2017**, *144*, 445–449. [CrossRef]

18.   Masekameni, M.D.; Moolla, R.; Gulumian, M.; Brouwer, D. Risk Assessment of Benzene, Toluene, Ethyl Benzene, and Xylene Concentrations from the Combustion of Coal in a Controlled Laboratory Environment. *Int. J. Environ. Res. Public Health* **2019**, *16*, 95. [CrossRef]

19.   McKenzie, L.M.; Witter, R.Z.; Newman, L.S.; Adgate, J.L. Human health risk assessment of air emissions from development of unconventional natural gas resources. *Sci. Total Environ.* **2012**, *424*, 79–87. [CrossRef]

20.   Easter, E.; Lander, D.; Huston, T. Risk Assessment of Soils Identified on Firefighter Turnout Gear. *J. Occup. Environ. Hyg.* **2016**, *13*, 647–657. [CrossRef]

21.   SKC. Available online: www.skcltd.com/sorbent-tubes/9-uncategorised/177-anasorb-sorbent-tubes (accessed on 26 February 2020).

22.   Dehghani, M.; Fazlzadeh, M.; Sorooshian, A.; Tabatabaee, H.R.; Miri, M.; Baghani, A.N.; Delikhoon, M.; Mahvi, A.H.; Rashidi, M. Characteristics and health effects of BTEX in a hot spot for urban pollution. *Ecotoxicol. Environ. Saf.* **2018**, *15*, 133–143. [CrossRef]

23.   Zabiegała, B.; Urbanowicz, M.; Szymańska, K.; Namieśnik, J. Application of passive sampling technique for monitoring of BTEX concentration in urban air: Field comparison of different types of passive samplers. *J. Chromatogr. Sci.* **2010**, *48*, 167–175. [CrossRef] [PubMed]

24.   U.S. Environmental Protection Agency. *Risk Assessment Guidance for Superfund Volume I: Human Health Evaluation Manual—Part F, Supplemental Guidance for Inhalation Risk Assessment)*; Office of Superfund Remediation and Technology Innovation Environmental Protection Agency: Washington, DC, USA, 2009.

25.   Chaiklieng, S.; Suggaravetsiri, P.; Autrup, H. Risk assessment on benzene exposure among gasoline station workers. *Int. J. Environ. Res. Public Health* **2019**, *16*, 2545. [CrossRef] [PubMed]

26.   International Agency for Research on Cancer. *Painting, Firefighting, and Shiftwork. IARC Monographs on the Evaluation of Carcinogenic Risks to Humans*; IARC Monographs: Lyon, France, 2010.

27.   Polish State Fire Service. Available online: http://www.kmpsruda.pl/statystyki (accessed on 24 April 2020).

28.   Zabiegała, B.; Urbanowicz, M.; Namieśnik, J.; Gorecki, T. Spatial and seasonal patterns of benzene, toluene, ethylbenzene, and xylenes in the Gdańsk, Poland and surrounding areas determined using Radiello passive samplers. *J. Environ. Qual.* **2010**, *39*, 896–906. [CrossRef] [PubMed]

29.   Schneider, P.; Gebefügi, I.; Richter, K.; Wölke, G.; Schnelle, J.; Wichmann, H.E.; Heinrich, J. Indoor and outdoor BTX levels in German cities. *Sci. Total Environ.* **2001**, *267*, 41–51. [CrossRef]

30.   Parra, M.; González, L.; Elustondo, D.; Garrigó, J.; Bermejo, R.; Santamaría, J. Spatial and temporal trends of volatile organic compounds (VOC) in a rural area of northern Spain. *Sci. Total Environ.* **2006**, *370*, 157–167. [CrossRef] [PubMed]

31.   Rogula-Kozłowska, W.; Klejnowski, K.; Rogula-Kopiec, P.; Ośródka, L.; Krajny, E.; Błaszczak, B.; Mathews, B. Spatial and seasonal variability of the mass concentration and chemical composition of PM2.5 in Poland. *Air Qual. Atmos. Health* **2014**, *7*, 41–58. [CrossRef] [PubMed]

32.   Miller, L.; Xu, X.; Wheeler, A.; Atari, D.O.; Grgicak-Mannion, A.; Luginaah, I. Spatial variability and application of ratios between BTEX in two Canadian cities. *Sci. World J.* **2011**, *11*, 2536–2549. [CrossRef]

33.   Hoque, R.R.; Khillare, P.S.; Agarwal, T.; Shridhar, V.; Balachandran, S. Spatial and temporal variation of BTEX in the urban atmosphere of Delhi, India. *Sci. Total Environ.* **2008**, *392*, 30–40. [CrossRef]

34.   Olszowski, T. Concentrations and co-occurrence of volatile organic compounds (BTEX) in ambient air of rural area. *Proc. ECOpole* **2012**, *6*, 375–381. [CrossRef]

35.   Khoder, M.I. Ambient levels of volatile organic compounds in the atmosphere of Greater Cairo. *Atmos. Environ.* **2007**, *41*, 554–566. [CrossRef]

36.   Mainka, A.; Kozielska, B. Assessment of the BTEX concentrations and health risk in urban nursery schools in Gliwice, Poland. *AIMS Environ. Sci.* **2016**, *3*, 858–870. [CrossRef]

37.   U.S. Environmental Protection Agency. *Environmental Technology Protocol Verification Report Emissions of VOCs and Aldehydes from Commercial Furniture*; US EPA: Washington, DC, USA, 1999.

38.   Barboni, T.; Chiaramonti, N. BTEX emissions during prescribed burning in function of combustion stage and distance from flame front. *Combust. Sci. Technol.* **2010**, *182*, 1193–1200. [CrossRef]

39.   Fent, K.W.; Evans, D.E.; Couch, J. *Evaluation of Chemical and Particle Exposures during Vehicle Fire Suppression Training, Health Hazard Evaluation Report*; HETA 2008-0241-3113; Miami Township Fire and Rescue: Yellow Springs, OH, USA, 2010.

40. Guzewski, P.; Wróblewski, D.; Małozięć, D. *Red Book of Fires. Selected Problems of Fires and Their Effects*; CNBOP-PIB: Józefów, Poland, 2016; Volume 1.

41. Miranda, A.I.; Ferreira, J.; Valente, J.; Santos, P.; Amorim, J.H.; Borrego, C. Smoke measurements during Gestosa-2002 experimental field fires. *Int. J. Wildland Fire* **2005**, *14*, 107–116. [CrossRef]

42. Demirel, G.; Ozden, O.; Dogeroglu, T.; Gaga, E. Personal exposure of primary school children to BTEX, NO_2 and ozone in Eskisehir, Turkey: Relationship with indoor/outdoor concentrations and risk assessment. *Sci. Total Environ.* **2014**, *473–474*, 537–548. [CrossRef] [PubMed]

43. Huang, L.; Mo, J.; Sundell, J.; Fan, Z.; Zhang, Y. Health risk assessment of inhalation exposure to formaldehyde and benzene in newly remodeled buildings, Beijing. *PLoS ONE* **2013**, *8*, 11. [CrossRef]

44. De Donno, A.; De Giorgi, M.; Bagordo, F.; Grassi, T.; Idolo, A.; Serio, F.; Ceretti, E.; Feretti, D.; Villarini, M.; Moretti, M.; et al. Health risk associated with exposure to PM_{10} and benzene in three italian towns. *Int. J. Environ. Res. Public Health* **2018**, *15*, 1672. [CrossRef]

5

Personal Control of the Indoor Environment in Offices: Relations with Building Characteristics, Influence on Occupant Perception and Reported Symptoms Related to the Building—The Officair Project

Ioannis Sakellaris [1,*], Dikaia Saraga [1,2], Corinne Mandin [3], Yvonne de Kluizenaar [4],
Serena Fossati [5], Andrea Spinazzè [6], Andrea Cattaneo [6], Tamas Szigeti [7], Victor Mihucz [7],
Eduardo de Oliveira Fernandes [8], Krystallia Kalimeri [1], Paolo Carrer [9] and John Bartzis [1,*]

[1] Department of Mechanical Engineering, University of Western Macedonia, Sialvera & Bakola Str., 50100 Kozani, Greece
[2] Environmental Research Laboratory, INRASTES, National Center for Scientific Research "DEMOKRITOS", Aghia Paraskevi Attikis, P.O. Box 60228, 15310 Athens, Greece
[3] Université Paris Est, CSTB-Centre Scientifique et Technique du Bâtiment, 84 avenue Jean Jaurès, 77447 Marne-la-Vallée CEDEX 2, France
[4] The Netherlands Organization for Applied Scientific Research (TNO), P.O. Box 96800, 2509 JE The Hague, The Netherlands
[5] ISGlobal, Institute for Global Health, 08036 Barcelona, Spain
[6] Department of Science and High Technology, University of Insubria, Via Valleggio 11, 22100 Como, Italy
[7] Cooperative Research Centre for Environmental Sciences, Eötvös Loránd University, Pázmány Péter sétány 1/A, H-1117 Budapest, Hungary
[8] Institute of Science and Innovation in Mechanical Engineering and Industrial Management, Rua Dr. Roberto Frias s/n, 4200-465 Porto, Portugal
[9] Department of Biomedical and Clinical Sciences-Hospital "L. Sacco", University of Milan, via G.B. Grassi 74, 20157 Milano, Italy
* Correspondence: isakellaris@uowm.gr (I.S.); bartzis@uowm.gr (J.B.).

Abstract: Personal control over various indoor environment parameters, especially in the last decades, appear to have a significant role on occupants' comfort, health and productivity. To reveal this complex relationship, 7441 occupants of 167 recently built or retrofitted office buildings in eight European countries participated in an online survey about personal/health/work data as well as physical/psycho-social information. The relationship between the types of control available over indoor environments and the perceived personal control of the occupants was examined, as well as the combined effect of the control parameters on the perceived comfort using multilevel statistical models. The results indicated that most of the occupants have no or low control on noise. Half of the occupants declared no or low control on ventilation and temperature conditions. Almost one-third of them remarked that they do not have satisfactory levels of control for lighting and shading from sun conditions. The presence of operable windows was shown to influence occupants' control perception over temperature, ventilation, light and noise. General building characteristics, such as floor number and floor area, office type, etc., helped occupants associate freedom positively with control perception. Combined controlling parameters seem to have a strong relation with overall comfort, as well as with perception regarding amount of privacy, office layout and decoration satisfaction. The results also indicated that occupants with more personal control may have less building-related symptoms. Noise control parameter had the highest impact on the occupants' overall comfort.

Keywords: IEQ; perceived comfort; sick building syndrome; health effects

1. Introduction

Office employees spend a significant part of their time in modern office buildings that are characterized by sealed facades and complex building systems (e.g., mechanical, electrical, plumbing, controls and fire protection systems) designed to reduce energy costs through controlled indoor environment conditions. Beside central control, a wide range of degrees of personal control, such as local thermostats, windows, personal lights etc., over the indoor environment can be found in modern office buildings. Personal control has a crucial role in achieving a healthy [1–3], comfortable [4–7] and productive [8–11] environment, reducing energy consumption in buildings without sacrificing the comfort of occupants [12,13]. The effect of personal control on occupant satisfaction, especially with regards to providing well-being and comfort is an important area of study [14,15].

Several studies have focused on estimating occupants' comfort in offices and the ability to control indoor environment parameters, producing a better Indoor Environment Quality (IEQ). In 1990, Paciuk [16] studied if indoor environment comfort was affected by personal control, leading to a model using thermal control parameters. Three different types of control parameters were involved: available control (the degree and type of control made available by the environment), exercised control (the relative frequency in which occupants engage in several types of controls to obtain comfort) and perceived control (how the different degrees of available and exercised control interact to produce different levels of perceived control). The model was applied on data from ten offices (511 workstations). The results pointed out that the occupants' level of control perception at their workstation enhances their satisfaction in their working environment.

The possibility to control light satisfaction was surveyed by Collins et al. [17] in 13 office buildings. The occupants provided with a task light recorded in general a higher satisfaction with light comfort than those without a task light. In addition, both groups expressed improved light satisfaction when they had the possibility to control light conditions. In another study [18], the role of personal control in natural versus mechanical ventilated office buildings was investigated. Personal control on operable windows, electronic lightning and solar blinds in natural ventilated buildings resulted in higher levels of perceived control. It was also mentioned that control systems should be simple and in compliance with the building design, as well as with quick response to alleviate discomfort as soon as it is experienced. The level of perceived control by occupants had a small influence on the indoor environment satisfaction, as described by Haghighat and Donnini [19] through a survey in 12 office buildings. Satisfaction with respect to temperature, air quality, ventilation, air circulation and overall comfort showed a moderate correlation with perceived control. On the contrary, there was a decline in prevalence of health symptoms with an increasing amount of control over the indoor environment.

The association of temperature personal control and operable windows with reported health symptoms and complaints was indicated in the early 1990s [20]. Many years later, Toftum [21] examined occupants' comfort perception and symptoms prevalent in mechanical versus natural ventilated buildings in 24 office buildings located in Denmark. Multiple logistic regression analysis indicated that the perceived control was more important for the prevalence of symptoms and environment perception than the type of ventilation.

Zagreus et al. [22] remarked that occupants with a sense of high degree of control over environmental parameters such as temperature, air movement, air quality and noise, were more satisfied with the indoor environment. Boerstra et al. [23] examined the impact of perceived control and access to control options in occupants' health and comfort, through 64 office buildings in Europe. The link between perceived comfort and control parameters such as temperature, ventilation, shading from the sun, light, noise was investigated. The analysis showed that occupants feel more comfortable when the perceived control over temperature, ventilation and noise is high. No significant correlation was found between comfort and the different types of access to control like operable windows, type of thermostats, etc., except for solar radiation.

In a recent study, Kwon et al. [24] tried to identify the relationship between the level of personal control and users satisfaction within offices. They found that higher controllability leads to more thermal and visual satisfaction, while the results revealed the psychological impact on the users' satisfaction by indicating differences among the available control types. The psychological aspect in personal control was also raised by Luo and Cao [25] and Karjalainen [26].

The physiological and psychological aspect of IEQ satisfaction [27] and more specific thermal satisfaction makes it harder to control indoor environment conditions and provide optimal results for everyone in a given space. The ASHRAE standard 55 [27,28] or ISO 7730 [29,30] tries to give a solution in this complex relation by stating the appropriate conditions that should be met in order to establish comfort levels in offices. To achieve that, these standards consider both personal factors, such as metabolic rate and clothing level, and environmental factors, such as air temperature, mean radiant temperature, air speed and humidity. Apart from thermal comfort, other factors that can influence comfort levels are usability of a space, acoustics, ventilation, daylight and energy use in a building.

Literature review highlights the role of personal control on the IEQ in office buildings. The objective of this paper is to provide an updated overview of the personal control in office buildings and the association with occupants' perception. To the best of our knowledge, this is the first time that European employees have participated in a questionnaire survey covering simultaneously in detail records of comfort, control and health perception in office buildings. This large-scale survey was performed under the framework of the European FP7-funded project OFFICAIR [31] and included eight widely distributed across European countries (Finland, France, Greece, Hungary, Italy, Portugal, Spain, The Netherlands) with different characteristics (e.g., geographical location, climate, socio-economic status). More specifically, the aim of this study is threefold: (i) to describe the degree of personal control over indoor environments in office buildings as reported by occupants (perceived personal control) and the association with access to available controls; (ii) to investigate the associations between perceived control and building characteristics; (iii) to study the associations between perceived control and perceived comfort and health of the occupants.

2. Materials and Methods

2.1. Data Collection

This study is based on data collected between October 2011 and May 2012 in the OFFICAIR project in modern office buildings in eight European countries [32–40]. 'Modern' buildings, constructed during the last 10 years, are described by the presence of several sorts of new electronic equipment and ventilation, heating and cooling systems, making the indoor environment almost unaffected by local climate [34]. About 19 to 24 modern buildings, selected on a voluntary basis, were investigated per country, resulting in 167 office buildings.

The protocol for data collection is described elsewhere in detail [34]. After a preliminary inspection of each building, a checklist was filled in by a local investigator along with a building manager, gathering information about building characteristics (e.g., presence of solar devices, operable windows), mechanical systems (e.g., type of mechanical ventilation, heating, cooling), rooms and activities (e.g., type of work, cleaning schedules). An online survey for the building occupants was developed in the national language of the participating countries and included questions on personal control of the indoor environment as perceived by the occupant, perceived comfort and building-related symptoms. The survey was anonymous and the participants gave their consent prior to participation. The study was approved by the competent local/national ethics committees. In total, 26,735 email invitations were sent to the occupants with an average response rate of 41% across the buildings. Although the questionnaire was online and its length might have influenced the response rate, the participation rate can be considered satisfying, in line with other recent surveys [34]. The final database involved 7441 participants—52% were females and 48% were males with an average age of 41 years.

2.2. Characteristics of the Buildings and Access to Control

The available control types in each office building were obtained from the OFFICAIR checklist (details in Table 1). The checklist included information regarding the presence of solar shading devices (grouped as: not present, internal, external), type of solar shading device control, type of temperature control, presence of operable windows, type of main lights control and type of mechanical ventilation control.

Table 1. Checklist used to investigate the types of controls over the indoor environment available for occupants in the building in the OFFICAIR project.

Parameter	Items
Are there solar shading devices present? Which kind?	Not present South side only One or more other facades External vertical blinds External shutters External roller shutters External louvers External screens External window films External horizontal blinds External awnings/canopies External overhangs External vertical fins Blind between glazing Internal vertical blinds Internal louvers Atrium Double façade Other
How are the solar shading devices controlled?	No control (fixed) Individual Central down, individual up Automatic
How is the room temperature controlled?	Manual radiator valve Local thermostat at radiator/heating unit Local thermostat (e.g., on wall) Central sensor Façade sensor(s)—i.e., outside temperature Zone sensor(s) Manual control in room(s) According to occupancy Other
Are the windows operable?	Yes Yes, some Yes, but occupants are not allowed to open them No
How are main lights (e.g., ceiling or wall) controlled?	Automatic by time (building/floor/zone) Automatic with manual end control (building/floor/zone) Demand control: Daylight (photocells) Demand control: Occupants (motion sensors) Manual
What type of control system is there for mechanical ventilation?	Central—Manual (on/off) Central—Clock/Central—Demand control (temperature, CO_2, other pollutant, relative humidity) Local—Manual (on/off) Local—Clock/Local—Demand control (temperature, CO_2, other pollutant, relative humidity) Recirculation control

2.3. Perceived Control over the Indoor Environment

Five controlling parameters were set for defining the occupants' evaluation of the perceived control and IEQ: Personal control of the occupants over temperature, ventilation, shade from the sun, light and noise was investigated using the following question with a seven-point Likert-like scale answer (from "1, not at all" to "7, full control"): 'How much control do you personally have over the following aspects of your working environment?'. The combined control variables were introduced. Several combinations were used (e.g., perceived control over temperature and ventilation) and the overall combined control variable containing all the 5 parameters [23].

2.4. Personal Comfort, Reported Health Symptoms and Self-Assessed Productivity

The satisfaction of the occupants toward the following parameters was evaluated: overall comfort, temperature (overall, too hot/cold, variation), air movement, air quality (overall air quality satisfaction, humid or dry air, stuffy or fresh air, odor), light (overall light satisfaction, natural, artificial, glare), noise (overall noise satisfaction, outside noise, noise from building systems, noise within the building), vibration, amount of privacy, office layout, office decoration, and view from the windows. The following question was used: 'How would you describe the typical indoor conditions in your office environment during the past month?' or 'How would you describe the following in your office?' A seven-point Likert-like scale answer (from "1, dissatisfied" to "7, satisfied") was provided for most of the questions, except for those questions investigating two extreme conditions in contrast where a seven-point scale answer ranging from −3 to 3 was adopted (Table 2) and converted to a scale from 1 to 7 as follow: +/−3 = 1; +/−2 = 3; +/− 1 =5; 0 = 7.

Table 2. Questions used to investigate occupants' indoor environment quality perception in the OFFICAIR project.

Parameter	Sub-Parameters	Type of Answer
Overall Comfort		Seven-point Likert-like scale
Temperature	Overall temperature	Seven-point Likert-like scale
	Too hot/too cold temperature	From "−3, too hot" to "+3, too cold"
	Temperature variation	From "−3, too much" to "+3, not enough"
Air movement		From "−3, too draughty" to "+3, too still"
Air quality	Overall air quality	Seven-point Likert-like scale
	Humid/dry air	From "−3, too humid" to "+3, too dry"
	Stuffy or fresh air	Seven-point Likert-like scale
	Odor	Seven-point Likert-like scale
Light	Overall light	Seven-point Likert-like scale
	Natural	Seven-point Likert-like scale
	Artificial	Seven-point Likert-like scale
	Glare	Seven-point Likert-like scale
Noise	Overall noise	Seven-point Likert-like scale
	Outside noise	Seven-point Likert-like scale
	Noise from building systems	Seven-point Likert-like scale
	Noise within the building	Seven-point Likert-like scale
Vibration		Seven-point Likert-like scale
Amount of privacy		Seven-point Likert-like scale
Office layout		Seven-point Likert-like scale
Office decoration		Seven-point Likert-like scale
View from the windows		Seven-point Likert-like scale

The occupants were also requested to estimate their productivity at the workstation and in other locations inside the building, considering the influence of environmental conditions on a scale of 7 levels, from +30% to −30%. In addition, they were requested to record building-related health symptoms. The Personal Symptom Index-5 (PSI-5) was calculated based on the incidence of five health symptoms—dry eyes, blocked or stuffy nose, dry throat, headache, and tiredness—which are considered to be the fundamental components of sick building symptoms as mentioned by Raw et al. [41]. This indicator has a 0 to 5 score, according to the prevalence of the reported symptoms. The respective question was: 'Have you ever experienced any of the following symptoms while working in this building (or workstation) (including today)?'

2.5. Statistical Analysis

The statistical analysis of the dataset was performed in four steps (Figure 1):

i.   Descriptive results were obtained.

ii.   The correlation between the available control and the perceived control using the Kruskal Wallis analysis of variance test was investigated [23,42]. Groups with less than five individuals were not included in the analysis.

iii. The relationship between the occupants' overall combined perceived control and the general physical building characteristics as well as the occupant personal characteristics was examined by applying a multilevel model [43,44], to account for the three-level structure of our data (level 1-occupant, level 2-building, level 3-country). The ordered logistic regression analysis was applied using building and country as random effects and the covariates as fixed effects. Four step-by-step models were applied. The first was an empty model without any variable and with building and country variance only. In the second model, individual level variables were imported. The final version of the second model included variables, with p-value below 0.2, such as gender, age (in four groups, <35, 35–45, 46–55, >55), effort reward ratio, experience of negative events, use of air fresheners at home, type of job (managerial, professional, clerical/secretarial, other), type of job contract (full-time, part-time), and job contract duration (permanent, fixed-term). In the third model, building characteristics were imported iteratively on the second model to identify significant relations with the perceived control. Variables with a p-value below 0.2 were selected to be used in the next model. In the fourth model, both individual and building level variables were imported. The results of the fixed effects were reported in Odds Ratios (OR) and 95% Confidence Interval (CI). For the random effects, the explained variance is reported as well as the Proportional Change in Variance (PCV) between the null model and the final model with the variables.

iv.   The potential relations between the occupants' perceived control and perceived comfort and reported health symptoms (PSI5) were examined by using the spearman correlation [45]. This study focused on the correlation between the various combined control scores (e.g., perceived control over temperature and ventilation) and the perceived occupants' comfort. Additionally, the dependent variable (overall comfort satisfaction) and the response-variables were expressed in values on an ordinal scale; hence, ordinal regression analysis was employed to determine the impact of the controlling parameters on overall comfort. In the regression model, the response-dependent variable was the overall comfort satisfaction and the predictor-independent variables were the satisfaction for each personal controlling parameter as evaluated by the occupants. The results are presented in the format of OR and its CI95%. The ORs were used to rank the effect of the personal controlling parameters on overall comfort. p-values <0.05 were considered as statistically significant. The statistical package IBM SPSS Statistics [46] was used for the analysis.

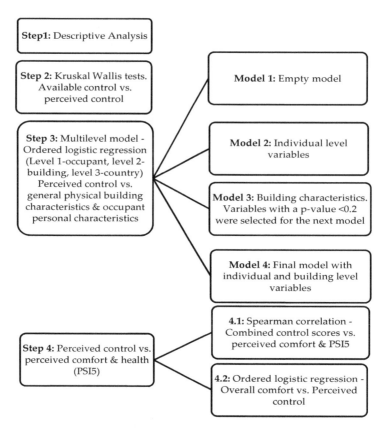

Figure 1. Schematic overview of the methodology used.

3. Results

3.1. Characteristics of Perceived Personal Control

The responses to the perceived controls (with a response rate above 99% for all control types) are presented in Figure 2. In general, noise, ventilation and temperature were perceived as poorly controlled. Regarding shading from the sun and lighting, the occupants reported the perceived control as moderate. A group of people (up to 35%) declared "no control" can be observed for all control parameters. For temperature and ventilation control, the rest of the occupants are equally distributed in the five levels of control degree. Very few occupants seem to be able to control the noise level at their workstation, where answers follow a descending rate (full control below 5%). On the contrary, regarding lighting and shading from the sun, after the distinctive "no control" group, the occupants' answers exhibited an ascending rate, indicating that they feel more able to control the light conditions (full control up to 20%).

Figure 3 presents the percentages of the office occupants who have no or low perceived control (values < 4) on their indoor working environment. The majority (63%) has no or low control on noise. Half of them declared no or low control on ventilation (53%) and temperature (47%). Almost one-third of them remarked that they do not have satisfactory levels of control for lighting and shading from sun conditions.

In Figure 4, the responses are presented based on the occupancy per room. Responses were categorized into five groups. In the first group, which is characterized by personal office rooms, occupants seem to have higher levels of control, except for ventilation. It is interesting to note that as the occupancy increases, the degree of personal control becomes lower. Offices with many occupants (30+), probably open space offices, show the lowest degrees of personal control for all parameters. Furthermore, in all groups, the occupants characterized noise control as the worst parameter, while lighting and shading from the sun control gathered higher degrees of control.

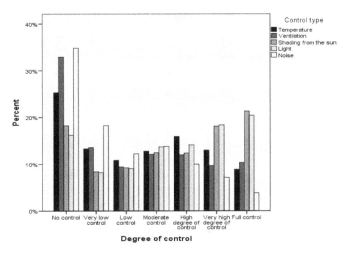

Figure 2. Percentage of occupants based on self-reported degree of control ($n = 7441$).

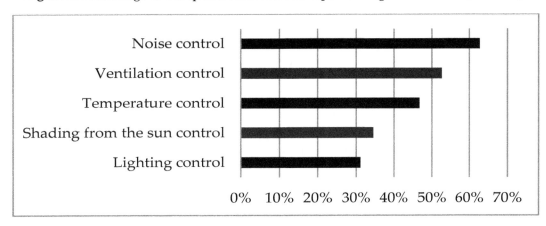

Figure 3. Percentage of occupants who reported no or low control.

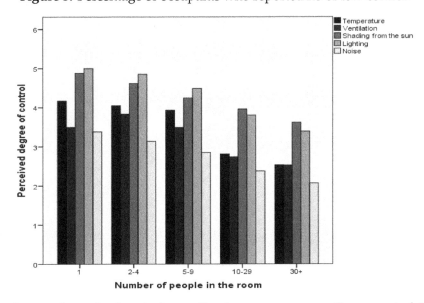

Figure 4. Mean degree of perceived control according to room occupancy (1 = no control, 7 = full control).

The satisfaction comfort towards IEQ parameters was also examined, with regard to their respective control options. The perceived satisfaction parameters were categorized in three levels of perceived control: no control (1), low control (2–3), high control (≥4) and are presented in Figure 5. For each IEQ parameter, the related personal control option was selected, e.g., for parameters about temperature satisfaction, temperature control was selected; for indoor air quality parameters, ventilation control was selected, etc. Occupants with high degrees of personal control over their working environment reported higher levels of satisfaction on average. In all cases, occupants with high perceived control reported higher satisfaction levels. Occupants without personal control evaluated satisfaction of the IEQ with lower levels. Only in some cases, 'none' and 'low personal' control options are reversed, e.g., in the case of satisfaction with odor and reflection.

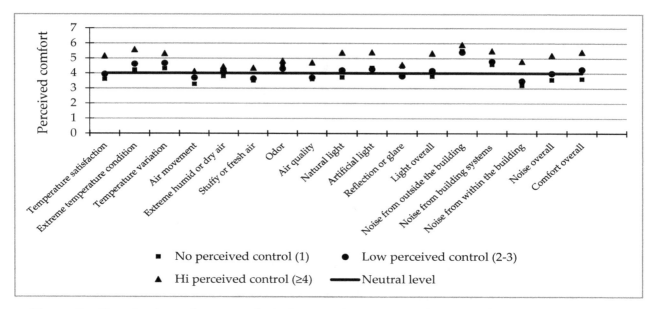

Figure 5. Perceived comfort towards indoor environment quality according to three levels of perceived control.

Figure 6 shows the degree of personal control vs. self-reported productivity at the workstation (n = 7289) and in other places in the building (n = 7154). In both cases, occupants with higher levels of personal control reported higher levels of productivity (Kruskal Wallis: $p < 0.001$). Regarding the workstations, a small increase in the low control area results in a clear increase in productivity.

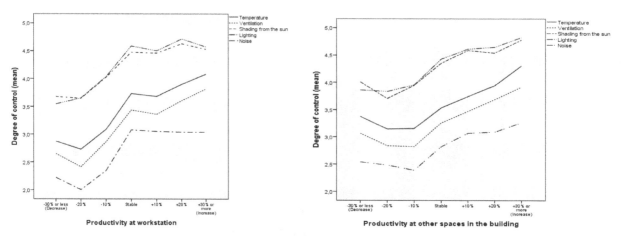

Figure 6. Self-reported productivity vs. degree of control.

3.2. Relationship between Perceived Control and Available Control

The statistically significant ($p < 0.05$) correlations between the available control and perceived control are presented in Figures 7–11.

Regarding the availability of operable windows (Figures 7–10), the scores of perceived control over temperature, ventilation, light and noise differ significantly ($p < 0.001$, $p < 0.001$, $p = 0.001$, $p < 0.001$, respectively) among buildings with operable windows, some operable windows, operable windows that people are not allowed to open, and no operable windows. The mean occupants' score over temperature in buildings with operable windows was 1.3 points higher than in buildings without operable windows ($p < 0.001$). The corresponding difference in the mean score for ventilation, light and noise was 2 ($p < 0.001$), 1 ($p < 0.001$) and 0.8 ($p = 0.001$), respectively.

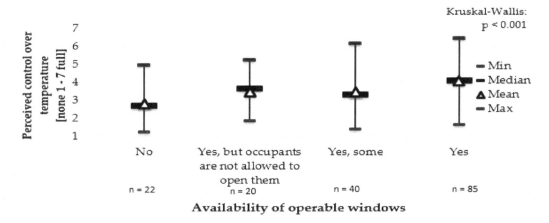

Figure 7. Perceived control over temperature vs. availability of operable windows (n is the number of office buildings).

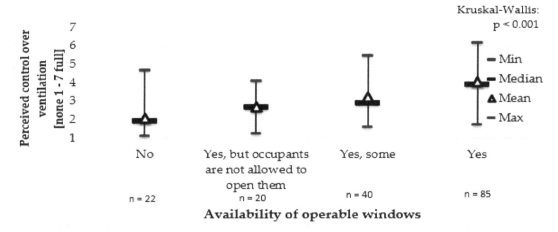

Figure 8. Perceived control over ventilation vs. availability of operable windows (n is the number of office buildings).

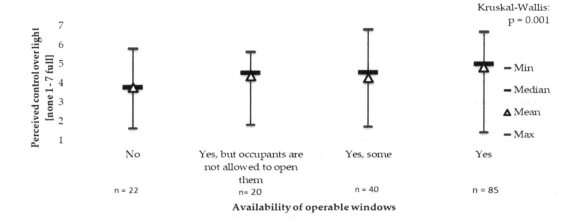

Figure 9. Perceived control over light vs. availability of operable windows (n is the number of office buildings).

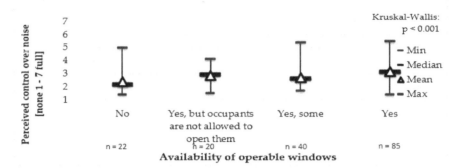

Figure 10. Perceived control over noise vs. availability of operable windows (n is the number of office buildings).

With regard to the presence of solar shading devices, the buildings were grouped into three categories based on the solar devices set up: internal, external, none (Figure 11). Perceived control over temperature varied significantly through the different types of the solar devices ($p = 0.028$). The highest score of perceived control over temperature was in buildings with no solar devices. This might be explained by the fact that the design of these office buildings provides adequate control over temperature without the use of solar shading devices; the multilevel regression analysis in the next step provides additional insights. No significant differences were observed between perceived control over shading and lighting in relation to the types of solar shading devices ($p = 0.635$ and $p = 0.255$, respectively, Table 3).

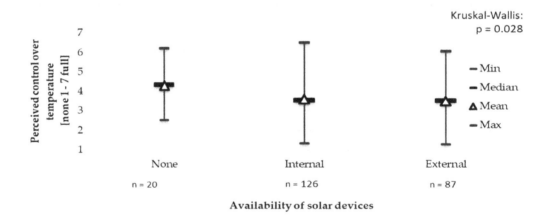

Figure 11. Perceived control over temperature vs. availability of solar shading devices (n is the number of office buildings).

Table 3 presents the remaining outcomes of the statistical tests performed between the available control and occupants' perceived control. Regarding temperature control, the analysis showed no significant differences ($p = 0.166$) between the perceived control over temperature and the various types of temperature controls (e.g., manual radiator valve and central sensor). The different types of operable windows did not affect the perceived control of the shading from the sun ($p = 0.100$). Subsequently, the type of control on the solar shading devices (fixed, individual, central control, automatic) was examined (Table 3). Unlike the various types of solar devices, analysis of the available different control types of the shading devices showed that there are no significant differences among the scores of the perceived control over shading, lighting and temperature ($p = 0.147$, $p = 0.710$ and $p = 0.755$, respectively). As far as the available controls of lights are concerned, no significant difference was found for perceived control over lighting ($p = 0.328$). Finally, the several types of mechanical ventilation control were examined. The occupants' perceived control over ventilation and temperature did not significantly differ ($p = 0.105$ and $p = 0.768$, respectively) from the available control types.

Table 3. Available control at building level and occupants' perceived control (from 1 = no control at all to 7 = full control) with $p > 0.05$.

	Perceived Control over Temperature	Perceived Control over Shading	Perceived Control over Light	Perceived Control over Ventilation
Type of Available Temperature Controls				
Manual radiator valve	3.38			
Local thermostat at radiator/heating unit	3.96			
Local thermostat (e.g., on wall)	3.53			
Central sensor	3.52			
Facade sensor(s)—i.e., outside temperature	4.12			
Zone sensor(s)	4.03			
Manual control in room(s)	3.70			
p-value (Kruskal-Wallis test)	0.166			
Availability of shading devices				
None		4.09	4.69	
Internal		4.43	4.43	
External		4.22	4.44	
p-value (Kruskal-Wallis test)		0.635	0.255	
Type of control of the available solar shading devices				
No control (fixed)		3.90	4.53	
Individual		4.43	4.43	
Central down, individual up		4.46	3.76	
Automatic		4.18	4.61	
p-value (Kruskal-Wallis test)		0.147	0.710	
Type of operable windows				
Yes		4.42		
Yes, some		4.15		
Yes, but occupants are not allowed to open them		4.81		
No		4.11		
p-value (Kruskal-Wallis test)		0.100		
Type of available light controls				
Manual			4.52	
Demand control: Occupants (motion sensors)			4.42	
Demand control: Daylight (photocells)			4.26	
Automatic with manual end control (building/floor/zone)			4.91	
Automatic by time (buildings/floor/zone)			4.54	
p-value (Kruskal-Wallis test)			0.328	
Type of available mechanical ventilation controls				
Central—Manual (on/off)				3.33
Central—Clock				3.23
Central—Demand control (temperature, CO_2, other pollutant, relative humidity)				3.27
Local—Manual (on/off)				4.27
Local—Clock				3.78
Local—Demand control (temperature, CO_2, other pollutant, relative humidity)				3.73
p-value (Kruskal-Wallis test)				0.105

3.3. Combined Perceived Control Versus Building Characteristics

The variances of country and building level according to Model 1 (the null model) are 0.38 and 1.17, respectively. The respective variances in Model 2, which includes the individual variables (gender, age groups, effort reward ratio, experience of negative events, use of air fresheners at home, type of job, type of job's contract, job's contract duration), were 0.52 and 0.94. The relations between the building characteristics, which were added in the model iteratively, and the combined perceived control with p-values < 0.2 are presented in Table 4 (Model 3). The strongest associations with a p-value < 0.001 were found for the building's location (suburban area), the maximum number of occupants, the documented complaints procedure, the number of people in office, as well as for the office type and availability of opening windows. Significant associations with a p-value < 0.05 were found for the number of adjacent facades with other buildings, glares from neighboring buildings, total floor area, pesticide treatment plan, smoking permission, use of portable air cleaner, floor of the workspace, partition in the offices, and noise source of occupants—distracting conversations and the location of air and exhaust devices.

Table 4. Relations between combined overall perceived control and building characteristics (Model 3).

Characteristics	n1/N1	n2/N2	OR (CI 95%)	p-Value
General building characteristics				
Location				
Mixed industrial/residential area (vs. industrial area)	696/7441	14/167	1.68 (0.76–3.69)	0.199
Commercial area (vs. industrial area)	789/7441	18/167	0.95 (0.46–1.95)	0.883
Mixed commercial/residential area (vs. industrial area)	2279/7441	50/167	1.71 (0.94–3.09)	0.076
City center, densely packed housing (vs. industrial area)	1344/7441	30/167	1.35 (0.7–2.63)	0.374
Town, with no or small gardens (vs. industrial area)	328/7441	8/167	**3.42 (1.33–8.75)**	**0.010**
Suburban, with larger gardens (vs. industrial area)	864/7441	22/167	**3.88 (1.92–7.82)**	**<0.001**
Village in a rural area (vs. industrial area)	24/7441	1/167	0.82 (0.09–7.46)	0.859
Rural area with no or few other homes nearby (vs. industrial area)	192/7441	6/167	1.27 (0.45–3.57)	0.646
Façades with adjacent buildings				
2 façades (vs. 1 façade)	1932/4664	37/102	**0.6 (0.38–0.95)**	**0.030**
3 façades (vs. 1 façade)	288/4664	9/102	1.48 (0.68–3.25)	0.322
Density of nearby obstructions				
Moderately dense (vs. very dense)	2610/7441	58/167	1.31 (0.69–2.5)	0.407
Few buildings (vs. very dense)	2992/7441	58/167	1.43 (0.74–2.76)	0.285
Free standing (vs. very dense)	1031/7441	26/167	1.95 (0.93–4.09)	0.076
Neighboring buildings with glass facades or light-colored facades causing glare				
Yes, in summer (vs. No)	215/7441	4/167	0.4 (0.13–1.27)	0.120
A little in summer (vs. No)	654/7441	12/167	**0.5 (0.25–1)**	**0.049**
Yes, in winter (vs. No)	102/7441	2/167	0.35 (0.07–1.71)	0.195
A little in winter (vs. No)	217/7441	5/167	**0.24 (0.09–0.68)**	**0.007**
Maximum number of occupants in the building				
Between 101 and 250 (vs. ≤100)	2288/7260	56/163	0.79 (0.52–1.2)	0.272
≥251 (vs. ≤100)	3544/7260	51/163	**0.39 (0.25–0.61)**	**<0.001**
Total floor area (m^2)				
Between 1441 and 3210 (≤1440)	2300/7234	53/160	0.78 (0.49–1.24)	0.295
≥3211 (≤1440)	3341/7234	54/160	**0.51 (0.31–0.83)**	**0.007**
Activities in the building besides office work				
Underground car park (vs. No)	2678/7441	44/167	0.73 (0.46–1.16)	0.178
Print shop (vs. No)	428/7441	7/167	1.92 (0.78–4.73)	0.156
Pesticide treatment plan (vs. No)	7025/7441	99/164	**0.59 (0.39–0.88)**	**0.010**
Documented complaints procedure for occupants (vs. No)	5112/7349	99/165	**0.41 (0.28–0.61)**	**<0.001**
Smoking permission				
Only outside the building (vs. No)	3916/7441	88/167	**0.55 (0.33–0.93)**	**0.026**
Only in separately ventilated rooms (vs. No)	973/7441	13/167	**0.29 (0.13–0.63)**	**0.002**
Percentage of office furniture is less than one year old and made of MDF	7025/7441	162/162	0.99 (0.99–1)	0.200
Portable air cleaner (vs. No)	225/7441	5/167	**0.33 (0.12–0.93)**	**0.035**

Table 4. *Cont.*

Characteristics	n1/N1	n2/N2	OR (CI 95%)	*p*-Value
Office characteristics				
Floor surface of the office	7410/7441	167/167	**1.05 (1.02–1.09)**	**0.001**
Number of people in the office on average				
Private (vs. 30+)	1501/7352	144/167	**7.65 (6.19–9.45)**	**<0.001**
2–4 (vs. 30+)	2134/7352	152/167	**5.15 (4.24–6.26)**	**<0.001**
5–9 (vs. 30+)	1019/7352	117/167	**2.62 (2.15–3.21)**	**<0.001**
10–29 (vs. 30+)	1492/7352	100/167	**1.48 (1.25–1.76)**	**<0.001**
Type of office				
Shared private office (vs. single person private office)	2236/7417	150/167	**0.56 (0.48–0.65)**	**<0.001**
Open space with partitions (vs. single person private office)	991/7417	115/167	0.21 (0.17–0.26)	**<0.001**
Open space without partitions (vs. single person private office)	2602/7441	139/167	**0.21 (0.17–0.25)**	**<0.001**
Other (vs. single person private office)	90/7417	60/167	**0.39 (0.24–0.65)**	**<0.001**
Partitions within the offices (vs. No)	4112/7441	91/167	**0.67 (0.45–0.99)**	**0.043**
Noise source of occupants–distracting conversations (vs. No)	2653/7441	64/167	**0.66 (0.45–0.98)**	**0.040**
PC or laptop monitor position				
In front of windows (vs. Not)	3018/7441	69/167	1.44 (1–2.08)	0.052
Printer/copy machines location				
In the offices (vs. on the corridor)	2976/7441	75/167	0.62 (0.39–1)	0.051
In a separate printing room (vs. on the corridor)	2501/7441	47/167	0.64 (0.39–1.05)	0.075
Operable windows				
Yes (vs. No)	3724/7441	85/163	**5.51 (3.14–9.67)**	**<0.001**
Yes, some (vs. No)	1769/7441	40/163	**3.08 (1.72–5.53)**	**<0.001**
Yes, but occupants are not allowed to open them (vs. No)	913/7441	20/163	1.54 (0.79–2.98)	0.204
Location of air supply devices inside offices—ceiling (vs. No air supply)	5783/7441	123/167	**0.63 (0.4–0.99)**	**0.046**
Location of air exhaust devices inside offices				
High (vs. None)	6189/7089	128/154	**0.46 (0.25–0.87)**	**0.017**
Low (vs. None)	382/7089	10/154	0.63 (0.24–1.63)	0.339

p-values in bold are significant at 5%. Adjusted for gender, age groups, effort reward ratio, experience of negative events, use of air fresheners at home, type of job, type of job's contract, job's contract duration. Characteristics with a p value lower than 0.20 are presented. Level 1—Occupant level, Level 2—Building level, Level 3—Country level. OR: Odd Ratio, CI: Confidence interval 95%, n1/N1: Occupants' answers/total number of occupants, n2/N2: building observations/total number of buildings.

All individual and building characteristics from Models 2 and 3 were imported to Model 4; the relevant results are presented in Table 5. The variance between buildings was equal to 0.14 in Model 4. PCV shows that 88% of the initial variance in overall perceived control was explained by the included variables. Buildings with larger total floor areas were positively associated with perceived overall control (OR 1.99, 95% CI: 1.22–3.25). The workspace floor was also positively associated with control perception (OR 1.06, 95% CI: 1.02–1.10). The existence of any kind of operable windows (OR 3.64 up to 6.53) and private and semi-private offices had the highest impact on the perceived overall control (OR 1.58 up to 3.73). On the other hand, the building's location (OR 0.24, 95% CI: 0.06–0.96) and the buildings' number of adjacent facades (OR 0.65, 95% CI: 0.44–0.97) with other buildings tended to negatively associate with the perceived control. Both high number of occupants in the building (OR 0.67, 95% CI: 0.46–0.99) and the type of office (OR from 0.23 to 0.36) had a significant negative effect on the perceived overall control. Indoor noise from distracting conversations was associated negatively with the perceived overall control (OR 0.69, 95% CI: 0.49–0.97).

Concerning individual characteristics, in addition to gender (OR 1.21, 95% CI: 1.06–1.38) and age group (OR 0.78, 95% CI: 0.61–0.99), strong association with the overall perceived control was found for the ERI (OR 0.67, 95% CI: 0.46–0.99) and the occupants' habit to use air fresheners at home (OR 1.26, 95% CI: 1.11–1.44).

Table 5. Associations between combined overall perceived control, building characteristics and individual characteristics (Model 4).

Factors	OR (CI 95%)	*p*-Value
Individual characteristics/personal activities		
Female (vs. Male)	**1.21 (1.06–1.38)**	**0.006**
Age		
<35 (vs. 55+)	0.89 (0.69–1.14)	0.345
35–45 (vs. 55+)	**0.78 (0.61–0.99)**	**0.040**
46–55 (vs. 55+)	0.95 (0.74–1.23)	0.720
Effort-reward ratio	**0.31 (0.25–0.38)**	**<0.001**
Experience of negative events (vs. No)	0.94 (0.83–1.07)	0.367
Type of job		
Managerial (vs. Other)	0.97 (0.75–1.27)	0.844
Professional (vs. Other)	0.98 (0.78–1.23)	0.873
Clerical-secretarial (vs. Other)	0.95 (0.77–1.17)	0.625
Type of job's contract–Full-time (vs. Part-time)	1.12 (0.89–1.41)	0.344
Job's contract duration–Permanent (vs. Fixed-term)	1.13 (0.91–1.39)	0.263
Air fresheners use in home (vs. No)	**1.26 (1.11–1.44)**	**<0.001**
Building characteristics		
Location		
Mixed industrial/residential area (vs. industrial area)	0.64 (0.31–1.29)	0.208
Commercial area (vs. industrial area)	1.59 (0.83–3.07)	0.164
Mixed commercial/residential area (vs. industrial area)	1.05 (0.61–1.83)	0.849
City Centre, densely packed housing (vs. industrial area)	1.15 (0.62–2.1)	0.660
Town, with no or small gardens (vs. industrial area)	2.98 (0.95–9.39)	0.062
Suburban, with larger gardens (vs. industrial area)	1.73 (0.63–4.72)	0.285
Village in a rural area (vs. industrial area)	**0.24 (0.06–0.96)**	**0.043**
Rural area with no or few other homes nearby (vs. industrial area)	1.02 (0.24–4.45)	0.974
Density of nearby obstructions		
Moderately dense (vs. Very dense)	1.11 (0.63–1.95)	0.729
Few buildings (vs. Very dense)	1.14 (0.65–2.02)	0.648
Free standing (vs. Very dense)	0.99 (0.44–2.19)	0.972
Maximum number of occupants in the building		
Between 101 and 250 (vs. ≤100)	**0.67 (0.46–0.99)**	**0.045**
≥251 (vs. ≤100)	0.92 (0.56–1.52)	0.745
Total floor area		
Between 1441 and 3210 (≤1440)	**1.99 (1.22–3.25)**	**0.006**
≥3211 (≤1440)	1.44 (0.81–2.57)	0.217
Façades with adjacent buildings		
2 façades (vs. 1 façade)	**0.65 (0.44–0.97)**	**0.033**
3 façades (vs. 1 façade)	1 (0.51–1.95)	0.992
Neighboring buildings with glass facades or light-colored facades causing glare		
Yes, in Summer (vs. No)	**3.16 (1.15–8.66)**	**0.025**
A little in Summer (vs. No)	1.52 (0.74–3.14)	0.256
Yes, in Winter (vs. No)	**0.24 (0.06–0.94)**	**0.040**
A little in Winter (vs. No)	0.49 (0.17–1.39)	0.180
Operable windows		
Yes (vs. No)	**4.81 (2.85–8.14)**	**<0.001**
Yes, some (vs. No)	**6.53 (3.76–11.34)**	**<0.001**
Yes, but occupants are not allowed to open them (vs. No)	**3.64 (1.82–7.27)**	**<0.001**
Activities in the building besides office work		
Underground car park (vs. No)	0.79 (0.52–1.19)	0.258
Print shop (vs. No)	1.4 (0.6–3.27)	0.435
Portable air cleaner (vs. No)	0.79 (0.31–2)	0.622
Pesticide treatment plan (vs. No)	1.33 (0.74–2.38)	0.336
Smoking permission		
Only outside the building (vs. No)	0.82 (0.46–1.46)	0.492
Only in separately ventilated rooms(vs. No)	0.66 (0.32–1.36)	0.260
Documented complaints procedure for occupants (vs. No)	0.77 (0.5–1.17)	0.221
Percentage of office furniture is less than one year old and made of MDF	1 (0.99–1.01)	0.800

Table 5. *Cont.*

Factors	OR (CI 95%)	*p*-Value
Office characteristics		
Floor surface of the office	**1.06 (1.02–1.1)**	**0.002**
Number of people in the office on average		
Private (vs. 30+)	**2.51 (1.59–3.97)**	**<0.001**
2–4 (vs. 30+)	**3.73 (2.78–4.99)**	**<0.001**
5–9 (vs. 30+)	**2.3 (1.79–2.96)**	**<0.001**
10–29 (vs. 30+)	**1.58 (1.29–1.93)**	**<0.001**
Type of office		
Shared private office (vs. Single person private office)	**0.36 (0.23–0.55)**	**<0.001**
Open space with partitions (vs. Single person private office)	**0.23 (0.14–0.36)**	**<0.001**
Open space without partitions (vs. Single person private office)	**0.26 (0.17–0.4)**	**<0.001**
Other (vs. Single person private office)	0.57 (0.25–1.29)	0.178
Partitions within the offices (vs. No)	1.02 (0.64–1.61)	0.935
Noise source of Occupants–distracting conversations (vs. No)	**0.69 (0.49–0.97)**	**0.034**
PC or laptop monitor position		
In front of windows (vs. Not)	1.32 (0.94–1.86)	0.103
Printer/copy machines location		
In the offices (vs. on the corridor)	0.99 (0.63–1.55)	0.956
In a separate printing room (vs. on the corridor)	1.23 (0.76–1.99)	0.396
Location of air supply devices inside offices—ceiling (vs. No air supply)	0.81 (0.51–1.28)	0.365
Location of air exhaust devices inside offices		
High (vs. None)	1.43 (0.61–3.34)	0.408
Low (vs. None)	0.93 (0.37–2.35)	0.879
County level σ2/PCV (%)	0.42/10	
Building level σ2/PCV (%)	0.14/88	

3.4. Impact of Perceived Control on Perceived Comfort and Health

3.4.1. Combined Perceived Control vs. Perceived Comfort and Health—Bivariate Analysis

The relationships between perceived control and perceived comfort are presented in Table 6. Perceived overall comfort correlated significantly and positively with perceived control over all control parameters. When the scores of the perceived control were combined, the correlation with the overall comfort increased, reaching the highest value of $r = 0.465$. The combined correlation values are stronger than the correlations between the single perceived control parameters. The strongest correlation for perceived overall temperature was indicated with control over temperature ($r = 0.420$) and not with the combined controls. The satisfaction with extreme hot or cold temperature conditions seems to have the strongest correlation only with the control over temperature ($r = 0.281$), while the combined perceived controls show equal effect. On the other hand, temperature variation was found to be more related to the combined control over temperature and shading from the sun controls ($r = 0.209$). Perceived air quality about dry or humid air and odor did not show strong correlations, either with single or with combined perceived control parameters, indicating that these parameters were more difficult to be controlled by the occupants. However, perceived overall air quality satisfaction and perceived satisfaction with fresh air were found to be more correlated with the combined control perception over temperature, ventilation and shading from the sun ($r = 0.380$ and 0.320, respectively).

Perceived comfort regarding natural light correlated positively with perceived control over shading from the sun and over light ($r = 0.370$). Moreover, artificial light perception showed correlation with the perceived control over light ($r = 0.377$). This means that occupants with higher degrees of shading and lighting controls feel more satisfied with the light levels in their offices. Perceived satisfaction with glare did not show strong correlation. In general, combined controls do not significantly affect light satisfaction.

Table 6. Correlations between combined perceived control and perceived comfort and health.

	Perceived Control over Temperature and Ventilation Combined [2: None at all-14: Full Control]	Perceived Control over Temperature and Shading from the Sun Combined [2: None at all-14: Full Control]	Perceived Control over Temperature, Ventilation and Shading from the Sun Combined [3: None at all-21: Full Control]	Perceived Control over Temperature, Ventilation, Shading from the Sun, Light and Noise Combined [5: None at all-35: Full Control]
Overall comfort [1: Unsatisfactory-7: Satisfactory]	0.367 <0.001	0.415 <0.001	0.415 <0.001	0.465 <0.001
Temperature [1: Too hot/cold- 7: Satisfactory]	0.258 <0.001	0.257 <0.001	0.255 <0.001	0.245 <0.001
Temperature [1: Varies too much/not enough variation-7: Satisfactory]	0.154 <0.001	0.209 <0.001	0.186 <0.001	0.188 <0.001
Temperature [1: Unsatisfactory-7: Satisfactory]	0.403 <0.001	0.414 <0.001	0.416 <0.001	0.43 <0.001
Air movement [1: Draughty/Still- 7: Satisfactory] Air movement	0.175 <0.001	0.164 <0.001	0.178 <0.001	0.182 <0.001
Air quality [1: Dry/humid-7: Satisfactory]	0.143 <0.001	0.142 <0.001	0.147 <0.001	0.162 <0.001
Air quality [1: Stuffy- 7: Fresh]	0.287 <0.001	0.312 <0.001	0.322 <0.001	0.342 <0.001
Air quality [1: Smelly-7: Odorless]	0.129 <0.001	0.183 <0.001	0.17 <0.001	0.195 <0.001
Air quality [1: Unsatisfactory-7: Satisfactory]	0.35 <0.001	0.363 <0.001	0.38 <0.001	0.405 <0.001
Natural light [1: Unsatisfactory-7: Satisfactory]	0.247 <0.001	0.356 <0.001	0.342 <0.001	0.361 <0.001
Artificial light [1: Unsatisfactory-7: Satisfactory]	0.245 <0.001	0.281 <0.001	0.284 <0.001	0.331 <0.001
Reflection or glare [1: Glare]-7: No glare]	0.123 <0.001	0.145 <0.001	0.142 <0.001	0.166 <0.001
Light overall [1: Unsatisfactory-7: Satisfactory]	0.295 <0.001	0.364 <0.001	0.361 <0.001	0.401 <0.001
Noise from outside the building [1: Unsatisfactory-7: Satisfactory]	-0.004 0.707	0.077 <0.001	0.048 <0.001	0.077 <0.001
Noise from building systems (e.g., heating, plumbing, ventilation, air conditioning) [1: Unsatisfactory-7: Satisfactory]	0.172 <0.001	0.191 <0.001	0.198 <0.001	0.226 <0.001
Noise from within the building other than from building systems (e.g., phone calls, colleagues chatting, photocopiers, etc.) [1: Unsatisfactory-7: Satisfactory]	0.28 <0.001	0.302 <0.001	0.304 <0.001	0.367 <0.001

Table 6. *Cont.*

	Perceived Control over Temperature and Ventilation Combined [2: None at all-14: Full Control]	Perceived Control over Temperature and Shading from the Sun Combined [2: None at all-14: Full Control]	Perceived Control over Temperature, Ventilation and Shading from the Sun Combined [3: None at all-21: Full Control]	Perceived Control over Temperature, Ventilation, Shading from the Sun, Light and Noise Combined [5: None at all-35: Full Control]
Noise overall [1: Unsatisfactory-7: Satisfactory]	0.286 **<0.001**	0.322 0.000	0.321 0.000	0.389 0.000
Vibration [1: Unsatisfactory-7: Satisfactory]	0.108 **<0.001**	0.177 **<0.001**	0.16 **<0.001**	0.193 **<0.001**
Amount of privacy [1: Unsatisfactory-7: Satisfactory]	0.333 **<0.001**	0.385 **<0.001**	0.379 **<0.001**	0.433 **<0.001**
Layout [1: Do not like at all-7: Like very much]	0.31 **<0.001**	0.367 **<0.001**	0.361 **<0.001**	0.403 **<0.001**
Decoration [1: Do not like at all-7: Like very much]	0.22 **<0.001**	0.295 **<0.001**	0.278 **<0.001**	0.315 **<0.001**
View from the windows [1: Do not like at all-7: Like very much]	0.147 **<0.001**	0.256 **<0.001**	0.233 **<0.001**	0.246 **<0.001**
PSI5	-0.249 **<0.001**	-0.251 **<0.001**	-0.263 **<0.001**	-0.289 **<0.001**

Rho and *p*-value of Spearman correlation. Significant at 5% is in bold.

Perceived comfort over outdoor noise seems to have low correlation with perceived controls. The occupants were not able to control the noise levels coming from outside probably due to inefficient available control types. Noise from building systems and noise within the building correlated positively with perceived control over noise ($r = 0.205$ and 0.377). Moreover, the occupants were more satisfied with the noise from building systems when they had higher degree of all combined controls ($r = 0.226$). The highest correlation was observed ($r = 0.408$) between the overall noise satisfaction and perceived noise control. This indicates that the occupants are more comfortable in buildings where they perceive high degree of control over noise.

The amount of privacy was found to have significant correlation with all control parameters, both single and combined, while the maximum correlation was recorded with all combined controls ($r = 0.433$). This implies that high degrees of personal control improve the privacy that the occupant needs. This positive relation was also observed for office layout and decoration perceived satisfaction with $r = 0.403$ and 0.315, respectively.

As far as perceived control and the presence of health symptoms assessed through the PSI5 are concerned, a significant negative correlation was found for all control parameters, both single and combined. The combined perceived control scores exhibited a stronger negative correlation with the PSI5.

3.4.2. Perceived Control vs. Perceived Overall Comfort—Regression Analysis

The relation between the overall comfort and five perceived control variables has been examined by applying the ordinal regression analysis. The results are presented in Table 7. The maximum OR value (1.28) corresponded to the perceived control over noise. The results showed that if the perceived control over noise increases by one unit in the 1–7 point scale, there is a 1.28 time likelihood that the overall comfort will increase by one unit. Perceived noise control was the parameter with the lowest score according to the occupants' recordings, as shown in Table 2. The impact on the overall comfort of the control over noise, lighting, temperature and shading from the sun, was almost equivalent (ORs: 1.16, 1.14 and 1.12, respectively). Lastly, the perceived control over ventilation, despite the fact that it was ranked as the second worst parameter, had the lowest impact on overall comfort (OR 1.03, 95% CI: 1.00–1.06).

Table 7. Relations between perceived control and perceived overall comfort.

Parameters	OR CI (95%)	p-Value
Perceived control over noise	1.28 (1.25–1.32)	$p < 0.001$
Perceived control over lighting	1.16 (1.13–1.19)	$p < 0.001$
Perceived control over temperature	1.14 (1.11–1.18)	$p < 0.001$
Perceived control over shade	1.12 (1.10–1.15)	$p < 0.001$
Perceived control over ventilation	1.03 (1.00–1.06)	$p < 0.05$

Moreover, occupants were separated in two groups: those who had no or low control and those who had high control for the sum of the five control parameters. The results are presented in Table 8. The ranking remains the same as in the previous case, but two remarks can be highlighted. First, in the "low control" group, occupants were less sensitive to controlling parameters, with lower OR values. In addition, the perceived control over ventilation was not significant ($p > 0.05$). Second, in the "high control" group, occupants were more sensitive to all controlling parameters with increased OR values. The perceived control over lighting showed an almost equal impact on overall comfort as the perceived control over noise.

Table 8. Relations between perceived control in low and high control groups and perceived overall comfort.

Parameters— Low Control Group	OR CI (95%)	p-Value	Parameters— High Control Group	OR CI (95%)	CI (95%)
Perceived control over noise	1.21 (1.15–1.27)	$p < 0.001$	Perceived control over noise	1.37 (1.32–1.43)	$p < 0.001$
Perceived control over lighting	1.14 (1.11–1.17)	$p < 0.001$	Perceived control over lighting	1.32 (1.25–1.39)	$p < 0.001$
Perceived control over temperature	1.11 (1.06–1.15)	$p < 0.001$	Perceived control over temperature	1.26 (1.20–1.31)	$p < 0.001$
Perceived control over shade	1.10 (1.07–1.13)	$p < 0.001$	Perceived control over shade	1.19 (1.14–1.25)	$p < 0.001$
Perceived control over ventilation	0.99 (0.95–1.04)	$p > 0.05$	Perceived control over ventilation	1.08 (1.04–1.12)	$p < 0.001$

4. Discussion

4.1. Occupants' Personal Control

The current study analyzed data from 7441 occupants in 167 European modern office buildings, sustaining and reinforcing findings from previous studies, revealing the importance of the perceived personal control over indoor environment parameters.

The overall occupants' comfort should be examined by investigating the role of every aspect of personal control. Paciuk [16] highlighted three aspects of personal control; (i) available control, (ii) exercised control; (iii) perceived control. Available control can be described by the degree and type of control made available by the environment. It can be defined by 'the degree of manipulation of thermostats and other manual controls as well as the existence of operable windows, blinds, sunshades, ventilation vanes, doors and HVAC system components'. Exercised control is defined by 'the relative frequency in which occupants engage in several types of thermal-related behaviors in order to obtain thermal comfort when needed'. The occupants' perceived control is being produced as the interaction of the different degrees of available (building controls) and exercised control (frequency of use). OFFICAIR database does not include information about the occupants' exercise control so in that study this aspect was not examined. The main idea is that personal control, both available and perceived, works as a moderator affecting the occupants' satisfaction on IEQ parameters and on the overall comfort.

Paciuk [16] examined the perceived control over temperature, ventilation, shading from the sun, lighting and noise. The results showed that occupants have moderate or low control on these parameters with a significant group of occupants, who declared that they had no access to the control of their environment. This was also remarked by Haghighat and Donnini [19], where almost 60% of the responders had no control access. Furthermore, around 30% of the participants were very dissatisfied with the level of control. An interesting relationship between the perceived productivity in offices and perceived control has emerged, where higher productivity levels were recorded when occupants perceived higher degrees of personal control. Boardass et al. [18] also pointed out the issue of productivity versus personal control, concluding in the same finding.

In this analysis, only a limited number of available control types were found to be associated with the perceived control. Perceived control over temperature, ventilation, light and noise was found to be associated only with the presence of operable windows. The effect of operable windows in the indoor environment on occupants' comfort was also raised by Brager et al. [47]. The scores of perceived temperature control differ significantly with the different types of solar shading devices. However, after importing these available control types along with various general building characteristics in the multilevel model, only operable windows remained significant. Similar findings were also reported in another study [48] where opening windows were the most frequent behavior of controlling thermal conditions. The limited effect of the available control on the perceived control was pointed out in previous studies [23,24], which found no association between available and perceived control. This could be explained by the fact that occupants could identify the available controls, but they were not allowed to use them or did not know how to set them up or the control types could not respond rapidly. The multilevel model, containing both individual and building characteristics (both general and available control types), showcased 88% variance in perceived control between buildings. A Building's location and the distance with adjacent facades from other neighboring buildings seem to be significant. The area and number of floors were also found to affect occupant control perception. The number of occupants inside the building showed strong effect, especially inside the office, as well as the type of office (private or open space with partitions). Gou et al. [49] remarked that the provision of personal control in open-plan work environments is an important means to alleviating adverse perceptions. Noise due to conversations and phone calls negatively affected the controlling perception, which was expected as people cannot control themselves the level of noise. Around one-third of the occupants reported negative perception of lighting control, which is similar to findings by Moore et al. [50]. Boardass et al. [18] and Toftum [21] also indicated that the perceived degree of controlling parameters (temperature, ventilation, lighting) decreases with the increased number of occupants inside offices. To some extent, it seems that occupants perceive controlling capabilities better when providing them with the feeling of freedom inside the office buildings. Effort-Reward Imbalance (ERI), which is a critical psychological parameter in office environments, significantly affected control perception and the occupants' perceived comfort inside offices [36]. Moreover, occupants who habitually use air fresheners in their home seem to be more sensitive to adjusting their indoor work environment. The psychological aspect of perceived control is reinforced, as observed by Luo et al. [25].

A significant positive but weak correlation was observed between perceived control and perceived comfort. This finding is in compliance with Boerstra et al. [23], Haghighat and Donnini [19] and Roulet et al. [51]. This work includes a deeper analysis of the appreciation of the perceived control. It should be noted that the analysis using the combined control parameters is not widely used in this type of studies. Combined scores of the controlling parameters were positively correlated with IEQ satisfaction, as Boerstra et al. [23] also found. However, in our study, the combined control perception affected with a stronger correlation the overall comfort, in contrast to the findings of Boerstra et al. [23]. This could be explained by the fact that occupants with higher combined control perception perceive overall comfort with more satisfaction. Combined control perception also resulted in occupants being more satisfied with the overall air quality as well as with the amount of privacy and office layout. In general, occupants who are more able to adjust their environment feel more satisfied.

Regarding building-related health symptoms, the combined scores unveil higher negative correlations with the PSI5. Office occupants reported less health symptoms related to buildings where they perceived higher degree of personal controls.

4.2. Strengths and Limitations

This study has several strengths: a relatively large sample size, a survey performed in different geographical areas across Europe, and the use of standardized procedures (questionnaire and checklist). Data on socio-demographics, psycho-social work environment, and perceived environmental quality were collected by a validated questionnaire. IEQ was assessed using both crude IEQ items (satisfaction with perceived control over temperature, ventilation, shading from the sun, light and noise, as well as satisfaction with overall thermal comfort, noise, light, and indoor air quality), and with 14 detailed indoor environmental parameters (e.g., layout, noise within the building, noise from building systems, noise outside the building, air movement). Some limitations, however, should be noted. Caution is needed when interpreting results, because data on IEQ were self-reported. Consequently, a potential (recall) bias cannot be excluded and this type of surveys do not always capture IEQ issues. A combination of objective and subjective measurements would be useful for assessing IEQ. Another limitation is the cross-sectional study design. Therefore, no causality of the identified relations can be confirmed.

5. Conclusions

The following conclusions can be drawn from the present study:

1. In general, occupants have a moderate or low access in the control of their indoor working environment. They have rated with a moderate control perception the lighting and shading from the sun parameters. Noise was the worst control parameter, while temperature and ventilation control were rated in the low control scale. Among the five control parameters, there was a significant group of occupants who were not able to control their environment at all. Nearly all occupants have no or low control on noise. Half of the occupants declared no or low control on ventilation and temperature conditions. Almost one-third of them remarked that they do not have satisfactory levels of control for lighting and shading from sun conditions. It is noteworthy that as the occupancy within the offices increases, the degree of personal control becomes lower.

2. Occupants with a higher level of personal control were reported to be more productive within their working environment. Moreover, occupants who declared high degrees of personal control reported higher levels of IEQ satisfaction.

3. Some significant correlations were found between the available controls within the building and the perceived control. The availability of operable windows had a higher impact on the occupants' control perception over temperature, ventilation, light and noise than floor area or occupancy. Perceived control over temperature differed significantly with the different types of solar devices.

4. Occupants' perceived control is related to psychological aspects. General building characteristics, such as floor number and floor area, office type, operable windows, etc., which help occupants feel freedom were positively associated with the perceived control. ERI remains a significant parameter of the controlling perception in office environments.

5. Concerning the impact of perceived control on perceived comfort, the results of the present study are in general agreement with the literature. More specifically:

   - The perceived combined control over all studied parameters is positive on the overall comfort
   - The combined control over all parameters seems to have a stronger effect on overall comfort than the single controls
   - Temperature variation seems to be more related to the combined control of temperature and shading from the sun control
   - Combined control perception over temperature, ventilation and shading positively affects the occupants and gives them the feeling of fresher air and an overall satisfaction with air quality
   - Noise from the buildings systems seems to be more affected by the combined control perception of all parameters rather than single controls
   - The combined perceived control of all parameters was found to affect the occupants' privacy, office layout and decoration satisfaction more

6. Regarding reported health symptoms, office occupants with a higher degree of personal controls reported less building-related health symptoms.

Author Contributions: Conceptualization, I.S., D.S., C.M. and J.B.; data curation, I.S.; formal analysis, I.S.; funding acquisition, J.B.; investigation, I.S., D.S., C.M., Y.d.K., S.F., A.S., A.C., T.S. and V.M.; methodology, I.S., C.M., Y.d.K. and J.B.; project administration, J.B.; resources, K.K.; supervision, D.S. and J.B.; visualization, I.S.; writing-original draft, I.S.; writing-review & editing, D.S., C.M., Y.d.K., S.F., A.S., A.C., T.S., V.M., E.d.O.F., K.K., P.C. and J.B.

Acknowledgments: This work was supported by the project "OFFICAIR" (On the Reduction of Health Effects from Combined Exposure to Indoor Air Pollutants in Modern Offices).

References

1. Huang, Y.-H.; Robertson, M.M.; Chang, K.-I. The Role of Environmental Control on Environmental Satisfaction, Communication, and Psychological Stress: Effects of Office Ergonomics Training. *Environ. Behav.* **2004**, *36*, 617–637. [CrossRef]

2. Gee, G.C.; Payne-Sturges, D.C. Environmental Health Disparities: A Framework Integrating Psychosocial and Environmental Concepts. *Environ. Health Perspect.* **2004**, *112*, 1645–1653. [CrossRef]

3. Kalimeri, K.K.; Saraga, D.E.; Lazaridis, V.D.; Legkas, N.A.; Missia, D.A.; Tolis, E.I.; Bartzis, J.G. Indoor air quality investigation of the school environment and estimated health risks: Two-season measurements in primary schools in Kozani, Greece. *Atmos. Pollut. Res.* **2016**, *7*, 1128–1142. [CrossRef]

4. Lee, S.Y.; Brand, J.L. Effects of control over office workspace on perceptions of the work environment and work outcomes. *J. Environ. Psychol.* **2005**, *25*, 323–333. [CrossRef]

5. Huizenga, C.; Abbaszadeh, S.; Zagreus, L.; Arens, E.A. Air quality and thermal comfort in office buildings: Results of a large indoor environmental quality survey. *Proceeding Healthy Build.* **2006**, *3*, 393–397.

6. Brager, G.; Baker, L. Occupant satisfaction in mixed-mode buildings. *Build. Res. Inf.* **2009**, *37*, 369–380. [CrossRef]

7. Langevin, J.; Wen, J.; Gurian, P.L. Relating occupant perceived control and thermal comfort: Statistical analysis on the ASHRAE RP-884 database. *HVAC&R Res.* **2012**, *18*, 179–194.

8. Vine, E.; Lee, E.; Clear, R.; DiBartolomeo, D.; Selkowitz, S. Office worker response to an automated Venetian blind and electric lighting system: A pilot study. *Energy Build.* **1998**, *28*, 205–218. [CrossRef]

9. Lee, S.Y.; Brand, J.L. Can personal control over the physical environment ease distractions in office workplaces? *Ergonomics* **2010**, *53*, 324–335. [CrossRef]

10. Samani, S.A. The Impact of Personal Control over Office Workspace on Environmental Satisfaction and Performance. *J. Soc. Sci. Humanit.* **2015**, *1*, 163–172.

11. Xuan, X. Study of indoor environmental quality and occupant overall comfort and productivity in LEED- and non-LEED–certified healthcare settings. *Indoor Built Environ.* **2018**, *27*, 544–560. [CrossRef]

12. Yun, G.Y. Influences of perceived control on thermal comfort and energy use in buildings. *Energy Build.* **2018**, *158*, 822–830. [CrossRef]

13. Zhao, Z.; Amasyali, K.; Chamoun, R.; El-Gohary, N. Occupants' Perceptions about Indoor Environment Comfort and Energy Related Values in Commercial and Residential Buildings. *Procedia Environ. Sci.* **2016**, *34*, 631–640. [CrossRef]

14. Al horr, Y.; Arif, M.; Katafygiotou, M.; Mazroei, A.; Kaushik, A.; Elsarrag, E. Impact of indoor environmental quality on occupant well-being and comfort: A review of the literature. *Int. J. Sustain. Built Environ.* **2016**, *5*, 1–11. [CrossRef]

15. Frontczak, M.; Schiavon, S.; Goins, J.; Arens, E.; Zhang, H.; Wargocki, P. Quantitative relationships between occupant satisfaction and satisfaction aspects of indoor environmental quality and building design. *Indoor Air* **2012**, *22*, 119–131. [CrossRef]

16. Paciuk, M. *The Role of Personal Control of the Environment in Thermal Comfort and Satisfaction at the Workplace*; University of Wisconsin-Milwaukee: Milwaukee, WI, USA, 1989.

17. Collins, B.L.; Fisher, W.; Gillette, G.; Marans, R.W. Second-Level Post-Occupancy Evaluation Analysis. *J. Illum. Eng. Soc.* **1990**, *19*, 21–44. [CrossRef]

18. Bordass, B.; Bromley, K.; Leaman, A. *User and Occupant Controls in Office Buildings*; BRE: Brussels, Belgium, 1993; p. 13.

19. Haghighat, F.; Donnini, G. Impact of psycho-social factors on perception of the indoor air environment studies in 12 office buildings. *Build. Environ.* **1999**, *34*, 479–503. [CrossRef]

20. Zweers, T.; Preller, L.; Brunekreef, B.; Boleij, J.S.M. Health and Indoor Climate Complaints of 7043 office Workers in 61 Buildings in the Netherlands. *Indoor Air* **1992**, *2*, 127–136. [CrossRef]

21. Toftum, J. Central automatic control or distributed occupant control for better indoor environment quality in the future. *Build. Environ.* **2010**, *45*, 23–28. [CrossRef]

22. Zagreus, L.; Huizenga, C.; Arens, E.; Lehrer, D. Listening to the occupants: A Web-based indoor environmental quality survey. *Indoor Air* **2004**, *14*, 65–74. [CrossRef]
23. Boerstra, A.; Beuker, T.; Loomans, M.; Hensen, J. Impact of available and perceived control on comfort and health in European offices. *Archit. Sci. Rev.* **2013**, *56*, 30–41. [CrossRef]
24. Kwon, M.; Remøy, H.; van den Dobbelsteen, A.; Knaack, U. Personal control and environmental user satisfaction in office buildings: Results of case studies in the Netherlands. *Build. Environ.* **2019**, *149*, 428–435. [CrossRef]
25. Luo, M.; Cao, B.; Ji, W.; Ouyang, Q.; Lin, B.; Zhu, Y. The underlying linkage between personal control and thermal comfort: Psychological or physical effects? *Energy Build.* **2016**, *111*, 56–63. [CrossRef]
26. Karjalainen, S. Thermal comfort and use of thermostats in Finnish homes and offices. *Build. Environ.* **2009**, *44*, 1237–1245. [CrossRef]
27. De Dear, R.; Brager, G.S. Developing an Adaptive Model of Thermal Comfort and Preference. *ASHRAE Trans.* **1998**, *104*, 1–18.
28. *ASHRAE Standard 55 Thermal Environmental Conditions for Human Occupancy*; ANSI/ASHRAE: Atlanta, GA, USA, 2013.
29. International Organization for Standardization ISO 7730:2005. Available online: http://www.iso.org/cms/render/live/en/sites/isoorg/contents/data/standard/03/91/39155.html (accessed on 26 July 2019).
30. Olesen, B.W.; Parsons, K.C. Introduction to thermal comfort standards and to the proposed new version of EN ISO 7730. *Energy Build.* **2002**, *34*, 537–548. [CrossRef]
31. OFFICAIR Project. Available online: http://www.officair-project.eu/ (accessed on 6 August 2019).
32. Nørgaard, A.W.; Kofoed-Sørensen, V.; Mandin, C.; Ventura, G.; Mabilia, R.; Perreca, E.; Cattaneo, A.; Spinazzè, A.; Mihucz, V.G.; Szigeti, T.; et al. Ozone-initiated Terpene Reaction Products in Five European Offices: Replacement of a Floor Cleaning Agent. *Environ. Sci. Technol.* **2014**, *48*, 13331–13339. [CrossRef]
33. Mihucz, V.G.; Szigeti, T.; Dunster, C.; Giannoni, M.; de Kluizenaar, Y.; Cattaneo, A.; Mandin, C.; Bartzis, J.G.; Lucarelli, F.; Kelly, F.J.; et al. An integrated approach for the chemical characterization and oxidative potential assessment of indoor PM2.5. *Microchem. J.* **2015**, *119*, 22–29. [CrossRef]
34. Bluyssen, P.M.; Roda, C.; Mandin, C.; Fossati, S.; Carrer, P.; de Kluizenaar, Y.; Mihucz, V.G.; de Oliveira Fernandes, E.; Bartzis, J. Self-reported health and comfort in 'modern' office buildings: First results from the European OFFICAIR study. *Indoor Air* **2016**, *26*, 298–317. [CrossRef]
35. De Kluizenaar, Y.; Roda, C.; Dijkstra, N.E.; Fossati, S.; Mandin, C.; Mihucz, V.G.; Hänninen, O.; de Oliveira Fernandes, E.; Silva, G.V.; Carrer, P.; et al. Office characteristics and dry eye complaints in European workers-The OFFICAIR study. *Build. Environ.* **2016**, *102*, 54–63. [CrossRef]
36. Sakellaris, I.; Saraga, D.; Mandin, C.; Roda, C.; Fossati, S.; de Kluizenaar, Y.; Carrer, P.; Dimitroulopoulou, S.; Mihucz, V.; Szigeti, T.; et al. Perceived Indoor Environment and Occupants' Comfort in European "Modern" Office Buildings: The OFFICAIR Study. *Int. J. Environ. Res. Public Health* **2016**, *13*, 444. [CrossRef]
37. Szigeti, T.; Dunster, C.; Cattaneo, A.; Cavallo, D.; Spinazzè, A.; Saraga, D.E.; Sakellaris, I.A.; de Kluizenaar, Y.; Cornelissen, E.J.M.; Hänninen, O.; et al. Oxidative potential and chemical composition of PM2.5 in office buildings across Europe—The OFFICAIR study. *Environ. Int.* **2016**, *92–93*, 324–333. [CrossRef]
38. Campagnolo, D.; Saraga, D.E.; Cattaneo, A.; Spinazzè, A.; Mandin, C.; Mabilia, R.; Perreca, E.; Sakellaris, I.; Canha, N.; Mihucz, V.G.; et al. VOCs and aldehydes source identification in European office buildings—The OFFICAIR study. *Build. Environ.* **2017**, *115*, 18–24. [CrossRef]
39. Mandin, C.; Trantallidi, M.; Cattaneo, A.; Canha, N.; Mihucz, V.G.; Szigeti, T.; Mabilia, R.; Perreca, E.; Spinazzè, A.; Fossati, S.; et al. Assessment of indoor air quality in office buildings across Europe—The OFFICAIR study. *Sci. Total Environ.* **2017**, *579*, 169–178. [CrossRef]
40. Szigeti, T.; Dunster, C.; Cattaneo, A.; Spinazzè, A.; Mandin, C.; Le Ponner, E.; de Oliveira Fernandes, E.; Ventura, G.; Saraga, D.E.; Sakellaris, I.A.; et al. Spatial and temporal variation of particulate matter characteristics within office buildings—The OFFICAIR study. *Sci. Total Environ.* **2017**, *587–588*, 59–67. [CrossRef]
41. Raw, G.J.; Roys, M.S.; Whitehead, C.; Tong, D. Questionnaire design for sick building syndrome: An empirical comparison of options. *Environ. Int.* **1996**, *22*, 61–72. [CrossRef]
42. Theodorsson-Norheim, E. Kruskal-Wallis test: BASIC computer program to perform nonparametric one-way analysis of variance and multiple comparisons on ranks of several independent samples. *Comput. Methods Programs Biomed.* **1986**, *23*, 57–62. [CrossRef]

43. Gelman, A.; Hill, J. *Data Analysis Using Regression and Multilevel/Hierarchical Models*; Cambridge University Press: Cambridge, UK, 2007; ISBN 978-0-521-68689-1.

44. Goldstein, H. *Multilevel Statistical Models*; John Wiley & Sons: West Sussex, UK, 2011; ISBN 978-1-119-95682-2.

45. Mukaka, M. A guide to appropriate use of Correlation coefficient in medical research. *Malawi Med. J.* **2012**, *24*, 69–71.

46. *IBM SPSS Statistics for Windows*; Version 22.0; IBM: Armonk, NY, USA, 2013.

47. Brager, G.; Paliaga, G.; de Dear, R.; Olesen, B.; Wen, J.; Nicol, F.; Humphreys, M. Operable Windows, Personal Control, and Occupant Comfort. *ASHRAE Trans.* **2004**, *110*, 17–35.

48. Raja, I.A.; Nicol, J.F.; McCartney, K.J.; Humphreys, M.A. Thermal comfort: Use of controls in naturally ventilated buildings. *Energy Build.* **2001**, *33*, 235–244. [CrossRef]

49. Gou, Z.; Zhang, J.; Shutter, L. The Role of Personal Control in Alleviating Negative Perceptions in the Open-Plan Workplace. *Buildings* **2018**, *8*, 110. [CrossRef]

50. Moore, T.; Carter, D.; Slater, A. User attitudes toward occupant controlled office lighting. *Light. Res. Technol.* **2002**, *34*, 207–216. [CrossRef]

51. Roulet, C.-A.; Johner, N.; Foradini, F.; Bluyssen, P.; Cox, C.; De Oliveira Fernandes, E.; Müller, B.; Aizlewood, C. Perceived health and comfort in relation to energy use and building characteristics. *Build. Res. Inf.* **2006**, *34*, 467–474. [CrossRef]

6

Indoor Comfort and Symptomatology in Non-University Educational Buildings: Occupants' Perception

Miguel Ángel Campano-Laborda, Samuel Domínguez-Amarillo, Jesica Fernández-Agüera * and Ignacio Acosta

Instituto Universitario de Arquitectura y Ciencias de la Construcción, Escuela Técnica Superior de Arquitectura, Universidad de Sevilla, 41012 Seville, Spain; mcampano@us.es (M.Á.C.-L.); sdomin@us.es (S.D.-A.); iacosta@us.es (I.A.)
* Correspondence: jfernandezaguera@us.es.

Abstract: The indoor environment in non-university classrooms is one of the most analyzed problems in the thermal comfort and indoor air quality (IAQ) areas. Traditional schools in southern Europe are usually equipped with heating-only systems and naturally ventilated, but climate change processes are both progressively increasing average temperatures and lengthening the warm periods. In addition, air renewal is relayed in these buildings to uncontrolled infiltration and windows' operation, but urban environmental pollution is exacerbating allergies and respiratory conditions among the youth population. In this way, this exposure has a significant effect on both the academic performance and the general health of the users. Thus, the analysis of the occupants' noticed symptoms and their perception of the indoor environment is identified as a potential complementary tool to a more comprehensive indoor comfort assessment. The research presents an analysis based on environmental sensation votes, perception, and indoor-related symptoms described by students during lessons contrasted with physical and measured parameters and operational scenarios. This methodology is applied to 47 case studies in naturally ventilated classrooms in southern Europe. The main conclusions are related to the direct influence of windows' operation on symptoms like tiredness, as well as the low impact of CO_2 concentration variance on symptomatology because they usually exceeded recommended levels. In addition, this work found a relationship between symptoms under study with temperature values and the environmental perception votes, and the special impact of the lack of suitable ventilation and air purifier systems together with the inadequacy of current thermal systems.

Keywords: educational buildings; schools; field measurements; ventilation; indoor air quality (IAQ); thermal comfort; thermal perception; health symptoms; CO_2 concentration; air infiltration

1. Introduction

1.1. State of the Art

Non-university educational buildings are one of the most widespread building typologies, in which teenagers, a more sensitive population than adults and with specific different thermal preferences due to their different metabolic rate values [1–4], spend more than 25% of their day time during winter and midseasons. Thus, indoor environment in non-university classrooms is one of the most analyzed problems in the thermal comfort and indoor air quality (IAQ) areas [5], being widely studied for cold [6–10], mild [11–17], and warm climates [18].

Traditional schools in southern Europe solve thermal control basically by heating-only systems (without mechanical ventilation), relying on air renewal to uncontrolled infiltration and users' frequent windows' operation, much more than usually found in central and northern Europe. This develops a behavior that could be defined as hybrid or mixed mode, with thermal systems operated and with a significant part of the time with the windows open. In addition, climate change processes are progressively lengthening the warm periods with greater presence within the school season. In addition, urban environmental pollution and pollen are exacerbating allergies and respiratory conditions among the youth population [19,20], especially in the case of outdoor atmospheric particulate matter (PM) with a diameter of less than 2.5 micrometers (PM 2.5) [12,21,22]. This context generates a situation of specificity where further study is necessary, given the different exposure scenarios with a greater influx from the outside although varying over time.

Given that ventilation is one of the main variables which affects the degree of environmental comfort [23,24], the European ventilation standard EN 13779:2008 [25], through its Spanish transposition [26], establishes a minimum outdoor airflow to guarantee the adequate indoor air quality (IAQ) in non-residential buildings. Mainly, its focus is to control CO_2 concentration, pollutants, and suspended particles [27] to avoid the development of symptomatology and respiratory health related to prolonged periods of exposure [28]. According to the national regulation, this ventilation must be mechanically controlled since 2007, also including an air filtering system, to ensure this IAQ, but given that the adaptation could entail a huge investment and a higher energy consumption, several public institutions in Spain are imposing natural ventilation as the only system for IAQ control, against standards.

In this way, previous studies in classrooms of southern Spain [16,17], Portugal [12], France [29], Italy [30], and other south European locations [31] have shown poor indoor conditions, both thermal and clean air, which can relate to the appearance of symptoms like dizziness, dry skin, headache, or tiredness. This environmental exposure has a significant effect on both the academic performance [32–34], the general health of the users and their psychological and social development [35], existing evidences of poor indoor air quality in schools with correlation with negative effects on the students' health, which potentially can lead to asthma or allergic diseases [36], which are two of the most prevalent diseases in children and young people [37], and can be mainly related to the high values found in classrooms for bacteria and PM, given their pro-inflammatory role [38].

In this way, previous studies in European schools analyzed the link between the IAQ conditions, obtained through measurements of CO_2, PM, and volatile organic compounds (VOCs), with health questionnaires made by parents, spirometry, exhaled nitric oxide tests, and asthma tests with medical kits [29,38]. This approach required complex equipment and tests, and were not directly related to on-site symptomatology but to long-term symptom development, as it was gathered in housing studies [39]. Users' perception of environmentally related symptoms had a direct potential to draw an actual comfort situation, not only determined by room-physical conditions but to occupants' responses, as was shown in [40–43], also with the capacity to identify individual answers, such as those related to gender or emotional situation [44–46].

Thus, the analysis of the occupants' symptoms and their environmental perception was identified as a potentially affordable complementary tool to obtain a more accurate indoor comfort condition assessment with a high degree of widespread applicability, together with the widely accepted rational (RTC) [23,47,48] or adaptive (ATC) [49,50] thermal comfort indicators, especially those analyzed in the Mediterranean area including educational buildings [51–53] or in non-air conditioned buildings in warm climates [54].

1.2. Objectives

The first objective of this research was to present the physical and operational characterization of the indoor environment of a representative sample of multipurpose classrooms in a wide area of southern Spain, as well as the environmental perception votes, personal clothing, and symptoms expressed by the occupants (aged 12–17 years) exposed to this environment during the measurement campaigns.

The second objective of the study was to contrast environmental sensation votes, perception, and indoor-related symptoms described by students during lessons with physical and environmental parameters and operational scenarios (focusing on windows' and doors' operation), in order to evaluate the impact and relationship between them.

2. Methods and Materials

The acquisition of both the physical measurement data and the occupants' sensation votes during a normal school day was developed through the following phases:

(1) Definition of the study sample;
(2) Characterization of the airtightness of the samples;
(3) Field measurements; and
(4) Design and distribution of surveys.

The data collection was performed both in winter and midseason in two sets per day: One in the early morning, at the beginning of the first lesson, and another previous to the midmorning break.

2.1. Definition of the Study Sample

The study sample was composed of 47 multipurpose classrooms (for ages 12–17) from 8 educational buildings, selected from the most representative climate zones of the region of Andalusia according to the Spanish energy performance zoning [55–57] (zones A4, B4, C3, and C4), which include temperate to cold zones in winter (types A, B, or C), as well as average to warm summers (3 or 4). These zones can also be classed in the Köppen climate scale [58] as cold semi-arid climate (Köppen BSk) and hot summer Mediterranean climate (Köppen CSa), as it can be seen in Table 1.

Table 1. Study samples by location and climate zone.

Climate Zone		Educational Institution	Classrooms	Occupants
Köppen Climate Zone [58]	Spanish Energy Performance Zone			
BSk	A4	E1	4	92
CSa	B4	E2	3	54
CSa	B4	E3	8	192
CSa	B4	E4	11	186
CSa	C3	E5	2	45
CSa	C3	E6	12	270
CSa	C3	E7	3	60
CSa	C4	E8	4	78

These multipurpose classrooms followed the design standards established by the regional educational agency (Andalusian Agency of Public Education) [59], with classrooms measuring approximately 50 m^2 and 3 meters high for accommodating up to 30 students with their teacher. This standard also defined the common access corridor with the adjoining classrooms and the distribution of the furniture, as well as the location of the windows to the left of the occupants and the two entrance doors in the partition to the corridor, as it can be seen in Figure 1.

The most common composition of the external vertical wall was a brick masonry cavity wall with some kind of thermal insulation with a simple hollow brick wall with plaster setting in the inner surface. The internal partitions were usually composed of a half-brick wall with plaster on either side.

The regional standards established hot water (HW) radiators as the main heating system of schools, with no provision for cooling systems [59]. In this way, all the classrooms under study were equipped with this heating system and, in addition, two of the schools had some add-on split-systems for cooling. Ventilation is traditionally performed in the Mediterranean area by user windows' operation and uncontrolled infiltrations. Despite the current Spanish standard on thermal installations in buildings (RITE) [26] that establishes the mechanical ventilation as the only option for new non-residential buildings, these systems are not normally started up in order to save energy when the building is equipped with them.

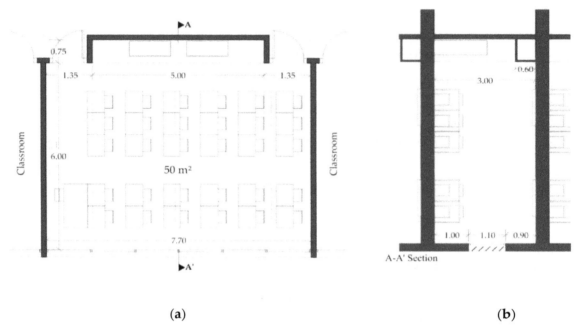

(a)   (b)

Figure 1. Multipurpose classroom according to Andalusian design standards: (**a**) Plant with windows, doors, and furniture standard distribution. (**b**) A-A' vertical section.

2.2. Characterization of the Airtightness of the Samples

The assessment of the infiltration level of the classrooms under study was performed by a series of airtightness tests (doors and windows closed) in order to obtain their expected average infiltrations rates (Figure 2). These tests consisted of decreasing the room pressure by using a fan, which extracted air until the indoor-outdoor differential pressure was stabilized. It was achieved by balancing the extracted airflow with the entering airflow through the envelope cracks. Then, the depressurization was decreased in steps by lowering the fan speed, in order to obtain the regression curve of the pressure/extracted airflow relation, which showed the entering airflow when the indoor pressure was equal to the atmospheric one.

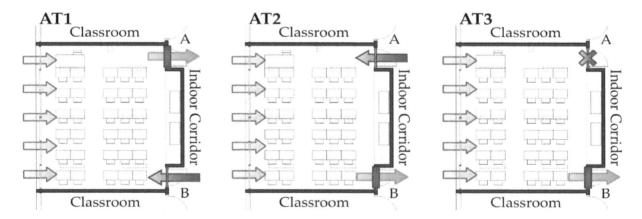

Figure 2. Protocols developed for the characterization of the airtightness of the classrooms.

These tests were performed by using enclosure pressurization-depressurization equipment or "blower door", as specified in the ISO standard 9972: 2015 [6], considering each classroom as a single zone to be analyzed. The specific model used was the Minneapolis Blower Door Model 4/230 V System, which was controlled by the TECTITE Express software.

The higher-pressure difference used to create this regression curve must be at least ± 50 Pa; in this study, it was reached until a ± 70 Pa differential pressure.

When the classroom had a single access point, the pressurization-depressurization test characterized the airflow that can pass through the envelope by sealing the corresponding door and the adjacent classrooms and common area. However, in most of the studied classrooms there were two access points, so it was necessary to perform three measurements in each classroom, changing the location of the blower door and sealing, or not, and the door in which the blower door was disposed. In this way, it was possible to determine the real airflow that entered the classroom during its normal operation. Adjacent classrooms and common area were sealed, too.

This protocol was designed for medium rooms with two access doors like the one under study, and required three different measurements (Figure 2):

- Airtight test 1 (AT1) to obtain $V_{50,P1}$: Blower door was placed in door A, and door B and windows were closed but not sealed.
- Airtight test 2 (AT2) to obtain $V_{50,P2}$: Blower door was placed in door B, and door A and windows were closed but not sealed.
- Airtight test 3 (AT3) to obtain $V_{50,P3}$: Blower door was placed in door B, and door A was sealed and windows were closed but not sealed.

where $V_{50\,P1}$ is the air leakage rate at 50 Pa in Protocol 1, $V_{50\,P2}$ is the air leakage rate at 50 Pa in Protocol 2, $V_{50\,P3}$ is the air leakage rate at 50 Pa in Protocol 3.

Infiltration values measured in each of these three ± 50 Pa depressurization test hypotheses, developed in each classroom, were obtained by the following expressions of the British Standard 5925 standard, obtained from a simplification of the "crack flow equation":

$$n_{50,AT1} = \frac{V_{50,DoorA} + V_{50,env}}{V} \tag{1}$$

$$n_{50,AT2} = \frac{V_{50,DoorB} + V_{50,env}}{V} \tag{2}$$

$$n_{50,AT3} = \frac{V_{50,env}}{V} \tag{3}$$

$$n_{50,t} = \frac{V_{50,DoorA} + V_{50,DoorB} + V_{50,env}}{V} \tag{4}$$

$$n_{50,t} = n_{50,P1} + n_{50,P2} - n_{50,P3} \tag{5}$$

where $n_{50,AT1}$ is the infiltration rate at 50 Pa in protocol 1, in h^{-1}; $n_{50,AT2}$ is the infiltration rate at 50 Pa in protocol 2, in h^{-1}; $n_{50,AT3}$ is the infiltration rate at 50 Pa in protocol 3, in h^{-1}; $n_{50,t}$ is the infiltration rate at 50 Pa through the envelope and doors of the room, in h^{-1}. $V_{50,DoorA}$ is the air leakage rate at 50 Pa which circulates through door A, in m^3/h; $V_{50,DoorB}$ is the air leakage rate at 50 Pa which circulates through door B, in m^3/h; $V_{50,env}$ is the air leakage rate at 50 Pa which circulates through the envelope, in m^3/h; V is the internal volume of the room, in m^3.

2.3. Field Measurements

The measurement campaign was developed in the selected classrooms according a data collection protocol [16,17,60] in which physical parameters relating to hygrothermal comfort, CO_2 concentration, and illuminance were obtained in a spatial matrix [61] previously and throughout the survey distribution period (30 min, twice per day), taking the average of the values obtained, both outdoor and indoor, as can be seen in Figure 3 and Table 2:

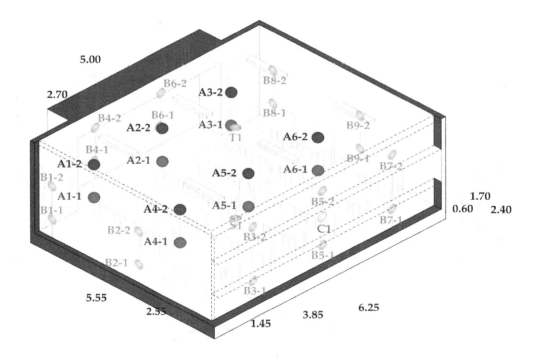

Figure 3. The 3-D array of measurement points superimposed in a multipurpose classroom. Red dots show values at 0.6 m, dark red dots at 1.7 m, orange dots represent measures in the room's envelope, and blue dot is the window.

Table 2. Acquisition points per physical parameters measured.

Parameter	Indoor Points	Outdoor Measurement
Air temperature (T_a)	All "A" points (0.60 and 1.70 m)	Yes
Surface temperature (t_s)	All "B" points (0.60 and 1.70 m), S1, T1, C1	No
Globe temperature (t_g)	A2-1, A6-1 (0.60 m)	No
Relative humidity (RH)	All "A" points (0.60 and 1.70 m)	Yes
Air velocity (V_a)	All "A" points (0.60 and 1.70 m)	Yes
CO_2 concentration (CO_2)	All "A" points (0.60 and 1.70 m)	Yes
Illuminance (E)	All "Ax-1" points (0.60 m)	No

During the measurements both the initial state and the operation of doors, windows, solar devices, heating systems, and electric lighting, as well as changes in the occupants' distribution, were collected. Performance and uncertainty of measurement instrumental are described in Table 3:

Table 3. Characteristics and uncertainty of the sensors used for on-site measurements.

Parameter	Sensor [1]	Units	Uncertainty
Air temperature (T_a)	Testo 0635.1535 (PT100)	°C	±0.3 °C
Surface temperature (t_s)	Testo 0602.0393 (Thermocouple type K)	°C	±0.3 °C
Mean radiant temperature (t_r)	Testo 0602.0743 (Globe probe, Thermocouple type K)	°C	±1.5 °C
Globe temperature (t_g)	Testo 0602 0743	°C	±1.5 °C
Relative humidity (RH)	Testo 0635.1535 (Capacitive)	%	±2%
Air velocity (V_a)	Testo 0635.1535 (Hot wire)	m/s	±0.03 m/s
CO_2 concentration (CO_2)	Testo 0632.1535	ppm	±50 ppm
Data acquisition system	Data Logger Testo 435-2	-	-
Illuminance (E)	PCE-134 lux-meter	lux	±5%

[1] The instruments for hygrothermal measurements listed comply with the requirements of ISO 7726 standard [62] for class C (comfort).

Mean radiant temperature ($\bar{t}_r$, in °C) was calculated using Equation (6) [63]:

$$\bar{t}_r = \left[\left(t_g + 273 \right)^4 + \frac{1.10 \cdot 10^8 \cdot v_a^{0.6}}{\varepsilon \cdot Dg^{0.4}} \cdot \left(t_g - T_a \right) \right]^{\frac{1}{4}} - 273, \tag{6}$$

where v_a is air velocity (in m/s), T_a is dry bulb air temperature (°C), t_g is black globe temperature (°C), ε is emissivity (dimensionless, 0.95 for black globe), and Dg is globe diameter (m).

The operative temperature (t_o, in °C) was obtained through Equations (7) and (8) [64]:

$$T_o = A \times T_a + (1 - A) \times \bar{t}_r \tag{7}$$

$$A = \begin{cases} 0.5 & \text{if } V_a < 0.2 \text{ m/s} \\ 0.6 & \text{if } 0.2 \text{ m/s} \le v_a < 0.6 \text{ m/s} \\ 0.7 & \text{if } 0.6 \text{ m/s} \le v_a < 1.0 \text{ m/s} \end{cases} \tag{8}$$

where v_a is air velocity (in m/s), T_a is dry bulb air temperature (°C), and t_g is black globe temperature (°C).

2.4. Design and Distribution of Surveys

The survey was designed to collect information and votes from occupants in order to comprehensively assess the environment in conjunction with the measurement of the physical parameters. The survey design was based on the experience of previous research [16,17,60,61,65] with the aim to collect data with an objective approach. The completion of the survey took around 20 minutes per classroom, being distributed during the measurement campaigns, and were performed twice per day both in a winter and a midseason day.

The survey distributed included questions about the following issues (the specific layout of the questionnaire is presented in Appendix A):

(1)   The occupant's age and sex.
(2)   The respondent's position inside the classroom.
(3)   The occupant's thermal vote [23] for:

- Sensation: Thermal sensation vote (TSV) using the 7-points ASHRAE scale.
- Preference: Thermal preference vote (TPV) using the 7-points ASHRAE scale.
- Acceptance: Thermal environment rejection percentage (PD_{acc}) from 0 (rejection) to 1 (acceptance).
- Level of comfort: Thermal comfort vote (TCV) from 4 (extremely uncomfortable) to 0 (comfortable).

(4)   The occupant's environmental perception vote (EPV) from 4 (repugnant odor) to 0 (without odor).
(5)   Symptoms and related health effects during the measurements:

- Difficulty concentrating (DC).
- Dry throat (DT).
- Dizziness (D).
- Itchiness (I).
- Dry skin (DS).
- Nausea (N).
- Nasal congestion (NC).
- Eye irritation (EI).
- Headache (H).
- Chest oppression (CO).
- Tiredness (T).

The perception of the hygrothermal environment was formulated to the occupants according the protocol established in the Spanish version of Standard ISO 10551 [66].

The clothing insulation values worn by the occupants were obtained from the surveys and subsequently quantified according to EN ISO 9920 [67] and EN ISO 7730 [23], considering the corrections proposed by Havenith et al. [68] for seated occupants, with a thermal insulation of clothing (I_{cl}) lower than 1.84 clo and air velocities under 0.15 m/s.

In addition, the protocol for the analysis of the surveys included a screening for the exclusion of the sample when the participant had previous and subsequent symptoms, related health problems, were developing some sickness, were taking medication for a long time, or when a strange answer was found for the multiple choices of a given question. In this way, during the measurements, a total number of 977 valid surveys was obtained (Table 4).

Table 4. Students participating in the survey campaign according to season and sex.

	Students	Average Value of Students per Classroom	Male Students	Female Students
All seasons	977	20.8	504	473
Winter season	693	20.4	364	329
Mid-seasons	284	21.9	140	144

3. Results

The results of the present study, part of a PhD dissertation [69], can be grouped into five subsections:

- Mean values of physical parameters;
- Mean values of airtightness of the samples;
- Mean values of occupants' votes;
- Mean values of occupants' clothing insulation; or
- Mean values of occupants' symptoms and related health effects.

These values were analyzed according to seasons (winter, W, and midseasons, MS) and windows' and doors' operation (open windows, OW, closed windows, CW, open door, OD). In this way, 26 classrooms (55% of the case studies) had the windows closed during the measurement period, with 23 of these during the winter session, and 21 had the windows open (45% of the case studies), with 11 of them during the winter period. No intervention by the researchers was made to modify classroom-state, allowing us to gather operational actual conditions.

3.1. Mean Values of Physical Parameters

The measured interior air temperature (Ta) ranged between 17.8 and 22.7 °C during winter season (Table 5), with the lowest mean temperature values obtained for the case studies with closed windows, especially when inner doors were open, with values of 20 °C. It can be related with the outdoor conditions, given that the lowest outdoor temperature values (Ta, outdoor) were measured for classrooms with closed windows and open inner doors. In addition, 8 of the case studies had the windows open during winter, which can mean that there was a bad regulation of the heating system and the heat excess had to be dissipated, or that the students considered that they had to ventilate the classroom due to a poor environment perception. Indoor air temperature in midseasons was oscillating around 22.4 °C, without a direct relation with window operation. Although winter time temperature expectations range between near 20–22 °C to 20.4–22.6 °C if windows are open (central quartile lower and upper values), this band nearly doubles in middle season, when temperatures from 20.6 to 24 °C may be expected (21.1 to 24.5 °C if windows are open). A quartile distribution plot for indoor thermal parameters, air temperature, and operative temperature is proposed in Figure 4. It is noteworthy to highlight that there was a statistical significance between seasons in a windows-state with independent behavior aspect that was verified through test of comparison of samples, F-test for the variance and a K-S (Kolmogórov-Smirnov) for the distributions of probability with p-values under 0.05 in all the cases.

Table 5. Mean values of environmental parameters obtained during the field measurements related to seasons and windows' and doors' operation.

		W	OW-W	CW-W	CW OD-W	MS	CW-MS	OW-MS
$T_{a,outdoor}$	Mean	9.9	11.4	9.1	8.3	18.5	18.5	18.4
(°C)	SD	4.4	2.8	4.9	5.1	5.3	3.1	5.9
T_a	Mean	21.1	21.5	21.0	20.0	22.4	22.3	22.5
(°C)	SD	1.4	1.3	1.5	0.8	2.1	1.6	2.3
$\overline{t_r}$	Mean	21.8	21.4	22.0	22.5	22.9	23.3	22.8
(°C)	SD	2.1	2.6	1.7	1.3	3.5	2.9	3.7
RH	Mean	51	49	51	55	45	44	46
(%)	SD	6	5	6	6	10	9	10
V_a	Mean	0.03	0.04	0.03	0.03	0.01	0.01	0.01
(m/s)	SD	0.03	0.04	0.02	0.01	0.01	0.01	0.01
CO_2 indoor	Mean	1951	1537	2164	1973	1267	1006	1354
(ppm)	SD	552	234	548	460	499	284	525
CO_2 outdoor	Mean	426	421	429	475	395	399	394
(ppm)	SD	41	47	38	5	18	15	19

W are measurements during winter, MS are measurements during midseasons, OW-W are measurements during winter with open windows, OW-MS are measurements during midseasons with open windows, CW-W are measurements during winter with closed windows, CW-MS are measurements during midseasons with closed windows, CW OD-W are measurements during winter with closed windows and open doors, SD are standard deviation.

Although average values of mean radiant temperature ($\overline{t_r}$) were within the recommended operating temperature ranges for classrooms according to ISO 7730 standard [23] (22.0 ± 2.0 °C for category B), there was a high dispersion of figures with a standard deviation (SD) between 1.7 °C in winter with closed windows and 3.7 °C in midseasons with open windows, which was due to the operation of HW radiator system, especially when windows were open, with $\overline{t_r}$ values of 27.2–28.0 °C with radiators on and values of 17.5–19.0 °C when radiators were turned off. This caused operative temperature to swing usually between 20 and 25 (central quartiles) during middle season, with typical values of 20.6 to 22.5 °C during winter, with a very similar band of 20.4 to 22.6 °C if windows were open, highlighting the effect of surface thermal control performed by the radiator heating system.

Relative humidity (RH) in winter was always over 40%, with a maximum value of 64% in the case of one of the classrooms with windows closed and inner doors opened. In midseasons, relative humidity was lower but with a higher oscillation, with a minimum value of 29%.

Air velocity (V_a) values were oscillating under 0.05 m/s, both in winter and midseasons, only exceeding the recommended design limit for comfort category B established by the ISO 7730 standard [23] of 0.16 m/s in one of the case studies with open windows, with a value of 0.18 m/s. In the case of closed windows, air velocity was always under 0.09 m/s. This showed poor air movement and limited air displacement potential.

Measurements of the CO_2 concentration usually show figures well above typical thresholds (Figure 5). The World Health Organization (WHO) recommends a limit for healthy indoor spaces of 1000 ppm [70]. In this way, the probability distribution derived from the measures showed that more than 92% of the distribution for closed windows was above this limit, while this only decreased to 88% of the time when windows were open. In addition, 47.5% of classrooms with windows closed exceeded the 2000 ppm threshold. The greatest relative effect of window operation was seen in the winter, when CO_2 concentration can be decreased by 25%, comparing median values. However, figures were above desirable levels, indicating the lack of capacity of the window operation to solve a suitable ventilation. In general, during the intermediate season, the operation of the windows did not provide a significant improvement of indoor air quality, which may be related to the lack of thermal differential between indoor and outdoor air, limiting the air exchange due to the absence of a thermodynamic effect (Figure 6).

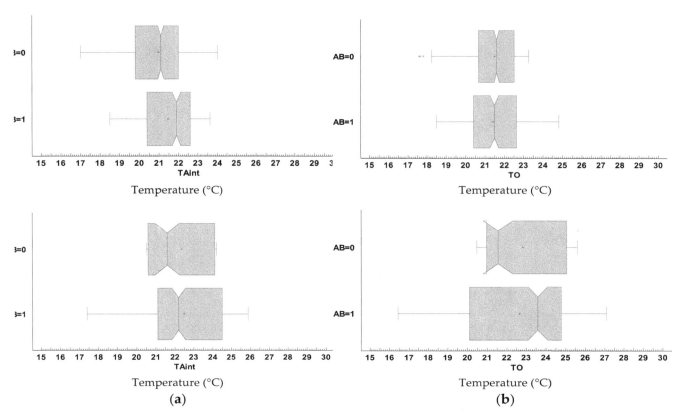

Figure 4. Quartile distribution for indoor air temperature and operative temperature. (**a**) Quartile distribution for indoor air temperature; (T_o) winter (top) and mid-season (down) with windows closed (0) and open (1). (**b**) Quartile distribution for operative temperature (T_o); winter (top) and mid-season (down) with windows closed (0) and open (1).

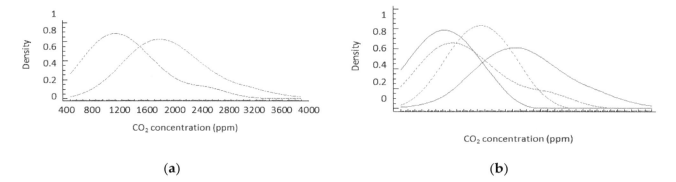

Figure 5. CO_2 concentration distribution. (**a**) General CO_2 concentration density trace for winter (blue) and mid-season (red). (**b**) Detailed CO_2 concentration density trace for winter (blue) and mid-season (red) with closed windows (continuous line) and windows open (dot line).

The median room mean illuminance (E) in the case studies oscillated between 461 and 560 lx (both cases with a SD of 222) according to the season, with an average lighting uniformity (U_o) of 0.48. However, although there seemed to be a greater illumination associated with the half-season period, it was not possible to rule out, without further measurements, the fact of being biased by the activities in execution during the measurements. The high SD was due both to the use of the projector and the solar protection devices (as low as 15 lx) and the lack of use of a solar protection device with direct solar radiation (figures as high as 1710 lx) (Figure 7). The homogeneity of the lighting solutions in almost all buildings generated visual fields with very similar characteristics, mainly dominated by the behavior of their electric lighting. The correlated color temperature was similar in all cases, varying from 3500 to 5500 K; hence, it can be considered that both the amount of light and hue did not affect the thermal perception of the participants, as exposed by Bellia et al. [71] and Acosta et al. [72].

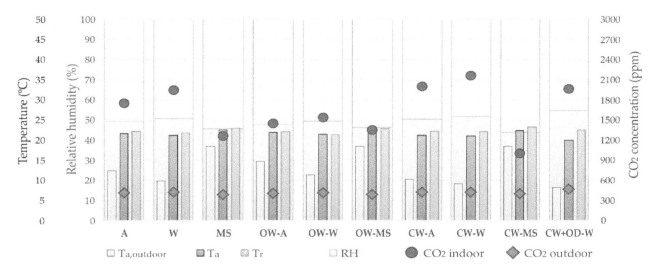

Figure 6. Indoor and outdoor air temperature values (T_a), mean radiant temperature values ($\bar{t}_r$), relative humidity (HR) values and indoor and outdoor CO_2 concentration values related to seasons and windows' and doors' operation.

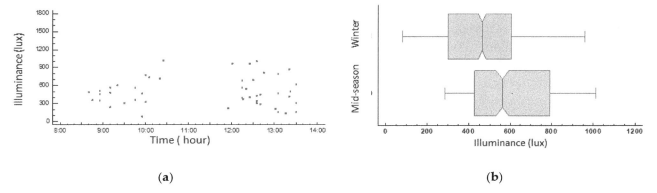

(a)   (b)

Figure 7. Room illuminance. (**a**) Hourly distribution for mean-room illuminance by season: winter (blue) and middle-season (red). (**b**) Quartile distribution of mean-room-illuminance by season (top) winter (down) mid-season.

3.2. Mean Values of Airtightness of the Samples

The values of airtightness of the classrooms under study with a difference of pressure between indoor and outdoor of 50 Pa (n_{50} range) varied from 2.6 h^{-1} to 10 h^{-1}, with an average value of n_{50} of 6.97 h^{-1} and a SD of 2.06 h^{-1}.

3.3. Mean Values of Occupants' Votes

The mean thermal sensation vote (TSV) of the students in both seasons was "slightly warm", with a value of +0.32 on the ASHRAE scale in winter and +0.38 in midseasons, having a SD of between 0.93 and 0.83, respectively. This can be identified as a common situation among poorly ventilated and crowded spaces. Even with open windows the actual air-removal capacity looked very limited as previously evaluated (Table 6). These thermal perceptions were higher (+0.10 points) when windows were open, highlighting an excess of heat release of the heating system due to inefficient regulation and the wish of the users of dissipation. In this case, the occupants' thermal preference vote (TPV) expressed was softer and closer to neutrality than the TSV, not fitting at all with the perceived thermal sensations ($R^2 = -0.47$, moderate correlation), as showed by Teli et al. [14,53].

Table 6. Mean values of occupants' votes obtained during the field measurements related to seasons and windows' and doors' operation.

		W	MS	OW-W	OW-MS	CW-W	CW-MS	CW OD-W
TSV	Mean	0.32	0.38	0.42	0.48	0.27	0.08	-0.22
(−3 to 3)	STD	0.93	0.83	0.91	0.80	0.93	0.84	0.88
TPV	Mean	0.06	−0.20	−0.06	−0.26	0.13	0.00	0.55
(−3 to 3)	STD	1.03	0.96	0.92	0.98	1.07	0.86	1.14
PD_{acc}	Mean	0.81	0.85	0.80	0.84	0.81	0.86	0.76
(0 to 1)	STD	0.39	0.36	0.40	0.37	0.39	0.35	0.43
TCV	Mean	0.39	0.30	0.44	0.32	0.36	0.25	0.38
(0 to 4)	STD	0.62	0.68	0.75	0.70	0.54	0.63	0.56
EPV	Mean	1.03	0.61	1.06	0.63	1.01	0.52	0.79
(0 to 4)	STD	0.93	0.68	0.97	0.71	0.92	0.61	0.77

TSV is the thermal sensation vote, TPV is the thermal preference vote, PD_{acc} is the thermal environment rejection percentage, TCV is the thermal comfort vote, and EPV is the environmental perception vote.

The average thermal environment rejection percentage (PD_{acc}) expressed by students, based on a scale from 1 (acceptance) to 0 (rejection), was low and homogeneous in both seasons, with a mean value of 0.81 in winter conditions and 0.85 for midseasons. In addition, thermal acceptance was, in general, slightly better in classrooms with closed windows in both seasons, but in the case of closed windows and open doors. The thermal comfort vote (TCV) allowed us to qualify this acceptance-rejection PD_{acc} index, given that less than 70% of students found "comfortable" the thermal environment in winter conditions compared to more than 80% in midseasons. This percentage increased to 96% for students with "comfortable" or "a bit uncomfortable" votes in winter conditions, but without reaching 92% in midseasons. By contrast, the number of users who, accepting a slight discomfort, considered the acceptable environment was superior in winter than in midseason, where the feeling of discomfort was slightly more marked, can be seen in Figure 8.

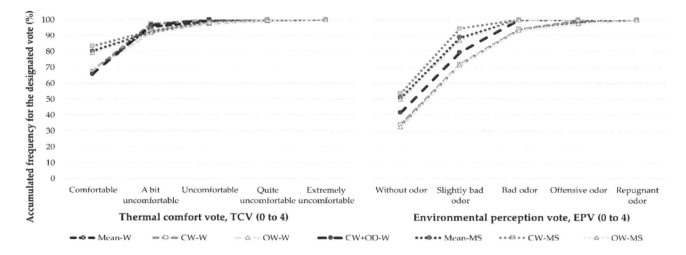

Figure 8. Accumulated frequency for the environmental perception and thermal comfort votes, related to seasons and windows' and doors' operation.

The mean environmental perception vote (EPV) showed during winter a 1.03 value (slightly bad odor), with low differences regarding windows' operation (less than 0.03 points); in midseasons, EPV was more favorable (0.61), with more than 0.10 points of difference regarding windows' operation. Figure 8 also shows the accumulated distribution of the EPV, in which less than 35% of students voted "without odor" in winter, while more than 50% voted it during midseasons. In addition, almost 30% of students perceived a slight odor or worse in winter in comparison to the midseasons, with 10%. Finally, around 7% of students voted "bad odor" or worse in winter, while there were no votes in this way during midseasons.

During the winter there is a more evident feeling of a poorly ventilated (not healthy) environment, in line with the measured CO_2 values acting as a token of the indoor ambient renovation state. The operation of windows produced little to no effect on the improvement of the environmental quality, especially during the winter. Although it was found that the opening of windows in this period generated noticeable dilution of the interior atmosphere, it was still insufficient to guarantee pleasant environments.

During midseason, although the ventilation mechanism was less effective (by means of a lack of thermal differential), the capability of diluting the indoor environment to threshold levels was perceived by the users as somewhat better. The assessment of these user perception-thresholds was a key aspect of research, since it will allow the design of more adequate and well-accepted spaces.

3.4. Mean Values of Occupants' Clothing Insulation

The occupants' clothing insulation (I_{cl}) showed two models of response linked to the season, as it can be seen in Table 7. Clothing distribution in winter was homogeneous, with a mean value of 0.90 clo and a SD of 0.19, common both for open and closed windows, and a minor divergence of 5% around 0.6–0.7 clo values related to windows' operation, showed in Figure 9. It should also be noted that the biggest slope of the insulation distribution was during winter with closed windows and inner doors open, which highlighted the smaller variation in clothing insulation of this group of case studies.

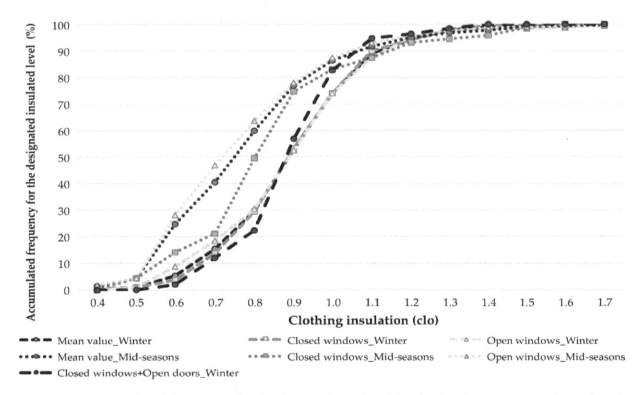

Figure 9. Accumulated frequency for the designed insulated level related to seasons and windows' and doors' operation.

Table 7. Mean values of occupants' clothing insulation obtained during the field measurements related to seasons and windows' and doors' operation.

		W	MS	OW-W	OW-MS	CW-W	CW-MS	CW OD-W
I_{cl}	Mean	0.90	0.78	0.90	0.76	0.90	0.84	0.89
(clo)	STD	0.19	0.23	0.21	0.23	0.18	0.23	0.14

I_{cl} is the clothing insulation level of the occupants.

In midseasons, the clothing insulation was lower and variable, with a SD of 0.23 and an asymmetrical distribution. There was a divergence of up to 25% in the frequency of the lowest levels of clothing insulation during midseasons regarding the windows' operation, coinciding both frequencies around the value of 0.90 clo (75–80% of the accumulated frequency).

3.5. Symptoms and Related Health Effects

The most commonly reported severe symptoms were headache and concentration difficulty (around 10%), followed by tiredness and a dry throat (under 10%), with a greater prevalence during wintertime and closed windows' operation. The action of the windows (Figure 10) was relatively weak, indicating the limited actual ventilation capacity of these spaces with only the opening of windows (reductions were around 25% less, in general). However, the perception of mild symptoms was very common in the classrooms, with tiredness, headache, and difficulty in concentration presenting a prevalence in the range of 40% to 50% for closed operation and slightly lower when windows were open (decreasing around 10–15 %), as shown Table 8.

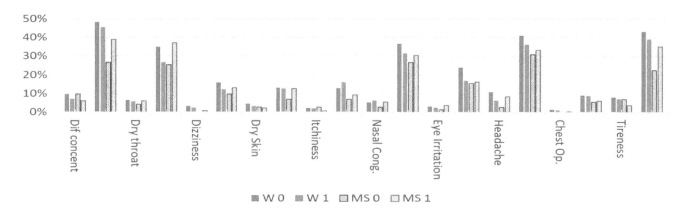

Figure 10. Average probability of reported symptoms by seasons (W is winter, MS is midseasons) and windows' operation (0 is closed, 1 is open) (severe, left group; light perception, right group for each symptom).

This situation changed in midseason, where the symptom report was lower, even for the situation of closed windows. However, unlike winter, symptomatic perception increased when the windows were open for both perceptions, severe and mild, especially for dry throat, itchiness, nasal congestion, and headache, which are symptoms that can be linked to the penetration of external species (in many cases aerobiological such as pollen [73–75]).

Aiming to evaluate the overall impact of the different perceptions of symptoms, while assuming the variability component of the subjective responses and different individual sensitivity to the environments, unlike the evaluation of physical parameters, users were asked to assess the intensity of the perception of discomfort on a scale of 0 to 1 (0 none and 1 maximum intensity). Although this was not a standardized parameter (it may vary between different users) it had a great potential to represent the importance that each user assigned to the nuisance and, therefore, to assess the actual perception of the indoor conditions. Similar subjective ratings in conjunction with objective environmental measures were used in relevant studies, such as [76–79]. An overall indicator was collected through the addition of the specific scores or valuations generated by the users of each symptom or condition.

This represented a global assessment of perceived impact, with a fundamentally qualitative character, since there was no univocal relationship but strong enough to highlight health discomfort ant to categorize best and worse indoor environments. The main values from the different classrooms are grouped by seasons and windows' situation in Table 9. This table contains the statistical summary for the data samples. Of particular interest are standardized bias and standardized kurtosis since, in all the cases (except the kurtosis of MS_1) these statistics were outside the range of −2 to +2 standard deviation, thus indicating significant deviations from normal.

Table 8. Relative probability of occupants' relating symptoms and health effects, from N (not perceived), to L (lightly perceived), and H (severe perception), with closed windows (0) and open (1).

	Season	Winter		Mid-Season	
	Windows	0	1	0	1
DC	H	9.83%	7.23%	9.86%	6.10%
	L	48.25%	45.53%	26.76%	38.97%
	N	41.92%	47.23%	63.38%	54.93%
DT	H	6.55%	5.96%	4.23%	6.10%
	L	34.93%	26.81%	25.35%	37.09%
	N	58.52%	67.23%	70.42%	56.81%
D	H	3.49%	2.55%		0.94%
	L	15.94%	12.34%	9.86%	13.15%
	N	80.57%	85.11%	90.14%	85.92%
DS	H	4.59%	3.40%	2.82%	2.35%
	L	13.10%	12.77%	7.04%	12.68%
	N	82.31%	83.83%	90.14%	84.98%
IT	H	2.40%	2.13%	2.82%	0.94%
	L	12.88%	16.17%	7.04%	9.39%
	N	84.72%	81.70%	90.14%	89.67%
NC	H	5.24%	6.38%	2.82%	5.63%
	L	36.68%	31.49%	26.76%	30.52%
	N	58.08%	62.13%	70.42%	63.85%
EI	H	3.28%	2.55%	1.41%	3.76%
	L	24.02%	17.02%	15.49%	16.43%
	N	72.71%	80.43%	83.10%	79.81%
H	H	10.92%	6.38%	2.82%	8.45%
	L	41.05%	36.17%	30.99%	33.33%
	N	48.03%	57.45%	66.20%	58.22%
CO	H	1.53%	1.28%		0.47%
	L	9.17%	8.94%	5.63%	6.10%
	N	89.30%	89.79%	94.37%	93.43%
T	H	8.08%	7.23%	7.04%	3.76%
	L	43.23%	39.15%	22.54%	35.21%
	N	48.69%	53.62%	70.42%	61.03%

DC is difficulty concentrating, DT is dry throat, D is dizziness, DS is dry skin, IT is itchiness, N is nausea, NC is nasal congestion, EI is eye irritation, H is headache, CO is chest oppression, and T is tiredness.

The distribution of symptoms' samples for each scenario (Figure 11) was asymmetrical, not normal (Shapiro–Wilk test with p-value less than 0.05 in all cases, so it can be ruled out with 95% confidence) with bias. Median values located between 1.4 as the lower impact case in half a season (closed windows) up to 2.10 for winter (also with closed windows). Although values concentrated around 2.00, there was a significant dispersion, reaching values of up to 11, which meant a maximum vote in practically all the symptoms. (This specific case must be understood as outlier). This highlighted that even in the best scenario analyzed, there was a significant perception of ambient-related symptoms and problems by the users. By contrast, there was also a non-negligible presence of users that did not reflect any discomfort or effects, especially in the midseason scenario with closed windows, with percentiles that stood at 39%, compared to lower values in the other states, where this group went from 6.1% to 16.4% (W0 to MS1). In this way, the low level of difference in the distribution according to windows' operation can also show that the ventilation airflow through windows was not enough to guarantee a noticeable reduction of the students' symptoms, although it can modify slightly the physical parameters of the interior environment. This aspect was of singular importance, since it indicated that the mere control of the usual environmental values did not guarantee satisfaction with the interior environment, at least with regard to the absence of bothersome symptoms. In the case of midseason, symptoms described with open windows can be due to the higher level of external aerobiological particles entering into the classrooms, such as pollen. That is why the appropriate ventilation to provide a perceptive reduction of the symptomatology should be done by means of fans with filter system.

Table 9. Statistics from symptoms' scores for individuals' response by season and windows' situation (MS, middle season; W, winter; 0, windows closed; and 1, open).

	MS_0	W_0	MS_1	W_1
Average	1.48	2.37	1.87	2.09
Median	1.40	2.10	1.80	2.00
STD	1.58	1.57	1.49	1.55
Min. Value	0.00	0.00	0.00	0.00
Max. Value	5.40	11.00	6.80	8.80
Stand. Bias	2.33543	10.2611	3.68023	7.34186
Stand. Kurt.	−1.32011	12.1936	0.342811	7.47186

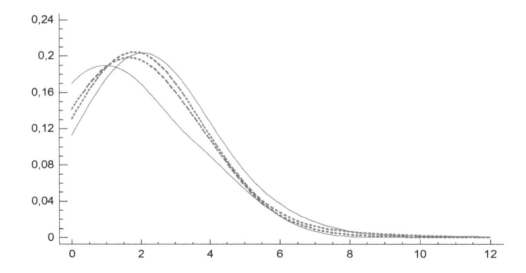

Figure 11. Probabilistic density trace distribution for individual symptoms' scores (winter, red; middle season, blue; windows closed, solid line; windows open, dashed line).

The probabilistic distributions of individual related symptoms' scores for the different scenarios showed some similarity in the global pattern response and central values, mainly for open windows, except for the MS_0 (closed windows). A set nonparametric contrast through K–S test (Kolmogorov–Smirnov for the global parameter) was developed to evaluate the pertinence to a common distribution. In all four cases, comparisons for accumulated distances of the samples showed statistically significant differences at 95% significance between the distributions (with all the cases with a p-value < 0.05 and DN values over $D_{crit.0.05}$), with DN around 0.122 to 0.148 for the samples with closest distribution (windows open winter vs. middle season and winter open vs. closed windows) and the greater DN value 0.380 for the furthest. So it can be established that there were different distributions for all the cases

3.6. Airtightness

The average value of the infiltration rate at 50 Pa (n_{50}) was 6.97 h^{-1}, with a standard deviation of 2.06 h^{-1}. Models with the lowest n_{50} values were those in the C3 climate zone, where the lowest average temperature values are recorded in the winter. The values of n_{50} ranged from 10 h^{-1} (maximum) to 2.6 h^{-1} (minimum), both recorded in the B4 climate zone (Table 10).

Table 10. Average values and standard deviation of n_{50}.

Climatic Zone	Mean n_{50}	Standard Deviation
A4	6.53	0.94
C3	6.12	1.67
B4	7.89	2.45
C4	7.6	0.56
Mean	6.97	2.06

4. Discussion

This section is focused on the analysis of the relationships between the symptoms described by occupants and the rest of the parameters under study (physical, building operation, and votes).

4.1. Relationship between Physical Parameters and Classroom Operation

It could be assumed that manual opening of windows in naturally ventilated buildings should depend on outdoor conditions, as this is the main element of control. However, it was observed that, despite the fact that in midseason windows remain open longer than in winter, no clear linear trend can be observed. In winter, the need for ventilation or indoor air changes is considered more important than the need to control the entry of outdoor cold air. Analysis by categories of the opening of windows (Figure 12) showed this occurs mostly in mid temperatures, although it was also observed in cooler conditions when necessary. Furthermore, no progressive growth was observed with the increased temperature, as could be expected. In midseason, it is more common to open windows, although there was no clear correlation with temperature, some of which was similar to winter, where more windows are opened in comparison.

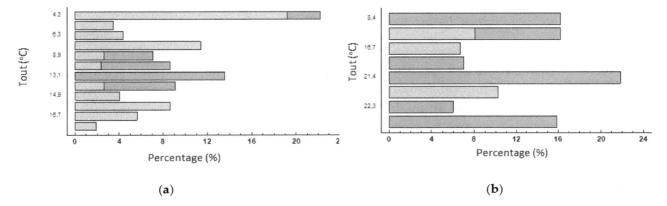

(a)   (b)

Figure 12. Cross-tabulation for windows' opening and outdoor air temperature. (**a**) Cross-tabulation for windows opening and outdoor air temperature in winter: windows closed (gray) and open (blue). (**b**) Cross-tabulation for windows opening and outdoor air temperature in mid-season: windows closed (gray) and open (blue).

It could be deduced that users are psychologically or culturally conditioned to some extent as to how and when they open windows. Although it would be preferable for the classroom windows to remain open, the act of opening was seen as a reaction to poorer indoor air quality, which was more noticeable for the same thermal conditions in spring. It, therefore, appears that there is an adaptation process.

As it can be seen in Figure 13, although there was statistical significance between the CO_2 concentration and the outdoor-indoor air temperature differential (p-value < 0.05), the correlation was somehow weak and more clear in winter time ($R^2 = 0.249$) than in midseason ($R^2 = 0.145$), with a better fit to a y-reciprocal relation. However, the predictive mathematical model lacked enough accuracy to be of utility to forecast actual situations. Besides the wide dispersion on values, there was a trend in the worsening of the indoor environment as DT increased, as can be usually expected, due to the lack of a controlled ventilation system.

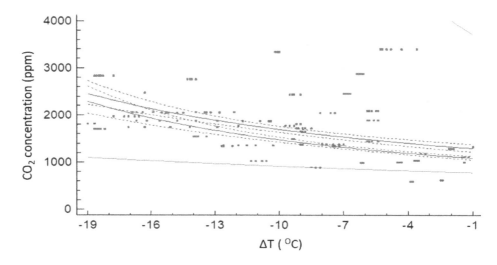

Figure 13. Fitted regression model plot for CO_2 indoor concentration related to indoor-outdoor air temperature differential, closed windows (red), and windows open (blue).

Is noteworthy to highlight that moderate DT winter and midseason trends were very similar, which matched with the foreseen windows' operation patterns, when most of the apertures occurred around cold-mild external temperatures.

4.2. Relationship between Physical Parameters and Symptoms Described

Some symptoms were more predominant when outdoor temperature was lower, although no clear linear relationship could be established. These symptoms were more frequent in winter when the thermal differential is at its highest, usually linked to a lack of ventilation at the time, as supported by the high CO_2 indices as a general air quality indicator.

Although the symptoms often appeared to be more evident when the windows were open, this should be seen as a consequence, not a cause, as user perception of the symptoms was generally clearer when opening the windows. This is interesting to note, as it could be due to a situation which exceeded the perception threshold. In the winter, it is more common to observe symptoms such as difficulty concentrating, dry throat, and tiredness. These are very closely linked to poor hygrothermal control, even with windows open, where temperature and relative humidity are far more important, especially with open windows, as well as increased indoor CO_2 linked to poor ventilation. In contrast, itchiness and chest tightness were barely noticeable.

The situation changed in midseason and symptoms, such as difficulty concentrating, tiredness, and nasal congestion, were less widely reported. However, symptoms less connected with the absence of hygrothermal regulation increased, while there was a greater presence of symptoms that may be linked to outdoor exposure.

When both lighting parameters, illuminance (E) and illuminance uniformity (Uo), were analyzed and referred to symptomatology, no clear correlation was obtained, as other previous studies showed for educational buildings [80]. This may be because illuminance values in the classrooms under study were generally over 350–400 lx with a uniformity of 0.40–0.50, so they were values good enough to not influence students at a symptomatic level.

The infiltration rate (n_{50}) and the symptoms related by occupants showed a very tenuous connection, with some weak trends in the case of tiredness, as well as difficulty on concentrating, dry throat, and headache. Given that the airtightness of the classrooms was, in general, adequate or even good, with an average value of 6.97 h^{-1} with a maximum value of 10 h^{-1}, its influence can be moderate due to its low impact on air renewal. It also indicates that other variables, like time spent inside the classroom or windows' and doors' operation, can have more importance than the airtightness of the room.

There was no clear linear correlation between the students' clothing insulation and the symptoms described during measurements. The possibility of freely varying the level of clothing insulation by the students, according to their individual thermal needs, may be a factor that influenced this lack of relationship between clothing and symptomatology, besides psychological factors linked to clothing.

When symptomatology was assessed as global, there were some trends that could be identified. If CO_2 was assumed as an overall indicator of indoor air renovation (not as a contaminant itself), the worsening of indoor environment linked with the increase of symptoms related. It can be approximated to a logarithmic regression relation (Figure 14), although a wide spread of values must be assumed. Different patterns for winter and midseason were described due to adaptation of users and the influence of outdoor species. Although this model presents some uncertainty for its use as a prediction tool, it does have the capacity to act as a qualitative indicator.

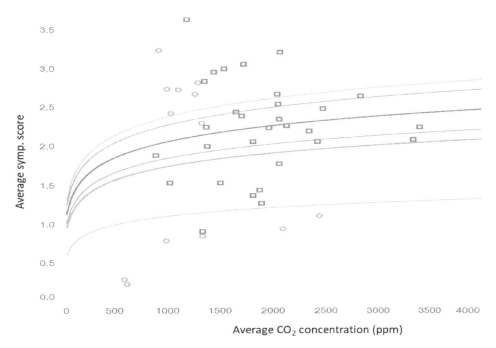

Figure 14. General average related symptoms' scores relation with indoor CO_2 grouped by measured classrooms (green for middle season and blue for winter).

A linear trend model was calculated for the average symptom score and given a record of the average indoor CO_2 (logarithmic fit). The model was statistically significant at $p < 0.05$, having a high correlation coefficient ($R^2 = 0.8833$) and a mean square error (MSE) of 0.6160.

A somewhat weaker linear relationship (logarithmic fit also) was seen ($R^2 = 0.509$ for midseason and $R^2 = 0.143$ for winter) but with statistical signification (p-value < 0.05 in both cases) and an error of MSE 0.425. Although dispersion was high, it was also a useful qualitative trend indicator, and was found between the overall perception of symptoms and the indoor operative temperature (Figure 15). In this case, it can be established that the symptoms tended to be more frequent when indoor temperatures increased, also with specific patterns for winter and middle season.

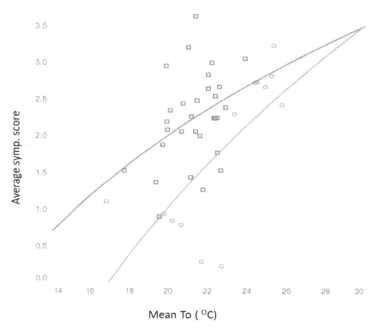

Figure 15. General average related symptoms' scores relation with indoor operative temperature grouped by measured classrooms (green for middle season and blue for winter).

5. Conclusions

A wide study sample of 47 naturally ventilated multipurpose classrooms of the most representative climate zones of southern Spain was characterized and analyzed through field measurements and surveys distributions, in order to contrast environmental sensation votes, perception, and indoor-related symptoms described by 977 students during lessons with physical and environmental parameters, as well as operational scenarios.

The main operational case to be analyzed, according to votes and symptoms, was the windows' operation. In this sense, the 61% of the case studies during winter season had the windows open, which can be related both to a bad regulation of the heating system (the slight heat excess had to be dissipated) as well as to a poor indoor environment perception. In this way, the case studies with open windows in winter had a higher mean indoor air temperature value (21.5 °C versus 21.0 °C) and higher standard deviation of the mean radiant temperature (2.6 °C versus 1.6 °C). The mean thermal perception of students in winter season with open windows reinforced this slight heat excess, given that it was in a comfort range but 0.15 points warmer than in the case of closed windows, also expressing a thermal preference of thermal neutrality-mild cold (−0.06 on the ASHRAE scale) with open windows in contrast to the preference for a warmer environment when the windows were closed (+0.13). The thermal assessment of the environment through the thermal comfort vote (TCV) also had a poorer value with open windows (−0.44 versus −0.35 from 0 to −4), also showing a higher deviation in the votes (0.75 versus 0.54) and a somewhat higher linear correlation with CO_2 concentration. Therefore, the architectural design should take into account to guarantee the air quality of the venue, as well as a comfortable heating system, in order to lead students to not open the windows uncontrollably, which produces, as explained above, a noticeable energy consumption and distorts interior comfort control.

The operation of windows during winter helps to decrease the mean value of CO_2 concentration, with 1537 ppm versus 2164 ppm with windows closed; but, in most of cases, this decrease was insufficient both to be within the standard recommendations for healthy environments and to reach threshold values of perceptions of the users. Given that the mean CO_2 concentration level was still high even when windows were open, the mean environmental perception of the students (EPV) was not strongly influenced by the opening of windows, with almost 30% of students expressing a certain level of annoying odor in both cases, but also having a moderate correlation between poor environment perception and CO_2 concentration just when windows were closed. Therefore, it can be stated that there was not a high correlation between the CO_2 value and the students' perceptions, mainly due to the olfactory adaptation phenomenon, irrespective of the need to provide a suitable air quality for healthiness purpose. In this way, when symptoms reported were added to this analysis, they presented a not-direct relationship with EPV, with the higher complaint values when windows were open. This odor perception was also somehow related with tiredness, difficulty on concentrating, eye irritation, headache, and dry throat.

In midseasons, windows' operation led to a greater variation of indoor thermal values, both air and radiant, also maintaining in general CO_2 levels over the WHO recommendations (mean vale of 1537 ppm). In addition, students' TSVs were higher with open windows, close to the thermal comfort limit by warmth. Furthermore, the odor perception (EPV) was also poorer (0.63 value versus 0.52) when windows were open in midseasons, reinforcing the finding that windows alone are not able to provide an adequate renewal capacity for the indoor environment.

The study of the symptoms reported during measurements showed that they were largely expressed by students, both for windows open and closed, particularly in the case of difficulty of concentrating (52%), headache and tiredness (46%), followed by dry throat and nasal congestion (39%), which also were the symptoms most frequently combined with the other symptoms. According to the studied scenario, without a mechanically controlled ventilation system, complaints were more often found during winter, especially when windows were closed. In midseason conditions, symptoms were somewhat less common, but students expressed more acute symptomatology when windows were open, especially for dry throat, itchiness, nasal congestion, and headache, which are symptoms that can be related to hypersensitivity to external agents such as allergies and other respiratory conditions. This conclusion states the clear need to provide a ventilation system with a suitable filtering.

Regarding the relationship with indoor temperature, it can also be established that the symptoms tended to be more frequent when indoor temperatures increased, also with specific patterns for winter and middle season, also related to the occupants' thermal perceptions.

Other operation factors, like illuminance and illuminance uniformity, as well as students' clothing insulation, were analyzed referred to this symptomatology, but no clear correlation was obtained. In the case of lighting parameters, almost all the classrooms under study were generally over 350–400 lx with a uniformity of 0.40–0.50, so they were values good enough to not influence students at a symptomatic level. The correlated color temperature was similar in all cases, varying from 3500 to 5500 K; hence, it can be considered that both the amount of light and hue did not affect the thermal perception of the participants. On the other hand, students had the possibility of freely varying the level of clothing insulation, according to their individual thermal needs, so its impact on symptomatology was diminished.

In conclusion, the findings of this study show that effectively controlled ventilation systems are needed to assure an actual indoor ambient renovation and clean air supply. The special sensibility to external species make it advisable to incorporate filtering and cleaning systems for outdoor air beyond the impact on investment costs and energy use that this may entail. In addition, the study of symptomatology suggests that CO_2 indicator should be complemented by other pollutants' measurements to assure a proper interpretation of data, given that they could not be correctly identified exclusively using this single CO_2 control parameter. As explained above, CO_2 levels have a fuzzy influence in the students' symptomatology; hence, the air quality should be complementarily assessed through other parameters, such as particle or VOCs' levels.

The following points can be established as key aspects:

The use of CO_2 as a standalone indicator of environmental quality, especially for the management of ventilation systems or driving the windows' opening, may be insufficient and can derivate in situations of increased user discomfort, alongside thermal-ambient disturbance. Although there was evidence that there is a relationship with the indoor CO_2 levels growing (assumed as general index) and the increase in reported global symptoms, this was not a direct link and tended to be asymptotic from certain threshold levels (around 2000 ppm).

In most cases, natural ventilation systems are not able to solve properly the removal of pollutants, generating situations with high rates of complaints even when windows are open, although they can mitigate the situations during indoor peak situations (such as produced in winter season). In many cases, windows' opening can be counterproductive, given that, although the classic indicators of the indoor environment valuation improve, the perception of the users was negative or, at least, worse than in situations with closed windows.

Assuming that indoor ambient is a complex and multifactor model, in the current state of the art of school buildings, the use of natural ventilation by itself (with the typical configuration of classrooms and enclosures of the buildings in the region) does not guarantee adequate control of the indoor environment, against popular assumption in the area, both by users and administrators. This aspect, although it was previously included in the text, has been emphasized.

This fact may be related to the need to review the classic indicators and parameters commonly used in the environmental management of these spaces. This research found situations of discomfort even within the ranges generally assumed as comfortable by the standards and design guides. Thus, it is necessary to develop complementary indicators based on the perception and the probability of developing symptoms that allow contributing to the correct valorization of the indoor environments from the users' points of view.

In this way, this analysis should also be complemented with corresponding measurements and surveys distributions in classrooms with mechanical ventilation systems in order to develop a comparison of results with adequate CO_2 levels, so further research on this field is required.

Author Contributions: Conceptualization, M.Á.C.-L., S.D.-A., J.F.-A., and I.A.; methodology, M.Á.C.-L., S.D.-A., J.F.-A., and I.A.; software, M.Á.C.-L., S.D.-A., J.F.-A., and I.A.; validation, M.Á.C.-L., S.D.-A., J.F.-A., and I.A.; formal analysis, M.Á.C.-L., S.D.-A., J.F.-A., and I.A.; investigation, M.Á.C.-L., S.D.-A., J.F.-A., and I.A.; resources, M.Á.C.-L., S.D.-A., J.F.-A., and I.A.; data curation, M.Á.C.-L., S.D.-A., J.F.-A., and I.A.; writing—original draft preparation, M.Á.C.-L., S.D.-A., J.F.-A., and I.A.; writing—review and editing, M.Á.C.-L., S.D.-A., J.F.-A., and I.A.; visualization, M.Á.C.-L., S.D.-A., J.F.-A., and I.A.; supervision, M.Á.C.-L., S.D.-A., J.F.-A., and I.A.; project administration, M.Á.C.-L., S.D.-A., J.F.-A., and I.A.; funding acquisition, M.Á.C.-L., S.D.-A., J.F.-A., and I.A. All authors have read and agreed to the published version of the manuscript.

Acknowledgments: The authors wish to express their gratitude to Jaime Costa-Luque for reviewing the manuscript and helping with several of the graphics, to Blas Lezo for the encouragement of this article, as well as to the students, teachers, and management team of the secondary schools which were part of this study. Finally, to the Public Entity "Agencia Pública Andaluza de Educación" from the Regional Government of Andalucía.

Appendix A

Table A1. Excerpt from the surveys distributed to students.

What is your perception of the environment at this precise moment□						
Hot □	Warm □	A bit warm □	Neutral □	A bit cool □	Cool □	Cold □
How do you feel at this precise moment□						
Comfortable □	A bit uncomfortable □		Uncomfortable □	Quite uncomfortable □	Extremely uncomfortable □	

How do you feel at this precise moment□

A lot warmer □	Warmer □		A bit warmer □	No change □	A bit colder □	Colder □	A lot colder □

Do you accept this thermal environment rather than reject it□   Yes □   No □

What does the air smell like at this precise moment□

Without odor □	Slightly bad odor □	Bad odor □	Offensive odor □	Repugnant odor □

Do you feel these symptoms at this precise moment□

				Difficulty concentrating	Quite □	A bit □	No □
Dry throat	Quite □	A bit □	No □	Dizziness	Quite □	A bit □	No □
Dry skin	Quite □	A bit □	No □	Itchiness	Quite □	A bit □	No □
Nausea	Quite □	A bit □	No □	Nasal congestion	Quite □	A bit □	No □
Eye irritation	Quite □	A bit □	No □	Headache	Quite □	A bit □	No □
Chest oppression	Quite □	A bit □	No □	Tiredness	Quite □	A bit □	No □

References

1.   Lane, W.R. *Education, Children and Comfort*; University of Iowa: Ames, IA, USA, 1965.
2.   Pepler, R.D. The thermal comfort of students in climate-controlled and non-climate-controlled schools. *Ashrae Trans.* **1972**, *78*, 97–109.
3.   Auliciems, A. Thermal sensations of secondary school children in summer time. *J. Hyg. (Lond)* **1973**, *71*, 453–458. [CrossRef] [PubMed]
4.   Humphreys, M. A study of the thermal comfort of primary school children in summer. *Build. Environ.* **1977**, *12*, 231–239. [CrossRef]
5.   Zomorodian, Z.S.; Tahsildoost, M.; Hafezi, M. Thermal comfort in educational buildings: A review article. *Renew. Sustain. Energy Rev.* **2016**, *59*, 895–906. [CrossRef]
6.   Karimipanah, T.; Awbi, H.B.; Sandberg, M.; Blomqvist, C. Investigation of air quality, comfort parameters and effectiveness for two floor-level air supply systems in classrooms. *Build. Environ.* **2007**, *42*, 647–655. [CrossRef]
7.   Shendell, D.G.; Prill, R.; Fisk, W.J.; Apte, M.G.; Blake, D.; Faulkner, D. Associations between classroom CO_2 concentrations and student attendance in Washington and Idaho. *Indoor Air* **2004**, *14*, 333–341. [CrossRef]
8.   Mors, S.T.; Hensen, J.L.; Loomans, M.G.; Boerstra, A.C. Adaptive thermal comfort in primary school classrooms: Creating and validating PMV-based comfort charts. *Build. Environ.* **2011**, *46*, 2454–2461. [CrossRef]
9.   Mainka, A.; Zajusz-Zubek, E.; Mainka, A.; Zajusz-Zubek, E. Indoor Air Quality in Urban and Rural Preschools in Upper Silesia, Poland: Particulate Matter and Carbon Dioxide. *Int. J. Environ. Res. Public Health* **2015**, *12*, 7697–7711. [CrossRef]
10. Hedge, A. *Ergonomic Workplace Design for Health, Wellness, and Productivity*; CRC Press: Boca Raton, FL, USA, 2016.
11. Corgnati, S.P.; Ansaldi, R.; Filippi, M. Thermal comfort in Italian classrooms under free running conditions during mid seasons: Assessment through objective and subjective approaches. *Build. Environ.* **2009**, *44*, 785–792. [CrossRef]
12. Almeida, S.M.; Canha, N.; Silva, N.; do Carmo Freitas, M.; Pegas, P.; Alves, C.; Evtyugina, M.; Adrião Pio, C. Children exposure to atmospheric particles in indoor of Lisbon primary schools. *Atmos. Environ.* **2011**, *45*, 7594–7599. [CrossRef]
13. de Giuli, V.; da Pos, O.; de Carli, M. Indoor environmental quality and pupil perception in Italian primary schools. *Build. Environ.* **2012**, *56*, 335–345. [CrossRef]
14. Teli, D.; James, P.A.B.; Jentsch, M.F. Thermal comfort in naturally ventilated primary school classrooms. *Build. Res. Inf.* **2013**, *41*, 301–316. [CrossRef]
15. de Giuli, V.; Zecchin, R.; Corain, L.; Salmaso, L. Measured and perceived environmental comfort: Field monitoring in an Italian school. *Appl. Ergon.* **2014**, *45*, 1035–1047. [CrossRef] [PubMed]

16. Campano, M.Á.; Domínguez-Amarillo, S.; Fernández-Agüera, J.; Sendra, J.J. Thermal Perception in Mild Climate: Adaptive Thermal Models for Schools. *Sustainability* **2019**, *11*, 3948. [CrossRef]

17. Fernández-Agüera, J.; Campano, M.A.; Domínguez-Amarillo, S.; Acosta, I.; Sendra, J.J. CO2 Concentration and Occupants' Symptoms in Naturally Ventilated Schools in Mediterranean Climate. *Buildings* **2019**, *9*, 197. [CrossRef]

18. Kalimeri, K.K.; Saraga, D.E.; Lazaridis, V.D.; Legkas, N.A.; Missia, D.A.; Tolis, E.I.; Bartzis, J.G. Indoor air quality investigation of the school environment and estimated health risks: Two-season measurements in primary schools in Kozani, Greece. *Atmos. Pollut. Researc* **2016**, *7*, 1128–1142. [CrossRef]

19. Hutter, H.P.; Haluza, D.; Piegler, K.; Hohenblum, P.; Fröhlich, M.; Scharf, S.; Uhl, M.; Damberger, B.; Tappler, P.; Kundi, M.; et al. Semivolatile compounds in schools and their influence on cognitive performance of children. *Int. J. Occup. Med. Environ. Health* **2013**. [CrossRef]

20. Sousa, S.I.V.; Ferraz, C.; Alvim-Ferraz, M.C.M.; Vaz, L.G.; Marques, A.J.; Martins, F.G. Indoor air pollution on nurseries and primary schools: Impact on childhood asthma - Study protocol. *Bmc Public Health* **2012**. [CrossRef]

21. Fromme, H.; Diemer, J.; Dietrich, S.; Cyrys, J.; Heinrich, J.; Lang, W.; Kiranoglu, M.; Twardella, D. Chemical and morphological properties of particulate matter (PM10, PM2.5) in school classrooms and outdoor air. *Atmos. Environ.* **2008**, *42*, 6597–6605. [CrossRef]

22. Koenig, J.Q.; Mar, T.F.; Allen, R.W.; Jansen, K.; Lumley, T.; Sullivan, J.H.; Trenga, C.A.; Larson, T.; Liu, L.J. Pulmonary effects of indoor- and outdoor-generated particles in children with asthma. *Environ. Health Perspect.* **2005**, *113*, 499–503. [CrossRef]

23. ISO. *ISO 7730:2005—Ergonomics of the Thermal Environment—Analytical Determination and Interpretation of Thermal Comfort Using Calculation of the PMV and PPD Indices and Local Thermal Comfort Criteria*; International Organization for Standardization: Geneva, Switzerland, 2005.

24. CEN—European Committee for Standardization. *CR 1752:1998—Ventilation for Buildings. Design Criteria for the Indoor Environment*; European Committee for Standardization: Brussels, Belgium, 2008.

25. CEN—European Committee for Standardization. *EN 13779:2008—Ventilation for non-Residential Buildings—Performance Requirements for Ventilation and Room-Conditioning Systems*; CEN—European Committee for Standardization: Brussels, Belgium, 2008.

26. Gobierno de España. Real Decreto 1826/2009, de 11 de diciembre, por el que se modifica el Reglamento de Instalaciones Térmicas en los Edificios, aprobado por el Real Decreto 1027/2007. *Boletín Estado* **2009**, *298*, 104924–104927.

27. AENOR—Asociación Española de Normalización y Certificación. *UNE-EN 15251:2008—Indoor Environmental Input Parameters for Design and Assessment of Energy Performance of Buildings Addressing Indoor Air Quality, Thermal Environment, Lighting and Acoustics*; AENOR—Asociación Española de Normalización y Certificación: Madrid, Spain, 2008.

28. Fisk, W.J. How home ventilation rates affect health: A literature review. *Indoor Air* **2018**, *28*, 473–487. [CrossRef] [PubMed]

29. Annesi-Maesano, I.; Hulin, M.; Lavaud, F.; Raherison, C.; Kopferschmitt, C.; de Blay, F.; André Charpin, D.; Denis, C. Poor air quality in classrooms related to asthma and rhinitis in primary schoolchildren of the French 6 Cities Study. *Thorax* **2012**, *67*, 682–688. [CrossRef] [PubMed]

30. Annesi-Maesano, I.; Baiz, N.; Banerjee, S.; Rudnai, P.; Rive, S. Indoor air quality and sources in schools and related health effects. *J. Toxicol. Environ. Heal. Part B Crit. Rev.* **2013**, *16*, 491–550. [CrossRef]

31. Regional Enviornmental Center. Schools Indoor Pollution and Health Observatory Network in Europe. *Eur. Union* **2014**.

32. Mohai, P.; Kweon, B.S.; Lee, S.; Ard, K. Air pollution around schools is linked to poorer student health and academic performance. *Health Aff.* **2011**, *30*, 852–862. [CrossRef]

33. Mendell, M.J.; Heath, G.A. Do indoor pollutants and thermal conditions in schools influence student performance? A critical review of the literature. *Indoor Air* **2005**, *15*, 27–52. [CrossRef]

34. Mendell, M.J.; Eliseeva, E.A.; Davies, M.M.; Lobscheid, A. Do classroom ventilation rates in California elementary schools influence standardized test scores? Results from a prospective study. *Indoor Air* **2016**, *26*, 546–557. [CrossRef]

35. Finell, E.; Tolvanen, A.; Haverinen-Shaughnessy, U.; Laaksonen, S.; Karvonen, S.; Sund, R.; Luopa, P.;

Pekkanen, J.; Ståhl, T. Indoor air problems and the perceived social climate in schools: A multilevel structural equation analysis. *Sci. Total Environ.* **2018**, *624*, 1504–1512. [CrossRef]

36. Chithra, S.V.; Nagendra, S.M.S. A review of scientific evidence on indoor air of school building: Pollutants, sources, health effects and management. *Asian J. Atmos. Environ.* **2018**, *12*, 87–108. [CrossRef]

37. Pearce, N.; Douwes, J.; Beasley, R. The rise and rise of asthma: A new paradigm for the new millennium? *J. Epidemiol. Biostat.* **2000**, *5*, 5–16. [PubMed]

38. Madureira, J.; Paciência, I.; Cavaleiro Rufo, J.; Ramos, E.; Barros, H.; Paulo Teixeira, J.; De Oliveira Fernandes, E. Indoor air quality in schools and its relationship with children's respiratory symptoms. *Atmos. Environ.* **2015**, *118*, 145–156. [CrossRef]

39. Silva, M.F.; Maas, S.; de Souza, H.A.; Gomes, A.P. Post-occupancy evaluation of residential buildings in Luxembourg with centralized and decentralized ventilation systems, focusing on indoor air quality (IAQ). Assessment by questionnaires and physical measurements. *Energy Build.* **2017**, *148*, 119–127. [CrossRef]

40. Hernandez, R.; Bassett, S.M.; Boughton, S.W.; Schuette, S.A.; Shiu, E.W.; Moskowitz, J.T. Psychological Well-Being and Physical Health: Associations, Mechanisms, and Future Directions. *Emot. Rev.* **2018**, *10*, 18–29. [CrossRef]

41. Sakellaris, I.A.; Saraga, D.E.; Mandin, C.; Roda, C.; Fossati, S.; de Kluizenaar, Y.; Carrer, P.; Dimitroulopoulou, S.; Mihucz, V.G.; Szigeti, T.; et al. Perceived indoor environment and occupants' comfort in European 'Modern' office buildings: The OFFICAIR Study. *Int. J. Environ. Res. Public Health* **2016**, *13*, 444. [CrossRef]

42. al horr, Y.; Arif, M.; Katafygiotou, M.; Mazroei, A.; Kaushik, A.; Elsarrag, E. Impact of indoor environmental quality on occupant well-being and comfort: A review of the literature. *Int. J. Sustain. Built Environ.* **2016**, *5*, 1–11. [CrossRef]

43. Wargocki, P.; Frontczak, M.; Schiavon, S.; Goins, J.; Arens, E.; Zhang, H. Satisfaction and self-estimated performance in relation to indoor environmental parameters and building features. In Proceedings of the 10th International Conference on Healthy Buildings, Brisbane, Australia, 8–12 July 2012.

44. Kraus, M.; Novakova, P. Gender Differences in Perception of Indoor Environmental Quality (IEQ). *Iop Conf. Ser. Mater. Sci. Eng.* **2019**, 603. [CrossRef]

45. Rupp, R.F.; Vásquez, N.G.; Lamberts, R. A review of human thermal comfort in the built environment. *Energy Build.* **2015**. [CrossRef]

46. Kim, J.; de Dear, R. The effects of contextual differences on office workers' perception of indoor environment. In Proceedings of the Indoor Air 2014—13th International Conference on Indoor Air Quality and Climate, Hong Kong, China, 8–12 July 2014.

47. Fanger, P. *Thermal Comfort: Analysis and Applications in Environmental Engineering*; Copenhagen Danish Technical Press: Copenhagen, Denmark, 1970.

48. d'Ambrosio, F.R.; Olesen, B.W.; Palella, B.I.; Povl, O. Fanger's impact ten years later. *Energy Build.* **2017**, *152*, 243–249. [CrossRef]

49. McCartney, K.J.; Nicol, J.F. Developing an adaptive control algorithm for Europe. *Energy Build.* **2002**, *34*, 623–635. [CrossRef]

50. Nicol, J.F.; Humphreys, M.A. Derivation of the adaptive equations for thermal comfort in free-running buildings in European standard EN15251. *Build. Environ.* **2010**, *45*, 11–17. [CrossRef]

51. Guedes, M.C.; Matias, L.; Santos, C.P. Thermal comfort criteria and building design: Field work in Portugal. *Renew. Energy* **2009**, *34*, 2357–2361. [CrossRef]

52. Matias, L. *Desenvolvimento de um Modelo Adaptativo Para Definição das Condições de Conforto Térmico em Portugal*; Universidade técnica de Lisboa: Lisboa, Portugal, 2010.

53. Teli, D.; Jentsch, M.F.; James, P.A. Naturally ventilated classrooms: An assessment of existing comfort models for predicting the thermal sensation and preference of primary school children. *Energy Build.* **2012**, *53*, 166–182. [CrossRef]

54. Fanger, P.O.; Toftum, J. Extension of the PMV model to non-air-conditioned buildings in warm climates. *Energy Build.* **2002**, *34*, 533–536. [CrossRef]

55. Junta de Andalucía. Consejería de Innovación. Orden VIV/1744/2008 de 9 de junio, por la que se aprueba la 'Zonificación Climática de Andalucía por Municipios para su uso en el Código Técnico de la Edificación en su sección de Ahorro de Energía apartado de Limitación de Demanda Energética (CTE-HE1). *Boletín Of. la Junta Andalucía* **2008**, *254*, 4216.

56. Ministerio de Fomento del Gobierno de España. Código Técnico de la Edificación. 2006. Available online: http://www.codigotecnico.org (accessed on 20 April 2019).

57. de la Flor, F.J.S.; Domínguez, S.Á.; Félix, J.L.M.; Falcón, R.G. Climatic zoning and its application to Spanish building energy performance regulations. *Energy Build.* **2008**, *40*, 1984–1990. [CrossRef]

58. Kottek, M.; Grieser, J.; Beck, C.; Rudolf, B.; Rubel, F. World Map of the Köppen-Geiger climate classification updated. *Meteorol. Z.* **2006**, *15*, 259–263. [CrossRef]

59. Junta de Andalucía. Orden de 24 de enero de 2003 de la Consejería de Educación y Ciencia de la Junta de Andalucía por la que se aprueban las 'Normas de diseño y constructivas para los edificios de uso docente. *Boletín Of. La Junta Andal.* **2003**, *43*, 4669.

60. Campano, M.A.; Pinto, A.; Acosta, I.; Muñoz-González, C. Analysis of the Perceived Thermal Comfort in a Portuguese Secondary School: Methodology and Results. *Int. J. Eng. Technol.* **2017**, *9*, 448–455. [CrossRef]

61. Campano, M.A.; Sendra, J.J.; Domínguez, S. Analysis of thermal emissions from radiators in classrooms in Mediterranean climates. *Procedia Eng.* **2011**, *21*, 106–113. [CrossRef]

62. AENOR. *UNE-EN ISO 7726 - Ergonomics of the Thermal Environment. Instruments for Measuring Physical Quantities*; AENOR - Asociación Española de Normalización y Certificación: Madrid, Spain, 2003.

63. American Society of Heating Refrigerating and Air Conditioning Engineers. *ASHRAE Handbook-Fundamentals*; ASHRAE Handbook Committee: Atlanta, GA, USA, 2017.

64. Winslow, C.E.A.; Herrington, L.P.; Gagge, A.P. Physiological reactions to environmental temperature. *Am. J. Physiol.* **1937**, *120*, 1–22. [CrossRef]

65. Campano, M.A.; Pinto, A.; Acosta, I.; Sendra, J.J. Validation of a dynamic simulation of a classroom HVAC system by comparison with a real model. *Sustain. Dev. Renov. Archit. Urban. Eng.* **2017**, 381–392.

66. International Organization for Standardization. *ISO 10551:1995—Ergonomics of the Physical Environment—Subjective Judgement Scales for Assessing Physical Environments*; International Organization for Standardization: Geneva, Switzerland, 1995.

67. International Organization for Standardization. *ISO 9920:2007—Ergonomics of the Thermal Environment e Estimation of the Thermal Insulation and Evaporative Resistance of a Clothing Ensemble*; International Organization for Standardization: Geneva, Switzerland, 2007.

68. Havenith, G.; Holmér, I.; Parsons, K. Personal factors in thermal comfort assessment: Clothing properties and metabolic heat production. *Energy Build.* **2002**, *34*, 581–591. [CrossRef]

69. Campano, M.A. Confort Térmico y Eficiencia Energética en Espacios con Alta Carga Interna Climatizados: Aplicación a Espacios Docentes no Universitarios en Andalucía. Ph.D. Thesis, Universidad de Sevilla, Seville, Spain, 2015.

70. Penney, D.; Adamkiewicz, D.; Ross, H.; Kenichi, A.; Vernon, A.; Benignus, P.; Hyunok, C.; Cohen, A.; Däumling, C.; Delgado, J.M.; et al. *WHO Guidelines for Indoor Air Quality: Selected Pollutants*; World Health Organization Regional Office for Europe: Copenhagen, Denmark, 2010; Volume 9.

71. Bellia, L.; Bisegna, F.; Spada, G. Lighting in indoor environments: Visual and non-visual effects of light sources with different spectral power distributions. *Build. Environ.* **2011**, *46*, 1984–1992. [CrossRef]

72. Acosta, I.; Campano, M.A.; Leslie, R.; Radetski, L. Daylighting design for healthy environments: Analysis of educational spaces for optimal circadian stimulus. *Sol. Energy* **2019**, *193*, 584–596. [CrossRef]

73. Cariñanos, P.; Galan, C.; Alcázar, P.; Domínguez, E. Airborne pollen records response to climatic conditions in arid areas of the Iberian Peninsula. *Environ. Exp. Bot.* **2004**. [CrossRef]

74. Galán, C.; García-Mozo, H.; Cariñanos, P.; Alcázar, P.; Domínguez-Vilches, E. The role of temperature in the onset of the Olea europaea L. pollen season in southwestern Spain. *Int. J. Biometeorol.* **2001**. [CrossRef]

75. Annesi-Maesano, I.; Moreau, D.; Caillaud, D.; Lavaud, F.; Le Moullec, Y.; Taytard, A.; Pauli, G.; Charpin, D. Residential proximity fine particles related to allergic sensitisation and asthma in primary school children. *Respir. Med.* **2007**, *101*, 1721–1729. [CrossRef]

76. Knasko, S.C. Ambient odor's effect on creativity, mood, and perceived health. *Chem. Senses* **1992**. [CrossRef]

77. Hoehner, C.M.; Ramirez, L.K.B.; Elliott, M.B.; Handy, S.L.; Brownson, R.C. Perceived and objective environmental measures and physical activity among urban adults. *Am. J. Prev. Med.* **2005**. [CrossRef]

78. Zhang, X.; Wargocki, P.; Lian, Z.; Thyregod, C. Effects of exposure to carbon dioxide and bioeffluents on perceived air quality, self-assessed acute health symptoms, and cognitive performance. *Indoor Air* **2017**. [CrossRef] [PubMed]

79. Fan, X.; Liu, W.; Wargocki, P. Physiological and psychological reactions of sub-tropically acclimatized subjects exposed to different indoor temperatures at a relative humidity of 70%. *Indoor Air* **2019**. [CrossRef] [PubMed]
80. Pastore, L.; Andersen, M. Building energy certification versus user satisfaction with the indoor environment: Findings from a multi-site post-occupancy evaluation (POE) in Switzerland. *Build. Environ.* **2019**, *150*, 60–74. [CrossRef]

7

How Working Tasks Influence Biocontamination in an Animal Facility

Anna M. Marcelloni [1], **Alessandra Chiominto** [1], **Simona Di Renzi** [1], **Paola Melis** [1], **Annarita Wirz** [2], **Maria C. Riviello** [2,3], **Stefania Massari** [4], **Renata Sisto** [1], **Maria C. D'Ovidio** [1] and **Emilia Paba** [1,*]

[1] Department of Occupational and Environmental Medicine, Epidemiology and Hygiene,
    Italian Workers' Compensation Authority (INAIL), Via Fontana Candida 1, Monte Porzio Catone,
    00078 Rome, Italy; a.marcelloni@inail.it (A.M.M.); a.chiominto@inail.it (A.C.); s.direnzi@inail.it (S.D.R.);
    p.melis@inail.it (P.M.); r.sisto@inail.it (R.S.); m.dovidio@inail.it (M.C.D.O.)
[2] Santa Lucia Foundation IRCCS, Via Ardeatina 306, 00142 Rome, Italy; a.wirz@hsantalucia.it
[3] Institute of Cell Biology and Neurobiology, National Research Council (CNR), Via E. Ramarini 32,
    Monterotondo, 00015 Rome, Italy; cristina.riviello@cnr.it
[4] Department of Occupational and Environmental Medicine, Epidemiology and Hygiene,
    Italian Workers' Compensation Authority (INAIL), Via Stefano Gradi 55, 00143 Rome, Italy; s.massari@inail.it
* Correspondence: e.paba@inail.it.

Abstract: The exposure to biocontaminants in animal facilities represents a risk for developing infectious, allergic and toxic diseases. The aim of this study was to determine what factors could be associated with a high level of exposure to biological agents through the measure and characterization of airborne fungi, bacteria, endotoxin, $(1,3)$-β-$\textsc{d}$-glucan and animal allergens. Airborne microorganisms were collected with an air sampler and identified by microscopic and biochemical methods. Endotoxin, $(1,3)$-β-$\textsc{d}$-glucan, Mus m 1, Rat n 1, Can f 1, Fel d 1, Equ c 4 allergens were detected on inhalable dust samples by Kinetic LAL, Glucatell, and ELISA assays, respectively. Our data evidenced that changing cages is a determinant factor in increasing the concentration of the airborne biocontaminants; the preparation of bedding and distribution of feed, performed in the storage area, is another critical working task in terms of exposure to endotoxins (210.7 EU/m^3) and $(1,3)$-β-$\textsc{d}$-glucans (4.3 ng/m^3). The highest concentration of Mus m 1 allergen (61.5 ng/m^3) was observed in the dirty washing area. The detection of expositive peaks at risk of sensitization (>2 µg/g) by Fel d 1 in animal rooms shows passive transport by operators themselves, highlighting their role as vehicle between occupational and living environments.

Keywords: allergens; endotoxin; biological agents; laboratory animal allergy; environmental monitoring; occupational exposure

1. Introduction

The exposure to biocontaminants is well documented and studied in several occupational settings, and few data are available in animal facilities, although more attention has been addressed to animal allergens. These biological agents are aeroallergens, mainly lipocalins, derived by different biological fluids and/or tissue (saliva, serum, urine, dander, hair, fur) that remain in suspension for different times in relation to meteorological conditions and factors influencing their dispersion [1,2]. In recent years, the topic regarding animal allergen exposure has been treated both in pets [3–5] and experimental animals [6–9], and attention has been also addressed to co-exposure of allergens and endotoxins in these workplaces, although in a more restricted way [10–15]. Exposure to biocontaminants in animal facilities in USA and in the United Kingdom has received great attention [16–19] as demonstrated by the NIOSH publication in 1998 [20], where the so-called LAA (Laboratory Animal Allergy) was considered an occupational risk, and by Gordon in 2001 [21]. Various papers have been published in scientific literature both as a review [22–24] and experimental studies [25–27] while in Italy the topic of LAA is not studied carefully yet.

Health effects reported on laboratory animal workers (LAWs) are mainly represented by asthma, rhinitis, conjunctivitis, dermatitis and anaphylaxis [8,19,28]. Discordant percentages of LAA have been reported in the LAWs and these differences could be attributed to different evaluation methods (questionnaires, analytical test, etc.). In this context a relevant aspect is represented by environmental exposure to animal allergens, and although a lot of papers report the monitoring of these biocontaminants, there is some difficulty in comparing results because of the differences in sampling and analytical methods [29–32]. In addition to allergens, there are several other biological agents in animal facilities with possible health consequences such as airborne bacteria, fungi and microbial indicators (endotoxin and (1,3)-β-D-glucan) which may originate from animals (fur, epidermal material), their waste (feces and urine) and materials used in their maintenance (food, bedding) [33]. Airborne endotoxin, an outer membrane component of Gram-negative bacteria, has been identified as the major risk factor for nasal, chest and skin symptoms also contributing to allergy diseases [11,34]. Some authors report that fecal bacteria in soiled bedding may produce increased concentrations of endotoxin compared to other settings and this may be an occupational health concern. The authors concluded that the percentage of dust and endotoxin in different types of rodent bedding could be an important factor affecting the occupational exposure of personnel working with laboratory animals causing airway inflammation, hypersensitivity pneumonitis, organic dust toxic syndrome (ODTS), chronic obstructive pulmonary disease (COPD) and asthma-like syndrome, as well as having harmful effects on the animals themselves [33].

Pathogenic moulds, yeast and their products may also exist in the air of animal units, causing further risk for personnel working there. In particular, (1,3)-β-D-glucans are non-allergenic and hardly soluble in water glucose polymers, which consist of a part of the cellular wall of most fungi, but also many plants and some bacteria. Due to their presence in both viable and dead cells, they may be considered a good indicator of exposure to fungi [35]. Although little information is available on their health effects in animal facilities, however, their association with dry cough, cough associated with phlegm, hoarseness, and atopy, has been reported in different studies [36–38]. Very few and not recent investigations dealing with airborne biological agents in laboratory animal settings, other than allergens, can be found [12–15,39] to date. These issues have received little attention in our country. The objective of this study was to determine factors (working tasks, cage changing frequency and animal strains) were associated with the greatest level of exposure to biological agents of the personnel working with rodents through the measure and characterization of airborne viable fungi, bacteria, endotoxin, (1,3)-β-D-glucan and animal allergens.

2. Materials and Methods

2.1. Animal Facility

The study was conducted in a Biomedical Research Institute where animals are maintained in agreement with Italian Legislative decree 26/2014 [40]. The facility includes 14 conventional animal rooms (3 housing rats and 11 housing mice) where animals are housed in stainless steel wire cages (on average 2–3 animals/cage) with perforated floor and trays underneath; filters on the top prevent the distribution of airborne particles. Each cage is stacked in a rack that can hold up to 30 cages for mice and 24 for rats. The facility houses about 7000 rodents.

The racks are washed every 4 months. The cages and their equipment (grids containing bottles of water and feed, the bottles and soiled bedding) are changed once or twice per week depending on experimental protocols. When cages are changed, a clean rack is brought into the room, animals and accessories are moved to new cages and dirty cages are moved to the cage-washing area where they are emptied and washed (dirty washing area). Subsequently, the cages are allowed to dry and restacked for future use (clean washing area). There is also a storage area where preparation of the bedding and distribution of feed are carried out; these activities are performed only in the afternoon. The staff is divided into laboratory technicians and researchers; laboratory technicians perform shifts of 8 h and 30 min, and are engaged in changing cages, bedding and feeding animals, cleaning rooms, washing cages; researchers are engaged in several experimental procedures such as collecting blood and urine specimens, surgery and sacrifice. Personal protective equipment is required for all personnel working in the facility and includes surgical cap, mask, gloves, shoe cover, and disposable coat.

Animal rooms and the storage area have a ventilation system separate from the rest of the facility. The outside air is collected and treated by an Air Handling Unit (AHU) located on the coverage plan. Another AHU treats the air entering the two washing areas. In each work environment the air enters through vents located on the ceiling (two in each room), while the exhaust air is removed through grilles located on the walls near the floor (four in each rooms) or on the ceiling in the dirty and clean washing areas. In animal rooms 15–20 air changes per hour are guaranteed; the values of temperature and relative humidity are maintained in a controlled range (T: 20 °C ± 1 °C; RH: 45–55%).

2.2. Sampling and Analysis of Airborne Bacteria and Fungi

Triplicate air samples were collected using a portable microbiological sampler (SAS Super ISO, PBI International, Milan, Italy) at a flow rate of 100 L min^{-1}, from 10 workplaces: 7 animal rooms randomly selected (3 housing rats and 4 housing mice) (see Table 1), 2 washing areas (dirty and clean), and the storage area. Two offices were investigated: the technician's and manager's offices. All measurements were taken in the middle. In order to evaluate the influence of changing cages on environmental contamination, in animal rooms air samples were taken over three consecutive days: the day of changing cages, the one before, and the one after. Regarding the storage area, air sampling was carried out in the morning (without activity) and in the afternoon.

Table 1. Animals housed in each investigated room during the study and frequency of changing cages.

Rooms	Strains	Total Number	Male	Female	Pups	Changing Cages
A	Rats	149	63	86	0	Bi-weekly
B	Mice	494	328	166	0	Weekly
C	Mice	682	311	267	104	Weekly
D	Mice	613	225	283	105	Bi-weekly
E	Rats	155	85	60	10	Weekly
F	Mice	723	316	380	27	Bi-weekly
G	Rats	377	37	85	255	Bi-weekly

Air volumes of 300 L were sampled. Total cultivable and Gram-negative bacteria were impacted on Tryptone Soy Agar (TSA) and Mac Conkey 3 (McC$_3$) plates, respectively, fungi on Malt Extract Agar (MEA) and Dichloran Glycerol Agar (DG18) (all from Oxoid S.p.A, Milan, Italy). TSA and McC$_3$ plates were subsequently incubated for 1–2 days at 37 °C (mesophilic bacteria), DG18 and MEA at 25 °C for 7–10 days.

The bacterial isolates were identified with microscopic and biochemical methods by API 20E and NE tests (bioMerieux, Marcy l'Etoile, France) and Microstation ID instrument (Biolog Inc., Hayward, CA, USA). Identification of moulds and yeasts were accomplished via macroscopic and microscopic examination, referring to the manual Medically Important Fungi: A guide to Identification [41], while biochemical identifications were performed using the Microstation ID instrument. Data are expressed as CFU/m^3. The limit of detection (LOD) was 2 CFU/m^3. The uncertainty assessment results are presented in Table 2.

2.3. Inhalable Dust Sampling and Analysis of Endotoxin, (1,3)-β-D-glucan and Allergens

Stationary inhalable dust samples were collected using airChek2000 pumps (SKC Inc., Eighty Four, PA, USA), at a sampling flow rate of 2 L·min^{-1}, equipped with stainless IOM sampler and glass (GF, pore size 1.6 μm) and polycarbonate (PC, pore size 0.8 μm) for respectively endotoxin and (1,3)-β-D-glucan analysis.

Closed-face cassette with support pad and mixed cellulose ester (MCE, diameter 33 mm, 0.8 μm porosity), at a flow rate of 2.5 L·min^{-1}, were used for allergen sampling. Samplers were placed in the middle of the room at 1.5 m above the floor to simulate human breathing zone. A field blank was deployed on each sampling. The sampling strategy was the same as environmental monitoring (three consecutive days in animal rooms, in the morning and in the afternoon inside the storage area). The sampling time was 3 h.

GF extracts (5 mL of 0,05% Tween 20 in Pyrogen Free Water) were centrifuged at 1000× g rpm for 20 min. and analysed in duplicate with Kinetic LAL method (QCL-LAL assay Lonza Walkersville, MD, USA). Concentrations are reported in EU/m^3. The LOD was 0.005 EU/mL.

For (1,3)-β-D-glucan, PC extracts (5 mL of 0.85% NaCl and 0.05% tween 80) were vigorously shaken for 30' at room temperature (250 rpm) and analyzed with the (1,3)-β-D-glucan specific Kinetic Chromogenic LAL assay (Glucatell, Associates of Cape Cod), as previously described [42,43]. (1,3)-β-D-glucan concentrations are presented in ng/m^3. The LOD in suspension was 2.53 pg/mL.

MCE filters were extracted by submerging in 1 mL phosphate saline buffer (PBS plus 0.01% Tween 20) in a 15 mL centrifuge tube, vortexed at room temperature and shaken for 2 h. The eluate was collected by compressing the filter in a plastic syringe, 1% of human serum albumin was added for protein stabilization and stored at −20 °C until assay. ELISA kits for Mus m 1, Rat n 1, Can f 1, Fel d 1, Equ c 4 were performed following the protocols of analysis (Indoor Biotechnologies Ltd., Manchester, UK.) with read of the plates at 405 nm optical density. Results are expressed as ng/m^3. The uncertainty assessment results are reported in Table 2.

(*) Type A standard uncertainty with a confidence level 1-α > 0.95 using the Student's T to find the appropriate coverage factor based on degrees of freedom. (**) Measures are detected in the morning (without activity) and in the afternoon (during activity).

Table 2. Uncertainty measurement (*) for sampled biological agents by working areas.

Workplace	Mesophilic Bacteria (UFC/m^3)	Fungi (UFC/m^3)	Endotoxin (EU/m^3)	(1,3)-β-D-glucan (ng/m^3)	Rat n 1 (ng/m^3)	Mus m 1 (ng/m^3)
Animal rooms	±52.26	±0.58	±1.35	±0.59	±0.16	±4.56
Washing area clean	±21.18	±16.91	±9.25			
Washing area dirty	±434.69	±66.18	±6.54			
Storage area (**)	±44.77	±94.89	±163.54	±4.57	±0.00	±8.32

(*) Type A standard uncertainty with a confidence level 1-α > 0.95 using the Student's T to find the appropriate coverage factor based on degrees of freedom. (**) Measures are detected in the morning (without activity) and in the afternoon (during activity).

2.4. Statistical Analysis

One way ANOVA and Students't-test were considered to evaluate significant differences in environmental biocontaminant concentrations in various work tasks and areas. Univariate and multivariate, both fixed and mixed effect linear regression models were used, in which the concentration of the different biological agents acts as outcome variable.

Data were analyzed using the statistical software R (R Foundation for Statistical Computing, Vienna, Austria—ISBN 3-900051-07-0.).

3. Results

Table 1 shows strains and number of animals (divided into males, females and pups) housed in investigated rooms (A–G) and the frequency of changing cages. Concentrations (mean value and standard deviation) of mesophilic bacteria, fungi, endotoxin, $(1,3)$-β-D-glucan, Rat n 1 and Mus m 1 detected in each working areas are reported in Table 3. Regarding animal rooms, data are also presented by sampling room and the number of rodents is indicated.

3.1. Airborne Bacteria and Fungi

A total of 306 air samples were analysed. Mean mesophilic bacteria levels ranged from 33.8 to 480 CFU/m^3 with a peak value of 885 CFU/m^3 found in an animal room during the changing cages. Gram-negative concentrations were below the LOD (2 CFU/m^3) except for animal rooms, during the changing cages, and for the storage area where a mean value of 3 CFU/m^3 was recorded. Among a total of 18 genera and species, the most frequently isolated were *Staphylococcus warneri* (23.9%), *Staphylococcus xylosus* (21.4%) *Micrococcus* spp. (14.5%), *Acinetobacter schindleri* (10.4%).

Colonies of *Escherichia coli*, *Enterobacter cloacae* and *Enterococcus faecalis* (risk group 2, according to Annex XLVI of the Italian Legislative Decrees [44,45], were isolated in four animal rooms during the changing cages.

Mean concentrations of moulds and yeasts were low (flora was dominated by moulds such as *Alternaria*, *Ulocladium*, *Cladosporium*, *Penicillium*, *Eurotium*, *Aspergillus*, *Scopulariopsis*, *Phoma*, with occasional isolates of yeasts. No pathogen was identified.

3.2. Endotoxin, $(1,3)$-β-D-glucan and Allergens

The mean concentrations of endotoxin, $(1,3)$-β-D-glucan and allergens (See Table 3) were obtained on 89 samples taken in different occupational environments. Statistically significant differences were found between offices and animal rooms, these last having a higher concentration of endotoxin ($p = 0.00616$) and between the animal rooms and the storage area in the concentration of $(1,3)$-β-D-glucan ($p = 0.038$).

The evaluation of allergens was performed on 181 air samples for ELISA tests. Mus m 1 allergens were detected in all working areas except for offices, with the highest value (61.5 ng/m^3) in the dirty washing area while Rat n 1 allergens were absent except for in animal rooms. Can f 1 and Equ c 4 allergens were absent while Fel d 1 allergen was detected in rat rooms with the highest levels during the changing cages activity (range 0.4–3 ng/m^3).

Table 3. Concentrations (mean value and standard deviation) of mesophilic bacteria, fungi, endotoxin, $(1,3)$-β-D-glucan, Rat n 1 and Mus m 1 by working areas (number of animals is indicated for each room).

Workplace	Mesophilic Bacteria (UFC/m^3)	Fungi (UFC/m^3)	Endotoxin (EU/m^3)	$(1,3)$-β-D-glucan (ng/m^3)	Rat n 1 (ng/m^3)	Mus m 1 (ng/m^3)
Animal rooms	130.7 ± 207.5	1.6 ± 3.04	5.9 ± 3.6	1.5 ± 1.3	0.1 ± 0.4	4.3 ± 11.0
Room A (149)	63.1 ± 92.7	2.1 ± 4.4	7.8 ± 0.7	1.8 ± 1.2	0.0 ± 0.0	0.0 ± 0.1
Room B (494)	122.9 ± 110.2	2.0 ± 1.7	7.2 ± 4.2	3.6 ± 2.3	0.0 ± 0.0	6.1 ± 9.9
Room C (682)	112.4 ± 115.7	1.1 ± 2.4	3.6 ± 3.5	0.9 ± 0.6	0.1 ± 0.3	2.7 ± 4.2
Room D (613)	64.8 ± 83.5	1.3 ± 2.7	0.1 ± 0.0	1.9 ± 0.8	0.5 ± 0.9	0.3 ± 0.2
Room E (155)	317.0 ± 427.0	3.3 ± 5.3	7.3 ± 3.2	0.1 ± 0.0	0.3 ± 0.5	0.1 ± 0.2
Room F (723)	66.7 ± 75.8	0.4 ± 1.1	11.1 ± 4.8	1.3 ± 0.8	0.0 ± 0.0	20.6 ± 24.1
Room G (377)	220.5 ± 232.5	1.7 ± 1.7	5.2 ± 2.6	0.9 ± 0.3	0.0 ± 0.0	0.1 ± 0.2
Washing area clean	78.3 ± 2.4	19.3 ± 13.6	7.4 ± 1.8	1.3 ± 0.7		0.4 ± 0.0
Washing area dirty	211.1 ± 175.0	107.8 ± 63.1	16.7 ± 4.7	1.6 ± 0.9		61.5 ± 0.0
Storage area *	266.7 ±36.1	112.5 ± 76.4	124.2 ± 102.8	3.9 ± 1.7		1.5 ± 0.8
Offices	308.0 ± 91.2	60.0 ± 25.9	2.4 ± 0.8	1.4 ± 1.1		

* Mean concentrations measured in the storage area refer to samplings performed the morning and in the afternoon.

3.3. Effect of the Working Activities

One of our main questions was whether or not specific working tasks may cause an increased risk of biological exposure for personnel working in animal facilities. The hypothesis that the airborne concentration of biocontaminants could increase during the changing cages was tested. To point out the increase in level of $(1,3)$-β-D-glucan the time was considered a two-level factor: sampling during the changing cages was compared to the sampling before and after. The β-glucans concentration "during" increases on average by a factor of 1.3 with respect to the concentration "before" and "after" and the difference is statistically significant ($p = 0.04$). For endotoxins, a mixed effect linear regression model was used in which animal room was introduced as a random effect variable. The concentrations increased on average by a factor of 1.8 with respect to the levels before or after ($p = 0.04$). The highest mean concentrations of endotoxin (210.7 EU/m^3) and (1,3)-β-D-glucan (4.3 ng/m^3) were measured in the afternoon during the preparation of bedding and distribution of feed (storage area) (See Table 4).

The allergenic concentration was treated as a continuous variable. As in the case of endotoxin, to increase the statistics, the time was considered a two-level factor and a mixed effect linear regression model was used with the animal room treated as a random effect variable. The concentration level of the mouse allergenic contaminant increases on average by a factor of 9.3 with respect to the levels before or after the changing cages ($p = 0.037$).

Table 4. Concentrations (mean value and standard deviation) of inhalable endotoxin, (1,3)-β-D-glucan, Rat n 1 and Mus m 1 by working activities.

Workplace	Endotoxin (EU/m^3)	(1,3)-β-D-glucan (ng/m^3)	Rat n 1 (ng/m^3)	Mus m 1 (ng/m^3)
Animal rooms				
Before	5.2 ± 3.7	0.8 ± 0.5		0.1 ± 0.1
During	7.1 ± 3.8	2.4 ± 1.9	0.4 ± 0.6	10.5 ± 17.4
After	5.4 ± 3.5	1.4 ± 0.6		2.2 ± 5.3
Storage area				
Morning	37.8 ± 37.3	3.5 ± 2.4		0.9 ± 0.0
Afternoon	210.7 ± 20.0	4.3 ± 1.6		2.2 ± 0.0

3.4. Effect of Animal Strains and Changing Cages Frequency

In order to determine the effect of the animal strains on environmental biocontamination, the concentration measured in animal rooms was normalized according to the number of animals. This normalized quantity was related to the strain treated as a two level factor. In the case of the mesophilic bacteria and endotoxin, the average level was lowered in the case of mice with respect to the case of rats by a factor of -0.5070 ($p = 0.020$) and by a factor of -0.032 ($p = 1.5 \cdot 10-8$) respectively. The result was not statistically significant for Gram-negative bacteria, fungi and (1,3)-β-D-glucan.

To test if animal species have an effect on the allergenic pollutants concentration, a Pearson's Chi-squared with Yates' continuity correction test was applied. The test was statistically significant in the case of the Mus m 1 and Fel d 1 ($p = 0.004$ and $p = 0.045$ respectively). The positive result of the Mus m 1 is statistically associated with the mouse species whilst the Fel d 1 is statistically associated with the rat.

In regards to the cage changing frequency, the concentrations normalized to the number of animals have been compared with once per week and twice per week cage changing frequency, treated as a two-level factor. A statistically significant difference has been found only in the case of mesophilic bacteria concentration. The twice per week frequency has on average a bacteria concentration reduced by a factor of 0.13 with respect to the level of rooms with a once per week frequency ($p = 0.019$).

4. Discussion

To the best of our knowledge, this is the first Italian study which assesses the exposure to biological agents in a conventional animal facility, taking into account the biocontamination of different occupational areas in conjunction with the effect of specific working tasks, cage changing frequency and animal strains on it.

The most frequent genera identified in this study are common skin inhabitants and are thus easily shed into the environment on desquamated epithelial skin cells. Some bacterial species have been reported to cause a variety of diseases, ranging from pneumonia (*Acinetobacter*) to intestinal infections (*E. coli*) (https://www.cdc.gov). Other microorganisms, such as *Enterobacter* and *Pseudomonas*, exhibit a high degree of biological activity, are producers of endotoxin, and often show strong allergenic properties that increase the risk of disease in workers [46,47]. However, colonies of bacterial species belonging to risk group 2, according to the Italian Legislative Decrees 81/2008 and 106/2009 [44,45], have been isolated in some animal rooms during the changing cages. With regard to endotoxin, no regulatory value is currently available, but various recommendations are available. Among them, Dutch Expert Committee on Occupational Standards recommends an occupational exposure limit of 90 EU/m^3 [48]. Endotoxin concentrations measured in this study were below this standard in all settings except for the storage area during activities where a mean value of 210.6 EU/m^3 was observed. Our data confirm what was reported by Kaliste et al., 2002 [33], that bedding material could be a significant source of accumulation and release of bacterial endotoxins. This evidence was demonstrated also by Whiteside et al., 2010 [49] who measured a range of endotoxins between 3121 and 5401 EU/g in hardwood rodent bedding.

Endotoxin concentrations measured in this study are higher than those reported by Hwang et al., 2016 [50], who found a mean value of 0.14 EU/m^3 (range = 0.03–0.60 EU/m^3) in an animal laboratory housing mice, where the main tasks were feeding/weighing with air exchange rates comparable to ours, and then those reported by Pacheco et al., 2006 [13] who measured a mean endotoxin concentration of 315 pg/m^3 (about 3.15 EU/m^3) with a peak value of 678 pg/m^3 (corresponding about 6.78 EU/m^3) during the changing cages in a mouse facility.

Our data are also slightly higher than those reported by Ooms et al., 2008 [39] (mean value = 4.7 EU/m^3; range = 3.4–6.3 EU/m^3) in rabbit rooms; this difference may be attributed to higher mechanical air-change rates (30.3 air changes/h) in the animal housing space. Finally, results of the present study are quite lower than those shown by Lieutier-Colas et al., 2001 [15], who found the highest concentration (15.4 ng/m^3 = 154.7 EU/m^3) associated with cage cleaning and feeding in animal rooms housing rats.

The detection of endotoxin in the technician's office (mean value = 2.3 EU/m^3) suggests that these components may be carried on clothes and shoes or on airborne particles like allergens.

The highest levels of (1,3)-β-ᴅ-glucan were found in the storage area, indicating that activities performed in this environment were also associated with an increased exposure to fungi, some of which belong to strongly allergenic and mycotoxin producers genera (*Alternaria* spp., *Penicillium* spp.).

Regarding animal allergens, no occupational environmental limits are available, partly due to the lack of standardized measurement methods and partly because of complex exposure-response relationships and the influence of genetic susceptibility; it is important to keep the allergen exposure as low as possible.

We found the highest concentration for Mus m 1 in the dirty washing area (mean value = 61.5 ng/m^3) and its concentrations increased significantly during the changing cages on average by a factor of 9.3 with respect to the levels before or after (p = 0.037). Straumfors et al., [16] in a recent paper report that cage emptying and cage washing in the cage washroom represented the highest exposure with a median value of 3.0 ng/m^3 for Mus m 1. The lower concentration found by these authors could be explained by the use of individually ventilated cages in respect to cage-rack systems and by the different sampling methodologies (stationary versus personal monitoring).

Our data regarding the highest values of Mus m 1 in respect to Rat n 1 in animal facilities are in agreement with the ones reported in literature (range: 17–564 ng/m^3 for Mus m 1; 0.43–27.36 ng/m^3 for Rat n 1) [1,15,51].

For cat allergens (Fel d 1), limit values between 2 and 8 μg/g Fel d 1 for sensitization and above 8 μg/g for the development of acute asthma attacks, respectively, have been proposed [52]. In this work we measured concentrations at risk of sensitization (>2 μg/g) in two rat rooms.

The detection of cat allergens (Fel d 1) in animal rooms is unexpected, along with the presence of rat and mouse allergens (Figure 1), showing a cross-contamination likely due to the passive transport by operators themselves; this highlights their role as vehicle between occupational areas and living environments [53–56].

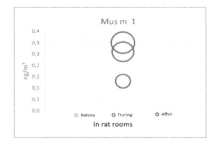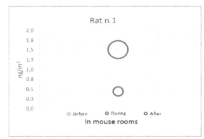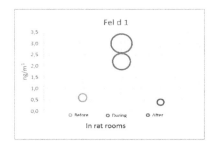

Figure 1. Aeroallergen levels over three consecutive days in mouse and rat rooms: before, during and after changing cages (Each circle represents the observed value and dimension/diameter indicates the concentration).

5. Conclusions

The environmental monitoring of airborne biocontaminants has been confirmed as a valid tool to identify the working tasks more critical in terms of exposure to biological agents in animal facilities.

Statistical analysis proves that the changing of cages is a determinant factor in increasing the concentration of all the airborne biocontaminants and in releasing biological agents that could risk the health of workers. It can also be pointed out that the concentrations of the biocontaminants increase significantly during this task with respect to the other days, but return to the background levels indicating the effectiveness of the ventilation system. Our data also identify the preparation of bedding and distribution of feed performed in the storage area as critical working tasks in terms of exposure to endotoxins, fungi and (1,3)-β-ᴅ-glucans.

The frequency of the changing of cages does not seem to affect the microbiological contamination; a statistical difference has been found only in the case of mesophilic bacteria levels. In regard to the animal strains, rats seem to influence only mesophilic bacteria and endotoxin levels.

Our data can be particularly useful for increasing knowledge about the biological risk in this occupational sector and to support all actors of prevention, employers, health and safety representatives and workers in all phases of risk management.

Simple preventive measures must be adopted to control the biological risk in this occupational setting.

Author Contributions: Environmental monitoring and laboratory procedures, A.M.M. and A.C.; Laboratory procedures, S.D.R.; Data set built, P.M.; Logistics support A.W. and M.C.R.; Statistical analysis, writing-review and editing, S.M. and R.S.; Conceived and designed the research topic, writing original draft, revising, editing and supervision, M.C.D. and E.P.

Acknowledgments: The authors are grateful to the workers who collaborated in our experimental phases of the study.

References

1.   Zahradnik, E.; Raulf, M. Respiratory allergens from furred mammals: Environmental and occupational exposure. *Vet. Sci.* **2017**, *4*, 38. [CrossRef] [PubMed]

2.   Zahradnik, E.; Raulf, M. Animal allergens and their presence in the environment. *Front. Immunol.* **2014**, *3*, 76. [CrossRef] [PubMed]

3.   Liccardi, G.; Calzetta, L.; Baldi, G.; Berra, A.; Billeri, L.; Caminati, M.; Capano, P.; Carpentieri, E.; Ciccarelli, A.; Crivellaro, M.A.; et al. Italian Allergic Respiratory Diseases Task ForceAllergic sensitization to common pets (cats/dogs) according to different possible modalities of exposure: An Italian Multicenter Study. *Clin. Mol. Allergy* **2018**, *16*, 3. [CrossRef]

4. Liccardi, G.; Salzillo, A.; Piccolo, A.; Calzetta, L.; Rogliani, P. Dysfunction of small airways and prevalence, airway responsiveness and inflammation in asthma: Much more than small particle size of pet animal allergens. *Ups. J. Med. Sci.* **2016**, *121*, 196–197. [CrossRef] [PubMed]

5. Patelis, A.; Dosanjh, A.; Gunnbjörnsdottir, M.; Borres, M.P.; Högman, M.; Alving, K.; Janson, C.; Malinovschi, A. New data analysis in a population study raises the hypothesis that particle size contributes to the pro-asthmatic potential of small pet animal allergens. *Ups. J. Med. Sci.* **2016**, *121*, 25–32. [CrossRef]

6. Stave, G.M. Occupational animal allergy. *Curr. Allergy Asthma Rep.* **2018**, *18*, 11. [CrossRef] [PubMed]

7. Mason, H.J.; Willerton, L. Airborne exposure to laboratory animal allergens. *AIMS Allergy Immunol.* **2017**, *1*, 78–88. [CrossRef]

8. Simoneti, C.S.; Ferraz, E.; Menezes, M.B.; Bagatin, E.; Arruda, L.K.; Vianna, E.O. Allergic sensitization to laboratory animals is more associated with asthma, rhinitis, and skin symptoms than sensitization to common allergens. *Clin. Exp. Allergy* **2017**, *47*, 1436–1444. [CrossRef]

9. Bush, R.K.; Stave, G.M. Laboratory animal allergy: An update. *ILAR J.* **2003**, *44*, 28–51. [CrossRef]

10. Lai, P.S.; Allen, J.G.; Hutchinson, D.S.; Ajami, N.J.; Petrosino, J.F.; Winters, T.; Hug, C.; Wartenberg, G.R.; Vallarino, J.; Christiani, D.C. Impact of environmental microbiota on human microbiota of workers in academic mouse research facilities: An observational study. *PLoS ONE* **2017**, *12*, e0180969. [CrossRef] [PubMed]

11. Oppliger, A.; Barresi, F.; Maggi, M.; Schmid-Grendelmeier, P.; Huaux, F.; Hotz, P.; Dressel, H. Association of endotoxin and allergens with respiratory and skin symptoms: A descriptive study in laboratory animal workers. *Ann. Work. Expo. Health* **2017**, *61*, 822–835. [CrossRef] [PubMed]

12. Samadi, S.; Heederik, D.J.; Krop, E.J.; Jamshidifard, A.R.; Willemse, T.; Wouters, I.M. Allergen and endotoxin exposure in a companion animal hospital. *Occup. Environ. Med.* **2010**, *67*, 486–492. [CrossRef] [PubMed]

13. Pacheco, K.A.; McCammon, C.; Thorne, P.S.; O'Neill, M.E.; Liu, A.H.; Martyny, J.W.; Vandyke, M.; Newman, L.S.; Rose, C.S. Characterization of endotoxin and mouse allergen exposure in mouse facilities and research laboratories. *Am. Occup. Hyg.* **2006**, *50*, 563–572.

14. Platts-Mills, J.; Custis, N.; Kenney, A.; Tsay, A.; Chapman, M.; Feldman, S.; Platts-Mills, T. The effects of cage design on airborne allergens and endotoxin in animal rooms: High-volume measurements with an ion-charging device. *Contemp. Top. Lab. Anim. Sci.* **2005**, *44*, 12–16. [PubMed]

15. Lieutier-Colas, F.; Meyer, P.; Larsson, P.; Malmberg, P.; Frossard, N.; Pauli, G.; de Blay, F. Difference in exposure to airborne major rat allergen (Rat n 1) and to endotoxin in rat quartes according to tasks. *Clin. Exp. Allergy* **2001**, *31*, 1449–1456. [CrossRef] [PubMed]

16. Straumfors, A.; Eduard, W.; Andresen, K.; Sjaastad, A.K. Predictors for increased and reduced rat and mouse allergen exposure in laboratory animal facilities. *Ann. Work Expo. Health* **2018**, *62*, 953–965. [CrossRef]

17. Feary, J.; Cullinan, P. Laboratory animal allergy: A new world. *Curr. Opin. Allergy Clin. Immunol.* **2016**, *16*, 107–112. [CrossRef]

18. Simoneti, C.S.; Freitas, A.S.; Barbosa, M.C.; Ferraz, E.; de Menezes, M.B.; Bagatin, E.; Arruda, L.K.; Vianna, E.O. Study of risk factors for atopic sensitization, asthma, and bronchial hyperresponsiveness in animal laboratory workers. *J. Occup. Health* **2016**, *58*, 7–15. [CrossRef]

19. Kampitak, T.; Betschel, S.D. Anaphylaxis in laboratory workers because of rodent handling: Two case reports. *J. Occup. Health* **2016**, *58*, 381–383. [CrossRef]

20. National Institute for Occupational Safety and Health (NIOSH). *Preventing Asthma in Animal Handlers*; NIOSH: Washington, DC, USA, 1998; pp. 97–116.

21. Gordon, S. Laboratory animal allergy: A British perspective on a global problem. *ILAR J.* **2001**, *42*, 37–46. [CrossRef]

22. Corradi, M.; Ferdenzi, E.; Mutti, A. The characteristics, treatment and prevention of laboratory animal allergy. *Lab. Anim.* **2012**, *42*, 26–33. [CrossRef] [PubMed]

23. D'Ovidio, M.C.; Martini, A.; Melis, P.; Signorini, S. Value of the microarray for the study of laboratory animal allergy (LAA). *G. Ital. Med. Lav. Ergon.* **2011**, *33*, 109–116. [PubMed]

24. Folletti, I.; Forcina, A.; Marabini, A.; Bussetti, A.; Siracusa, A. Have the prevalence and incidence of occupational asthma and rhinitis because of laboratory animals declined in the last 25 years? *Allergy* **2008**, *63*, 834–841. [CrossRef]

25. D'Ovidio, M.C.; Wirz, A.; Zennaro, D.; Masssari, S.; Melis, P.; Peri, V.M.; Rafaiani, C.; Riviello, M.C.; Mari, A. Biological occupational allergy: Protein microarray for the study of laboratory animal allergy (LAA). *AIMS Public Health* **2018**, *5*, 352–365. [CrossRef] [PubMed]

26. Larese Filon, F.; Drusian, A.; Mauro, M.; Negro, C. Laboratory animal allergy reduction from 2001 to 2016: An intervention study. *Respir. Med.* **2018**, *136*, 71–76. [CrossRef] [PubMed]

27. Tafuro, F.; Selis, L.; Goldoni, M.; Stendardo, M.; Mozzoni, P.; Ridolo, E.; Boschetto, P.; Corradi, M. Biomarkers of respiratory allergy in laboratory animal care workers: An observational study. *Int. Arch. Occup. Environ. Health* **2018**, *91*, 735–744. [CrossRef]

28. Bhabha, F.K.; Nixon, R. Occupational exposure to laboratory animals causing a severe exacerbation of atopic eczema. *Austr. J. Dermatol.* **2012**, *53*, 155–156. [CrossRef] [PubMed]

29. Jones, M. Laboratory animal allergy in the modern era. *Curr. Allergy Asthma Rep.* **2015**, *15*, 73. [CrossRef]

30. Muzembo, B.A.; Eitoku, M.; Inaoka, Y.; Oogiku, M.; Kawakubo, M.; Tai, R.; Takechi, M.; Hirabayashi, K.; Yoshida, N.; Ngatu, N.R.; et al. Prevalence of occupational allergy in medical researchers exposed to laboratory animals. *Ind. Health* **2014**, *52*, 256–261. [CrossRef]

31. Acton, D.; McCauley, L. Laboratory animal allergy: An occupational hazard. *AAOHN J.* **2007**, *55*, 241–244. [CrossRef]

32. Elliott, L.; Heederik, D.; Marshall, S.; Peden, D.; Loomis, D. Incidence of allergy and allergy symptoms among workers exposed to laboratory animals. *Occup. Environ. Med.* **2005**, *62*, 766–771. [CrossRef] [PubMed]

33. Kaliste, E.; Linnainmaa, M.; Meklin, T.; Nevalainen, A. Airborne contaminants in conventional laboratory rabbit rooms. *Lab. Anim.* **2002**, *36*, 43–50. [CrossRef]

34. Freitas, A.S.; Simoneti, C.S.; Ferraz, E.; Bagatin, E.; Brandão, I.T.; Silva, C.L.; Borges, M.C.; Vianna, EO. Exposure to high endotoxin concentration increases wheezing prevalence among laboratory animal workers: A cross-sectional study. *BMC Pulm. Med.* **2016**, *16*, 69. [CrossRef] [PubMed]

35. Douwes, J. (1–>3)-Beta-D-glucans and respiratory health: A review of the scientific evidence. *Indoor Air* **2005**, *15*, 160–169. [CrossRef] [PubMed]

36. Paba, E.; Chiominto, A.; Marcelloni, A.M.; Proietto, A.; Sisto, R. Exposure to airborne culturable microorganisms and endotoxin in two Italian poultry slaughterhouses. *J. Occup. Environ. Hyg.* **2014**, *11*, 469–478. [CrossRef]

37. Rylander, R.; Lin, R.H. (1–>3)-beta-D-glucan-relationship to indoor air-related symptoms, allergy and asthma. *Toxicology* **2000**, *152*, 47–52. [CrossRef]

38. Rylander, R.; Norhall, M.; Engdahl, U.; Tunsater, A.; Holt, P.G. Airways inflammation, atopy, and (1–>3)-beta-D-glucan exposure in two schools. *Am. J. Respir. Crit. Care Med.* **1998**, *158*, 1685–1687. [CrossRef]

39. Ooms, T.G.; Artwohl, J.E.; Conroy, L.M.; Schoonover, T.D.; Fortman, J.D. Concentration and emission of airborne contaminants in a laboratory animal facility housing rabbits. *J. Am. Assoc. Lab. Anim. Sci.* **2008**, *47*, 39–48. [PubMed]

40. Implementation of the Directive 2010/63/EU on the protection of animals used for scientific purposes, Italian Legislative Decree 4 march 2014, n. 26. *Off. J.* **2014**, *60*.

41. Larone, D.H. *Medically Important Fungi: A guide to Identification*, 3rd ed.; ASM: Washington, DC, USA, 1995.

42. Lee, T.; Grinshpun, S.A.; Kim, K.Y.; Iossifova, Y.; Adhikari, A.; Reponen, T. Relationship between indoor and outdoor airborne fungal spores, pollen, and (1→3)-β-D-glucan in homes without visible mold growth. *Aerobiologia* **2006**, *22*, 227–236. [CrossRef]

43. Iossifova, Y.Y.; Reponen, T.; Bernstein, D.I.; Levin, L.; Kalra, H.; Campo, P.; Villareal, M.; Lockey, J.; Hershey, G.K.; LeMasters, G. House dust (1-3)-beta-D-glucan and wheezing in infants. *Allergy* **2007**, *62*, 504–513. [CrossRef]

44. Implementation of the article 1 of the law 3 august 2007, no. 123 concerning the protection of health and safety in the workplaces, Italian Legislative Decree no. 81/2008. *Ordinary Suppl. No.108 Off. J.* **2008**, *101*.

45. Supplementary and corrective provisions of the legislative decree 9 April 2008, no. 81, concerning the protection of health and safety in the workplaces, Italian Legislative Decree no. 106/2009. *Ordinary Suppl. No. 142 Off. J.* **2009**, *180*.

46. Duchaine, C.; Grimard, Y.; Cormier, Y. Influence of building maintenance, environmental factors, and seasons on airborne contaminants of swine. *Am. Ind. Hyg. Assoc. J.* **2000**, *61*, 56–63. [CrossRef]

47. Mandryk, J.; Alwis, K.U.; Hocking, A.D. Effects of personal exposures on pulmonary function and work-related symptoms among sawmill workers. *Ann. Occup. Hyg.* **2000**, *44*, 281–289. [CrossRef]

48. Health Council of the Netherlands. *Endotoxins-Health-Based Recommended Occupational Exposure Limit*; Publication no. 2010/04OSH. 010; Health Council of the Netherlands: The Hague, The Netherlands, 2010.

49. Whiteside, T.E.; Thigpen, J.E.; Kissling, M.G.; Grant, M.G.; Forsythe, D.B. Endotoxin, coliform, and dust levels in various types of rodent bedding. *J. Am. Assoc. Lab. Anim. Sci.* **2010**, *49*, 184–189. [PubMed]

50. Hwang, S.H.; Park, D.J.; Park, W.M.; Park, D.U.; Ahn, J.K.; Yoon, C.S. Seasonal variation in airborne endotoxin levels in indoor environments with different micro-environmental factors in Seoul, South Korea. *Environ. Res.* **2016**, *145*, 101–108. [CrossRef]

51. Ohman, J.L.; Hagberg, K.; MacDonald, M.R.; Jones, R.R.; Paigen, B.J.; Kacergis, J.B. Distribution of airborne mouse allergen in a major mouse breeding facility. J. Allergy. *Clin. Immunol.* **1994**, *94*, 810–817.

52. Liccardi, G.; Triggiani, M.; Piccolo, A.; Salzillo, A.; Parente, R.; Manzi, F.; Vatrella, A. Sensitization to common and uncommon pets or other furry animals: Which may be common mechanisms? *Transl. Med. UniSa* **2016**, *14*, 9–14.

53. Liccardi, G.; Salzillo, A.; Piccolo, A.; D'Amato, M.; D'Amato, G. Can the levels of Can f 1 in indoor environments be evaluated without considering passive transport of allergen indoors? *J. Allergy Clin. Immunol.* **2013**, *131*, 1258–1259. [CrossRef]

54. Krop, E.J.; Doekes, G.; Stone, M.J.; Alberse, R.C.; van der Zee, J.S. Spreading of occupational allergens: Laboratory animal allergens on hair-covering caps and in mattress dust of laboratory animal workers. *Occup. Environ. Med.* **2007**, *64*, 267–272. [CrossRef] [PubMed]

55. Pereira, F.L.; Silva, D.A.; Sopelete, M.C.; Sung, S.S.; Taketomi, E.A. Mite and cat allergen exposure in Brazilian public transport vehicles. *Ann. Allergy Asthma Immunol.* **2004**, *93*, 179–184. [CrossRef]

56. Ferrari, M.; Perfetti, L.; Moscato, G. A case of indirect exposure to cat at school. *Monaldi Arch. Chest. Dis.* **2003**, *59*, 169–170. [PubMed]

8

Exposure to Submicron Particles and Estimation of the Dose Received by Children in School and Non-School Environments

Antonio Pacitto [1], Luca Stabile [1,*], Stefania Russo [2] and Giorgio Buonanno [1,3]

[1] Department of Civil and Mechanical Engineering, University of Cassino and Southern Lazio, 03043 Cassino, Italy; alpacitto@unicas.it (A.P.); buonanno@unicas.it (G.B.)

[2] FIMP-Federazione Italiana Medici Pediatri, 00185 Roma, Italy; russostef.ped@tiscali.it

[3] International Laboratory for Air Quality and Health, Queensland University of Technology, Brisbane, QLD 4000, Australia

* Correspondence: l.stabile@unicas.it.

Abstract: In the present study, the daily dose in terms of submicron particle surface area received by children attending schools located in three different areas (rural, suburban, and urban), characterized by different outdoor concentrations, was evaluated. For this purpose, the exposure to submicron particle concentration levels of the children were measured through a direct exposure assessment approach. In particular, measurements of particle number and lung-deposited surface area concentrations at "personal scale" of 60 children were performed through a handheld particle counter to obtain exposure data in the different microenvironments they resided. Such data were combined with the time–activity pattern data, characteristics of each child, and inhalation rates (related to the activity performed) to obtain the total daily dose in terms of particle surface area. The highest daily dose was estimated for children attending the schools located in the urban and suburban areas (>1000 mm^2), whereas the lowest value was estimated for children attending the school located in a rural area (646 mm^2). Non-school indoor environments were recognized as the most influential in terms of children's exposure and, thus, of received dose ($>70\%$), whereas school environments contribute not significantly to the children daily dose, with dose fractions of 15–19% for schools located in urban and suburban areas and just 6% for the rural one. Therefore, the study clearly demonstrates that, whatever the school location, the children daily dose cannot be determined on the basis of the exposures in outdoor or school environments, but a direct assessment able to investigate the exposure of children during indoor environment is essential.

Keywords: exposure assessment; school; children; number concentration; lung-deposited surface area; dose

1. Introduction

Many studies highlighted the link between the exposure to airborne particles and health effects, such as respiratory diseases and inflammation [1], cardiovascular diseases [2,3], diabetes [4], higher systolic blood pressure and pulse pressure [5], and decreased cognitive function in older men [6]; in particular, the World Health Organization (WHO) estimated that the overexposure to particulate matter (PM) causes about 4.2 million deaths per year worldwide [7]. Moreover, the WHO has recently classified PM, referred to as outdoor pollution, as a carcinogenic pollutant for humans (group 1) [8–10]. The harmful potential of airborne particles is related to their ability to penetrate and deposit in the deepest areas of human respiratory tract (i.e., alveolar region), causing irritation, inflammation and possible translocation into the blood system, carrying with them carcinogenic and toxic compounds [11–14]. The inhalation and consequent deposition of these compounds are strictly related to the size of the carrying particles: higher deposition fractions in the lungs are characteristics of submicron and ultrafine particles [15]. Moreover, smaller particles are also recognized to translocate from lungs to the cardiovascular system and from there to other organs (liver, spleen, kidneys, brain) [16–18].

In the last years, the attention of scientific studies has shifted from super-micron particles (whose contribution is expressed in terms of mass concentrations of particles smaller than 10 and 2.5 μm, i.e., PM_{10} and $PM_{2.5}$) [19,20] to submicron and ultrafine particles (UFPs, particles smaller than 100 nm) whose contribution is better related to particle number [21,22] and surface area concentrations [23,24] than mass concentration. In fact, many studies highlighted that dose-response correlation in terms of human health effects is better related to surface area of particles deposited in the lungs than other metrics of exposure. To summarize, particle surface area is the most relevant dose metric for acute submicron particle lung toxicity [1,25–32].

In light of this, to evaluate the health effect of the exposure to airborne particles, a critical factor that should be assessed and provided to medical experts is the dose of submicron particles received by individuals [33–35]. Moreover, the airborne particle dose is the main input data for human health risk model [36–39]. Airborne particle doses received by people can be evaluated on the basis of measurements obtained from ad-hoc exposure assessment research. Nonetheless, even though the scientific community is moving from particle mass-based (PM) to number- and surface area-based metrics (submicron particles), the current legislation is still limited to the outdoor concentration of PM_{10} and $PM_{2.5}$; such measurements are limited to some outdoor fixed sampling points (FSPs) placed in specific points classified as a function of the type of site (rural, urban, suburban) and the type of station, i.e., proximity to main sources (background, industrial, or traffic) [40–42]. Moreover, PM_{10} and $PM_{2.5}$ measurements at FSPs cannot be considered proxies for exposure to submicron and ultrafine particles since they present different dynamics (e.g., dilution, deposition) and origins/sources [43–50]. Indeed, differently from PM_{10} concentrations that are typically quite homogeneously distributed around the city, the concentrations of submicron particle metrics (number and surface area) are strongly affected by the proximity to the source [51,52]. Finally, the measurement at an outdoor FSP cannot take into account for the exposure in indoor environments; therefore, a proper evaluation of the overall human exposure to submicron and ultrafine particles can be only obtained through personal monitoring able to measure the exposure at a personal scale and also to include the exposure in indoor microenvironments [53–55].

One of the most vulnerable populations in terms of air pollution exposure is represented by children [56,57]. This is due, amongst other things, to their high inhalation rates, resulting in larger specific doses than adults [58–61]. Children use to spend a large part of their day in indoor environments, such as schools and homes. In our previous studies involving adults, we found that some environments and activities affect the total daily dose more than other ones: in particular, the indoor environments were recognized to contribute up to 90% of the total daily dose in terms of particle surface area, with cooking and eating activities alone accounting up to 50% [53,62,63]. Schools as well may be considered a critical indoor environment under certain circumstances, in fact, the long exposure time in schools (children spend from 175 to 220 days and from 5 to 8 hours at school [64]) could significantly affect the overall dose received by children. Actually, the exposure (and then the dose) in school environments is not affected by the presence of submicron particle sources (smoking is typically not allowed and cooking activities are in most of the cases no longer performed in the school) but mainly by the outdoor-to-indoor penetration of submicron particles produced outdoors, which depends on (i) airtightness of the building, (ii) type of ventilation and (iii) particle physical-chemical properties (e.g., size) [65–71]. Therefore, the location of the school, as highlighted in few previous studies [39,72–74], is the main parameter affecting the students' exposure to submicron particles leading to critical exposure scenarios for those attending schools located near highly trafficked urban roads. To the best of the authors' knowledge the dose of submicron particles received by children in school and non-school environments was investigated just in one (our) previous paper [55], but in this study the investigated schools (both in the rural and urban areas) were placed in the same city, thus the possible contribution of the outdoor concentration levels to the daily dose was not adequately deepened.

Within this context, the aim of the present research is to evaluate the actual exposure to submicron particles of children attending schools located in different urban contexts and cities (urban, suburban rural sites) and to estimate the corresponding doses received both in schools and in other non-school environments where they spend time. To this end, an extensive experimental campaign was performed by measuring the personal exposure of 60 children (for 48 h each) attending three different schools in Italy, characterized by different outdoor concentration levels, using wearable monitors able to measure particle number and lung-deposited surface area concentrations.

2. Methodology

2.1. Study Area and Monitoring Site

Children considered in the experimental campaign attended three naturally ventilated schools located in three different cities in Italy (Salerno, South of Italy, Roma, Central Italy, and Parma, North of Italy); the locations of the three schools within the urban contexts are completely different. In particular, the school in Salerno (S1) is placed in a suburban area as it is 1.6 km outside of the city centre, but quite close to a highway. The school in Rome (S2) is located in the urban area, and, in particular, in the proximity of highly trafficked roads, whereas, the school located in Parma (S3) is in the rural area, about 5 km from the city centre, and quite far from trafficked roads. The experimental campaigns in the three schools were performed from November 2018 to May 2019 for about two months in each school as summarized Table 1.

Table 1. School sites, sampling periods and summary of the meteo-climatic conditions (temperature, T, relative humidity, RH) and air quality parameters (NO_2, PM_{10} and $PM_{2.5}$) measured by the closest fixed sampling stations of the Italian environmental protection agency. The data related to every single period of the campaign are expressed as daily average values and their ranges (min–max).

School	City	Sampling Period	T (°C)	RH (%)	NO_2 ($\mu g \cdot m^{-3}$)	PM_{10} ($\mu g \cdot m^{-3}$)	$PM_{2.5}$ ($\mu g \cdot m^{-3}$)	Distance of the Closest Fsp to the School and Definition According to the Standard
S1	Salerno	November–December 2018	15 (13–17)	71 (n.a.)	34 (8–66)	21 (8–56)	13 (5–44)	Distance: <100 m type of site: suburban type of station: background
S2	Roma	February–March 2019	11 (6–16)	66 (n.a.)	42 (26–122)	29 (7–63)	16 (12–18)	Distance: 1 km type of site: urban type of station: traffic
S3	Parma	April–May 2019	15 (10–18)	71 (n.a.)	12 (10–42)	12 (<5–27)	7.5 (<5–21)	Distance: 12 km type of site: rural type of station: background

In order to better describe the three sampling sites in terms of outdoor air quality, in Table 1, the distance of the closest fixed sampling point (FPS) installed by the Italian environmental protection agency to the schools are reported, as well as its definition in terms of type of station (background, industrial, or traffic) and type of site (urban, suburban rural sites). The closest FPSs to school S1 (100 m), S2 (1 km), and S3 (12 km), are defined as suburban/background station, urban/traffic station, and rural/background station, respectively. The parameters measured by the three FSPs during the three different sampling periods (November–December 2018, February–March 2019, and April–May 2019 for school S1, S2 and S3, respectively) clearly highlight the different outdoor air quality of the locations investigated: indeed, the highest NO_2, PM_{10} and $PM_{2.5}$ values were measured by the FSP close to the school in Rome (S2) (average values of 42, 29, and 16 $\mu g \cdot m^{-3}$, respectively) whereas the lowest, as expected, were measured by the FPS close to the school in Parma (S3) (average values of 12, 12, and 7.5 $\mu g \cdot m^{-3}$ for NO_2, PM_{10} and $PM_{2.5}$, respectively). The authors, once again, point out that the concentration of the different PM fractions cannot be considered as a good proxy for ultrafine or submicron particles. Indeed, the latter, along with NO_2, are good markers of the tailpipe emissions of the vehicular traffic, whereas PM_{10} is only partially due to tailpipe emissions of vehicles (a significant fraction is due to the traffic-induced particle resuspension) and it is a good marker, amongst others, of biomass combustion for residential heating [75]. Therefore, an overall correlation between outdoor concentrations of PM_{10} and submicron particles can be found, but, in some conditions (e.g., co-presence of other sources) these two metrics could be poorly correlated. Actually, since the FSPs close to S1 and S2 are strongly affected by traffic sources, a good correlation between PM_{10} and submicron particles is somehow expected; this is partially confirmed by the fact that NO_2 and PM fractions data shown in Table 1 present very good correlations (linear regressions with r^2 equal to 0.95 and 0.99 for PM_{10} and $PM_{2.5}$, respectively). Finally, regarding the meteo-climatic parameters, temperature and relative humidity values were found to be roughly similar in the three sites during the three measurement periods. This is a not trivial aspect—indeed, generally, the time of the year (e.g., season) can affect the children's exposure and doses both in terms of time-activity patterns and ventilation of the microenvironments since warmer conditions would have increased the time spent outdoor and the manual ventilation in indoor environments (e.g., schools and homes). Thus, the similar outdoor meteo-climatic conditions had a relatively negligible effect on the time of the year on the results.

2.2. Study Design

To evaluate the surface area dose received by children attending the three schools considered in the present study, particle number (PN) and lung-deposited surface area (LDSA) concentrations and average particle sizes (D_p) were measured by means of a personal monitor, which is a handheld diffusion charger particle counter (NanoTracer, Philips). The children were equipped with the mobile monitor fixed to a belt at the hip for 48 h.

During the campaign, 20 children for each school (60 children in total) were monitored. In particular, children aged 6–10 years were monitored (both males and females). Measurements were performed only on school days; weekends were not considered in the study. The authors monitored such high number of children in each school in order to obtain sufficient data that could be representative of the exposure level in each microenvironment where they live/reside. Indeed, the exposure of the children in each microenvironment and during each activity was affected by several parameters, such as the outdoor concentration levels, the volume of the indoor environments, and the presence and the strength of indoor sources (e.g., cooking, smokers, incense, candles etc.). As an example, the children's exposure when they stay in the kitchen during parents' cooking activities is strongly affected by the kitchen volume and the different types of foods and stoves ([76–80]), thus having performed different measurements (on different children) allowed the authors averaging amongst all these influencing parameters. Similarly, the exposure during transport can vary significantly as a function of the transportation modes (i.e., car, walking, bus, etc.; [43,81]), thus, once again, multiple measurements allowed to take into account for all these conditions.

In order to estimate the dose, the children, with the support of their parents, were asked to fill in an activity diary to take note about the place, time, and activity performed. A pre-compiled form of the activity diary was prepared by the authors and given to the children along with the portable instrument; the form was prepared considering 15-min time slots (e.g., 00:00–00:15, 00:15–00:30, etc.) in order to make it easy to fill in the forms with the required information. The diary was then used during the data post-processing in order to evaluate the time spent in each activity (i.e., the time-activity pattern) and to determine the exposure during each activity and in each microenvironment. The daily dose of the children under investigation in terms of particle surface area in the tracheobronchial and alveolar regions of the lungs (δ), was calculated as sum of the dose received during the activities performed in the j microenvironments:

$$\delta = \sum\nolimits_{j=1}^{n} \left\{ IR_{activity,j} \cdot LDSA_j \cdot T_j \right\} \ (mm^2) \tag{1}$$

where $IR_{activity}$ ($m^3 \cdot h^{-1}$) is the inhalation rate of the child, LDSA is the Lung-Deposited Surface Area concentration ($\mu m^2 \cdot cm^{-3}$), and T_j (h) is time spent in each microenvironment. The $IR_{activity}$ is a function of the age and activity performed by the children; in particular, we have considered the IR data for 6–10-year-old children summarized in Buonanno et al., 2012 [63]. In Equation (1) the term "microenvironment" is used for the sake of simplicity: the activities performed by the children, obtained based on the time-activity patterns, were grouped in six main microenvironments, summarized in Table 2. Particular attention should be paid to the "Cooking & Eating" microenvironment; indeed, children do not perform cooking activities per se, thus, the exposure related to this microenvironment is due to cooking activities performed by the parents. To compare the received dose of the children in different microenvironments, the dose-intensity ratio (i_δ, $mm^2 \cdot min^{-1}$), i.e., the ratio between the daily dose fraction and the daily time fraction characteristics of each microenvironment, was also evaluated [53].

Table 2. Classification of the activities performed by the citizens in seven main microenvironments.

Microenvironment	Activities
Transportation	Trip and use of time not specified, round-trip to work
School	All type of activities performed in school environments
Cooking & eating	Cooking, eating and drinking
Outdoor day	Gardening and animal care, restoration, sport and outdoor activities, physical workout, Productive exercise, Sports-connected activities
Indoor day	Personal care, studying not specified, studying in the free time, activities for home and family not specified, housework, purchasing goods and services, helping adult family members, helping other family members, active activities, social activities and entertainment, social life, entertainment and culture, inactivity, hobbies and computer science, art and hobbies, computing, playing, media, reading, watching TV, DVD or videos, listening to the radio or recording
Sleeping	Sleeping

2.3. Instrumentation and Its Quality Assurance

As mentioned above particle number (PN) and lung-deposited surface area (LDSA) concentrations and average particle sizes (D_p) were measured by means of a hand-held diffusion charger particle counter (NanoTracer Philips). It measures the particle number concentration and the average particle size in the range 10–300 nm, with a sampling time of 10 seconds. The operating principle of this instrument is based on the diffusion charging technique. In particular, the sampled aerosol is charged in a standard positive unipolar diffusion charger imparting an average known charge on the particles that is approximately proportional to the particle diameter of the aerosol. The number of charges, and thus the number of particles, is then detected by an electrometer [82–84]. Since over 99% of total particle number concentrations in urban environments are due to particles below 300 nm in diameter [85,86], the instrument was considered adequate for the experimental campaign. Actually, the lung-deposited surface area (LDSA) concentration cannot be considered, strictly speaking, a direct measurement, since it is provided by the instrument on the basis of built-in semi-empiric relationships allowing calculating the particle surface area deposited in the alveolar and tracheobronchial through the PN concentration and average particle size (D_p) measured data as described in details in Marra, et al. [87] and Fierz, Houle, Steigmeier and Burtscher [82]. Then, the LDSA concentration was evaluated as sum of the alveolar- and tracheobronchial-deposited contributions. Nonetheless, in order to take into account for calibrated PN concentrations and D_p values, we have used the semi-empiric relationships to calculate the LDSA concentrations on the basis of the calibrated values. In particular, the calibration of the device was performed before and after each experimental campaign. To this end, both a Condensation Particle Counter (CPC 3775, TSI Inc., Shoreview, MN, USA) and a Scanning Mobility

Particle Sizer (SMPS 3936, TSI Inc.) were used to compare the devices in terms of number concentration and particle size, respectively. The SMPS consisted of an Electrostatic Classifier (EC 3080, TSI Inc.), a Differential Mobility Analyzer (DMA 3081, TSI Inc.), and a CPC 3775. The SMPS 3936 was used, with an aerosol/sheath flow ratio of 0.3/3.0 L·min^{-1}, thus measuring particle number distributions in the range 14–700 nm. The calibration was carried out at the European Accredited Laboratory of Industrial Measurements (LaMI) of the University of Cassino and Southern Lazio (Italy) in a 150 m^3-room, with a conventional mechanical ventilation system guaranteeing constant thermo-hygrometric conditions (20 ± 2 °C and 50 ± 5% RH). Comparisons were performed for two different aerosols: aged indoor aerosol and freshly emitted aerosol produced by incense burning. Tests were conducted for 2 h performing simultaneous measurements with the Nanotracer, the CPC 3775, and the SMPS 3936. CPC and SMPS sampling times were set at 1 s and 135 s, respectively. SMPS measurements were corrected for multiple charge and diffusion losses. The correction factors obtained by averaging the results of the two aerosols investigated before and after each experimental campaign were applied as correction factors for each campaign. The differences in correction factors measured before and after the campaigns were found lower than 10%.

2.4. Statistical Analysis of the Data

In order to perform a statistical analysis of the concentrations experienced by the children in the different microenvironments (in terms of PN and LDSA) a preliminary normality test (Shapiro–Wilk test) was performed to check for the statistical distribution of the data. Since the data did not meet the assumptions of Gaussian distribution, non-parametric tests and further post-hoc tests (Kruskal–Wallis test [88]) were considered in the analysis. The statistically significant result was referred to a significance level of 99% (a p-value < 0.01). In particular, the Kruskal–Wallis tests were performed (a) amongst the six different microenvironments for each group of children separately (S1, S2, and S3; thus 3 non-parametric tests and further post-hoc tests) and (b) amongst the three groups of children for each microenvironment separately (six microenvironments plus the whole day data, thus seven non-parametric tests and further post-hoc tests).

The PN and LDSA concentration data considered in the statistical analysis, and then shown in the result section, included all the data provided by the instrument (roughly 48 h of total sampling per each child with a sampling frequency of 10 s), thus, a huge number of values were available for each microenvironment of each children group.

On the contrary, the dose values reported and discussed in the results represented the median values (and corresponding ranges) obtained from the 20-dose data (i.e., 20 children) per each microenvironment per each children group. Thus, due to the limited number of dose data, the statistical analysis on such values was not performed as it could led to misleading results.

3. Results

3.1. Time Activity Patterns

In Table 3, data on time-activity patterns of the children under investigation are reported, which were obtained from the activity diaries filled in by children and parents during the measurements. The median data demonstrate that children spend the most significant time fraction performing indoor activities in indoor microenvironments: indeed, the median time spent by the children indoor, as sum of the microenvironments labelled as "sleeping", "indoor day", "cooking & eating", resulted equal to 68–69%, to which must be added the time spent at school (25%). On the contrary, the time fraction spent in "outdoor day" (2–3%) and "transport" (3–4%) microenvironments resulted very limited, likely due to the fact that just school days were included in the experimental analysis, thus, the time spent in "transport" microenvironment is mostly limited to the time to take children to school. The huge time spent in indoor environments is consistent with our previous studies analyzing western populations, in which emerged that also adults spend a significant time fraction (roughly 90%)

performing indoor activities [53,62,63]. Amongst the indoor activities, the time spent in "cooking & eating" microenvironment (here 8%) is of particular concern since these activities were recognized in our previous papers as the most influencing in terms of exposure and health risk [36,89].

Table 3. Time activity pattern, particle concentrations (PN and LDSA) and dose received by children of the three schools in the different microenvironments expressed as median values and range (5th and 95th percentile). Total daily doses as sum of the median doses received in the different microenvironments are also reported as well as daily dose fractions and intensity–dose ratios.

Microenvironment	School	Time (min)	Time Fraction	PN conc. (104 part. cm³)	LDSA conc. ($\mu m^2 \cdot cm^{-3}$)	δ (mm²)	Daily Dose Fraction	i_δ (mm²·min⁻¹)
Sleeping	S1	540 (480–590)	38%	1.11 (0.62–2.04)	66 (37–123)	182 (101–340)	17%	0.34
	S2	530 (485–560)	37%	1.12 (0.61–1.93)	66 (37–117)	180 (100–319)	15%	0.34
	S3	597 (540–620)	41%	0.62 (0.33–1.08)	36 (21–67)	111 (63–204)	17%	0.19
Indoor	S1	320 (135–365)	22%	1.85 (0.59–7.70)	84 (27–356)	407 (128–1714)	38%	1.27
	S2	345 (140–395)	24%	1.79 (0.58–7.36)	81 (26–342)	426 (136–1777)	36%	1.23
	S3	268 (110–320)	19%	1.32 (0.43–5.26)	60 (20–248)	245 (80–1002)	38%	0.91
Outdoor	S1	40 (18–68)	3%	1.91 (0.68–5.20)	79 (28–220)	48 (17–133)	5%	1.20
	S2	35 (13–58)	2%	2.58 (0.86–7.12)	106 (36–304)	57 (119–162)	5%	1.62
	S3	36 (15–60)	3%	0.42 (0.36–1.18)	17 (15–49)	10 (8–27)	2%	00.27
School	S1	360 (295–375)	25%	1.57 (0.54–3.38)	66 (22–144)	163 (56–362)	15%	0.45
	S2	360 (290–385)	25%	2.13 (0.76–4.77)	89 (33–205)	222 (81–515)	19%	0.62
	S3	360 (300–390)	25%	0.34 (0.28–0.76)	14 (12–33)	36 (28–82)	6%	0.10
Transport	S1	60 (16–138)	4%	2.38 (1.29–6.38)	106 (57–274)	62 (33–160)	6%	1.04
	S2	50 (12–106)	3%	1.93(0.91–5.55)	86 (41–254)	41 (20–122)	4%	0.82
	S3	62 (15–131)	4%	0.66 (0.36–1.84)	29 (17–84)	17 (10–50)	3%	0.28
Cooking & Eating	S1	120 (97–148)	8%	4.20 (1.44–15.3)	112 (38–412)	200 (69–744)	19%	1.66
	S2	120 (93–156)	8%	5.11 (1.69–16.6)	136 (46–453)	244 (83–823)	21%	2.03
	S3	117 (90–160)	8%	4.91 (1.67–19.3)	130 (45–525)	227 (80–923)	35%	1.94
Day	S1	1440		1.44 (0.61–5.99)	71 (28–248)	1062		0.74
	S2	1440		1.55 (0.65–6.33)	77 (34–261)	1169		0.81
	S3	1440		0.62 (0.30–5.07)	34 (12–169)	646		0.45

3.2. Exposure to Submicron Particles

In Table 3 and Figure 1, the submicron particle concentrations, in terms of particle number and lung-deposited surface area, to which the children attending the three different schools (S1, S2, and S3) were exposed to in the different microenvironments (sleeping, indoor day, outdoor day, school, transport, cooking & eating) are shown. In the box plots of Figure 1, exposure data not statistically different amongst the six different microenvironments for each group of children separately (S1, S2, S3) and amongst the three groups of children for each microenvironment separately are also indicated ($p > 0.01$) as resulting from the statistical analysis explained in Section 2.4 (Kruskal–Wallis test). Due to the huge amount of data available for each microenvironment, most of the exposure received in the six microenvironments by the same group of children as well as those received in the same microenvironment by the three groups of children resulted in statistically different results.

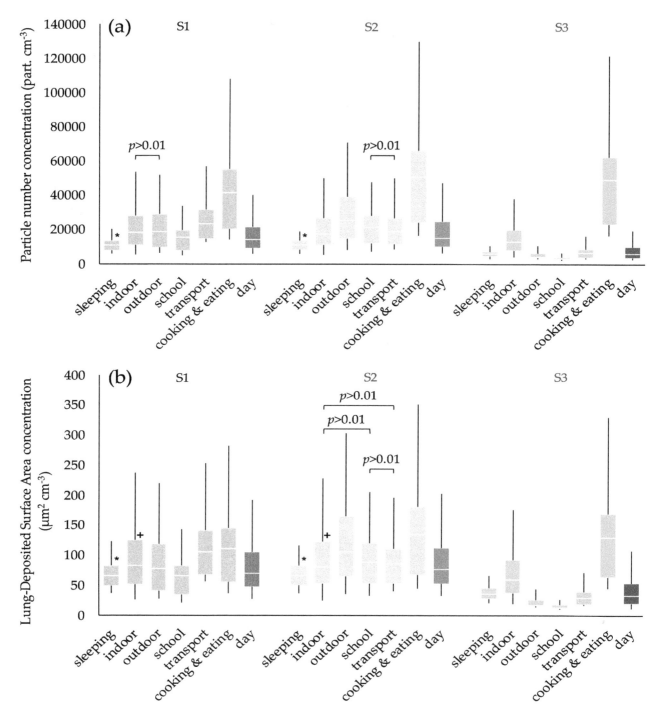

Figure 1. Statistics of (**a**) particle number and (**b**) lung-deposited surface area concentrations experienced by three groups of children (attending school S1, S2, and S3) in each microenvironment. Data was not statistically different within each group of children (S1, S2, S3) and amongst the same microenvironments of different groups (*,+) are also indicated ($p > 0.01$).

The children's exposure to submicron particles in the "school" microenvironment presents a significant deviation amongst the three schools. Indeed, children attending school S1, S2 and S3 were exposed to median PN and LDSA concentrations of 1.57×10^4 part. $cm^{-3}/66$ $\mu m^2 \cdot cm^{-3}$, 2.13×10^4 part. $cm^{-3}/89$ $\mu m^2 \cdot cm^{-3}$, and 3.39×10^3 part. $cm^{-3}/14$ $\mu m^2 \cdot cm^{-3}$, respectively. In particular, the concentration levels in the school S3 were much lower than S1 and S2 ones. This is due to the different outdoor concentrations, indeed, if no indoor submicron particle sources are in operation in the schools (as mentioned in the methodology section), the indoor concentrations are just affected by the outdoor-to-indoor penetration factors [65,90,91]. Thus, the low concentrations measured in school S3 are just related to the low outdoor concentrations typical of the rural site under investigation and discussed in the methodological section (Table 1). Indeed, the median particle number and lung-deposited surface area concentrations in the "outdoor day" microenvironment were equal to 1.91×10^4 part. $cm^{-3}/79$ $\mu m^2 \cdot cm^{-3}$, 2.58×10^4 part. $cm^{-3}/106$ $\mu m^2 \cdot cm^{-3}$, and 4.22×10^3 part. $cm^{-3}/17$ $\mu m^2 \cdot cm^{-3}$, for children attending school S1, S2 and S3, respectively. The resulting "school"/"outdoor day" concentration ratios (considering the median concentrations) were equal to 0.80–0.83 and 0.82–0.86 in terms of PN and LDSA concentrations, respectively, then consistent with the typical penetration factors reported in the scientific literature for naturally ventilated schools [65,90,91]. The location of the children's schools and homes is then the most influencing parameters in their exposure to submicron particles in "outdoor day" and "school" microenvironments, in fact the highest correlations between average outdoor NO_2 concentrations measured at the FSPs (Table 1) and PN concentrations measured during the experimental campaigns were determined for these two microenvironments (linear regressions with r^2 >0.99). The correlation between outdoor and indoor concentrations gets weaker when it comes to non-school environments, indeed, here the possible presence of indoor sources (cooking, incense, candles, heating systems) can lead to high indoor concentrations. In this context, as expected, the most critical microenvironment is "cooking & eating" which presents median values of PN and LDSA concentrations of 4.20×10^4 part. $cm^{-3}/112$ $\mu m^2 \cdot cm^{-3}$, 5.11×10^4 part. $cm^{-3}/136$ $\mu m^2 \cdot cm^{-3}$, and 4.91×10^4 part. $cm^{-3}/130$ $\mu m^2 \cdot cm^{-3}$, for children attending school S1, S2 and S3, respectively. The correlation with the average outdoor NO_2 concentrations measured by the FSPs barely doesn't exist, indeed the concentrations are much larger than the outdoor ones, and also children attending school S3 are exposed to very high submicron concentrations in "cooking & eating" microenvironment and roughly comparable to the S1 and S2 ones despite the much lower outdoor concentrations.

In regard to the other indoor environments labelled as "indoor day" microenvironment, the children's exposure resulted in lower statistical rates than the "cooking & eating" ones for all the three children groups. Nonetheless, the exposure in the "indoor day" microenvironment, when compared to the "outdoor day" one, varied amongst the different children groups. Indeed, the exposure in the "indoor day" microenvironment resulted statistically similar results, slightly lower, and much larger than the "outdoor day" environment for S1 (1.85×10^4 part. $cm^{-3}/84$ $\mu m^2 \cdot cm^{-3}$), S2 (1.79×10^4 part. $cm^{-3}/81$ $\mu m^2 \cdot cm^{-3}$), and S3 group of children (1.32×10^4 part. $cm^{-3}/60$ $\mu m^2 \cdot cm^{-3}$), respectively. The huge "indoor day"-"outdoor day" difference in the exposure detected for S3 group of children is related to the very low outdoor concentration level; thus, even a minor indoor source can easily increase the indoor concentration to values higher than the outdoor ones. Regarding the exposure in the "sleeping" microenvironment, the concentrations resulted in 0.5–0.6-fold of the "indoor day" microenvironment for all the three groups of children. Finally, during the "transport" microenvironment, higher concentrations were measured for children attending school S1 (2.38×10^4 part. $cm^{-3}/106$ $\mu m^2 \cdot cm^{-3}$) and S2 (1.93×10^4 part. $cm^{-3}/86$ $\mu m^2 \cdot cm^{-3}$), which are close to trafficked roads. On the contrary, children attending school S3 were exposed to quite low concentrations (6.58×10^3 part. $cm^{-3}/29$ $\mu m^2 \cdot cm^{-3}$), likely due to the location of the schools (rural area).

In summary, the daily exposure of the children is not only affected by the location of schools and homes, i.e., the proximity to outdoor sources, but also by the presence of indoor sources (mainly cooking); therefore, using outdoor concentration values as proxies of the daily exposure of the children could lead to serious under- or overestimation of the exposure. This is clearly highlighted by the daily median exposure data reported in Table 3; the concentrations, in terms of PN and LDSA, were equal to 1.44×10^4 part. $cm^{-3}/71$ $\mu m^2 \cdot cm^{-3}$, 1.55×10^4 part. $cm^{-3}/77$ $\mu m^2 \cdot cm$ and 0.62×10^4 part. $cm^{-3}/34$ $\mu m^2 \cdot cm^{-3}$, for children attending school S1, S2 and S3, respectively. Indeed, such values were 0.75-, 0.60-, and 1.48-fold the outdoor PN concentration values and 0.90-, 0.73-, and 2.00-fold the outdoor LDSA concentration values for S1, S2 and S3 children groups, respectively.

3.3. Particle Doses Received by Children

Median values (and corresponding 5th–95th percentile ranges) of particle surface area doses received by the three groups of children investigated (attending school S1, S2 and S3) in each microenvironment are shown in Table 3, here the daily doses are also reported. The doses received in the different microenvironments were calculated through Equation (1) considering the above mentioned and discussed (i) time–activity patterns and (ii) exposure data, as well as the (iii) inhalation rates characteristics of the children age and activity as resulting from the activity diaries, whereas the total daily doses here reported represent the sum of the median doses received in the different microenvironments.

The total daily doses for children attending school S1, S2 and S3 resulted equal to 1062, 1169 and 646 mm^2, respectively. The higher doses received by children of schools S1 and S2 are mostly due to their higher median daily exposures discussed in Section 3.1, while the time activity patterns (and then the inhalation rates) were quite similar amongst the three children groups.

The dose received in "school" microenvironment resulted equal to 163, 222 and 36 mm^2 for school S1, S2 and S3, respectively; with contributions of 15%, 19%, and 6% to daily dose. The dose received by children in school S3 is extremely low due to the low outdoor concentration of that rural area, whereas, the more polluted outdoor environments of S1 and S2 lead to higher doses. Anyway, such doses can be considered not extremely high if compared to the important time fraction of the day spent in such environments (25% of the day): this is clearly confirmed by the dose-intensity ratio (i_δ) summarized in Table 3; such ratios were lower than 1 for all the schools and, apart from "sleeping", they were the lowest values (0.45, 0.62, and 0.10 $mm^2 \cdot min^{-1}$ for S1, S2, and S3, respectively) amongst the microenvironments investigated.

Regarding the non-school environments, the contributions of "outdoor day" (2–5% of the daily dose) and "transportation" (3–6% of the daily dose) microenvironments are very limited due to the reduced time spent therein. As mentioned above, children attending school S1 and S2 were exposed to quite high concentrations in these two microenvironments then leading to dose-intensity ratios >1: this suggests that higher doses would be received in days and seasons characterized by different time-activity patterns with longer periods spent in such environments.

The main contribution to the daily dose is obviously received in non-school indoor environments, indeed summing up the doses received by children in "sleeping", "indoor day" and "cooking & eating" microenvironments, total contributions of 74%, 73%, and 90% were estimated for children attending school S1, S2 and S3, respectively. The most important contribution is due to the "indoor day" environment (36–38%) due to the both the significant time fraction (19–24%) and the possible presence of other sources leading to concentrations higher than the outdoor ones: indeed, dose-intensity ratios close or larger than 1 were measured for that environment. The contribution of the "sleeping" microenvironment is quite low (15–17%) if compared to the huge time spent in such activities (dose-intensity ratios extremely low), whereas an important dose fraction is received by children in "cooking & eating" microenvironments due to the high concentrations to which children are exposed to. Indeed, despite the time fraction spent in "cooking & eating" microenvironment is about 8% for all the three children groups, the contributions to the daily dose resulted equal to 19%, 21%, and 35% for children attending school S1, S2 and S3, respectively. In fact, such microenvironment resulted the one with the highest dose-intensity ratios (1.66, 2.03, 1.94), then consistently exceeding the "transportation" and "outdoor" microenvironments typically affected by outdoor sources.

In conclusion, the results on exposure levels in the different microenvironments confirm that indirect exposure assessments based on measurements at city scale or outdoor scale, typically adopted in cohort studies evaluating epidemiological effects on large populations [92,93] due to their easiness and cheapness, cannot provide a good estimate of the dose received by children whatever the location of their homes and schools. Thus, direct exposure assessment based on measurements at a personal scale, i.e., sampling aerosol from the breathing zone of the person using wearable instruments carried as personal monitors, is the only accurate experimental approach allowing proper dose estimates as it takes into account the different personal exposure of people moving between different microenvironments also including the indoor ones.

Regarding the exposure assessment results shown here, some broader implications can be drawn from the paper. In particular, concerning the exposure in outdoor-driven microenvironments (e.g., schools, outdoors), it can be reduced just building the schools and performing outdoor activities as far as possible from main outdoor sources (e.g., vehicular traffic). The reduction of the exposure (and then the dose) in indoor microenvironments can be reached (i) mitigating the particle sources (e.g., using ad-hoc hoods during kitchen activities, avoiding the use combustion sources such as biomass burning, candles, etc.) and/or (ii) reducing the exposure (e.g., increasing the air exchange rates through proper ventilation approaches, using air purifiers).

4. Conclusions

In the present study, an assessment of the total daily dose in terms of submicron particle surface area received by children living in different Italian areas and attending different schools located in different urban contexts (rural, suburban and urban area), was performed. The study aimed at investigating the children daily doses received in different microenvironments (both school and non-school environments) also taking into account the impact of the outdoor concentration levels on the received dose. To this end, an experimental analysis using portable instruments able to measure the concentrations at personal scale of the children was performed.

The findings of the study shown that the contribution of the school environment to the overall daily dose of the children is quite limited although they spent a significant time fraction of the day therein. Such dose is mainly affected by the outdoor concentrations; thus, schools placed close to main outdoor sources (e.g., trafficked roads) may results in higher rates of exposure and related doses then rural ones.

Outdoor and transport microenvironments present an almost negligible contribution to the children daily doses, whatever the investigated sites, due to the reduced exposure time in such environments. Therefore, a child's daily dose is mainly affected by indoor non-school environments, e.g., homes. In particular, the contribution of non-school indoor microenvironments to the children's daily dose account for more than 70% of the data from the children and school locations. Such a high contribution is led by "cooking & eating" and other "indoor day" microenvironments. Indeed, the "cooking & eating" microenvironment contributes up to 36% of the daily dose despite the reduced time spent therein: this is due to the high levels of exposure from high-emitting cooking activities.

In conclusion, the results of the study demonstrate that a proper evaluation of the submicron particle dose received by children cannot be performed only relying upon outdoor concentration data and that despite the location of the school and home, the contribution of indoor non-school environments is essential to properly assess the dose received by children.

Author Contributions: Conceptualization, G.B. and S.R.; methodology, A.P.; investigation, A.P. and L.S.; data curation, L.S and A.P.; writing—original draft preparation, A.P.; writing—review and editing, L.S and G.B.; funding acquisition, S.R. and G.B. All authors have read and agreed to the published version of the manuscript.

Acknowledgments: The authors want to thank the children and their families for their key contribution in the successful experimental campaign.

References

1. Schmid, O.; Möller, W.; Semmler-Behnke, M.; Ferron, G.A.; Karg, E.; Lipka, J.; Schulz, H.; Kreyling, W.G.; Stoeger, T. Dosimetry and toxicology of inhaled ultrafine particles. *Biomarkers* **2009**, *14*, 67–73. [CrossRef]

2. Buteau, S.; Goldberg, M.S. A structured review of panel studies used to investigate associations between ambient air pollution and heart rate variability. *Environ. Res.* **2016**, *148*, 207–247. [CrossRef]

3. Rizza, V.; Stabile, L.; Vistocco, D.; Russi, A.; Pardi, S.; Buonanno, G. Effects of the exposure to ultrafine particles on heart rate in a healthy population. *Sci. Total Environ.* **2019**, *650*, 2403–2410. [CrossRef]

4.	Brook, R.D.; Jerrett, M.; Brook, J.R.; Bard, R.L.; Finkelstein, M.M. The Relationship Between Diabetes Mellitus and Traffic-Related Air Pollution. *J. Occup. Environ. Med.* **2008**, *50*, 32–38. [CrossRef]

5.	Auchincloss, A.H.; Roux, A.V.D.; Dvonch, J.T.; Brown, P.L.; Barr, R.G.; Daviglus, M.L.; Goff, D.C.; Kaufman, J.D.; O'Neill, M.S. Associations between Recent Exposure to Ambient Fine Particulate Matter and Blood Pressure in the Multi-Ethnic Study of Atherosclerosis (MESA). *Environ. Health Perspect.* **2008**, *116*, 486–491. [CrossRef]

6.	Power, M.C.; Weisskopf, M.G.; Alexeeff, S.E.; Coull, B.A.; Spiro, A.; Schwartz, J. Traffic-Related Air Pollution and Cognitive Function in a Cohort of Older Men. *Environ. Health Perspect.* **2010**, *119*, 682–687. [CrossRef]

7.	World Health Organization. *Mortality and Burden of Disease from Ambient Air Pollution: Situation and Trends*; WHO: Geneva, Switzerland, 2014.

8.	Beelen, R.; Stafoggia, M.; Raaschou-Nielsen, O.; Andersen, Z.J.; Xun, W.W.; Katsouyanni, K.; Dimakopoulou, K.; Brunekreef, B.; Weinmayr, G.; Hoffmann, B.; et al. Long-term Exposure to Air Pollution and Cardiovascular Mortality. *Epidemiology* **2014**, *25*, 368–378. [CrossRef] [PubMed]

9.	International Agency for Research on Cancer. *Outdoor Air Pollution a Leading Environmental Cause of Cancer Deaths*; IARC: Lyon, Geneva, 2013. [CrossRef]

10.	Loomis, D.; Grosse, Y.; Lauby-Secretan, B.; El Ghissassi, F.; Bouvard, V.; Benbrahim-Tallaa, L.; Guha, N.; Baan, R.; Mattock, H.; Straif, K. The carcinogenicity of outdoor air pollution. *Lancet Oncol.* **2013**, *14*, 1262–1263. [CrossRef]

11.	Peters, A.; Von Klot, S.; Heier, M.; Trentinaglia, I.; Hörmann, A.; Wichmann, H.E.; Lowel, H. Exposure to Traffic and the Onset of Myocardial Infarction. *N. Engl. J. Med.* **2004**, *351*, 1721–1730. [CrossRef] [PubMed]

12.	Schins, R.P.; Lightbody, J.H.; Borm, P.J.; Shi, T.; Donaldson, K.; Stone, V. Inflammatory effects of coarse and fine particulate matter in relation to chemical and biological constituents. *Toxicol. Appl. Pharmacol.* **2004**, *195*, 1–11. [CrossRef]

13.	Weichenthal, S. Selected physiological effects of ultrafine particles in acute cardiovascular morbidity. *Environ. Res.* **2012**, *115*, 26–36. [CrossRef] [PubMed]

14.	Unfried, K.; Albrecht, C.; Klotz, L.-O.; Von Mikecz, A.; Grether-Beck, S.; Schins, R.P. Cellular responses to nanoparticles: Target structures and mechanisms. *Nanotoxicology* **2007**, *1*, 52–71. [CrossRef]

15.	International Commission on Radiological Protection. Human respiratory tract model for radiological protection. A report of a Task Group of the International Commission on Radiological Protection. *Ann. ICRP* **1994**, *24*, 1–482.

16.	Peters, A.; Veronesi, B.; Calderón-Garcidueñas, L.; Gehr, P.; Chen, L.C.; Geiser, M.; Reed, W.; Rothen-Rutishauser, B.; Schürch, S.; Schulz, H. Translocation and potential neurological effects of fine and ultrafine particles a critical update. *Part. Fibre Toxicol.* **2006**, *3*, 13. [CrossRef] [PubMed]

17.	Nakane, H. Translocation of particles deposited in the respiratory system: A systematic review and statistical analysis. *Environ. Health Prev. Med.* **2011**, *17*, 263–274. [CrossRef] [PubMed]

18.	Campagnolo, L.; Massimiani, M.; Vecchione, L.; Piccirilli, D.; Toschi, N.; Magrini, A.; Bonanno, E.; Scimeca, M.; Castagnozzi, L.; Buonanno, G.; et al. Silver nanoparticles inhaled during pregnancy reach and affect the placenta and the foetus. *Nanotoxicology* **2017**, *11*, 687–698. [CrossRef] [PubMed]

19.	Loomis, D. Sizing up air pollution research. *Epidemiology* **2000**, *11*, 2. [CrossRef]

20.	Pope, C.A. What do epidemiologic findings tell us about health effects of environmental aerosols? *J. Aerosol Med.* **2000**, *13*, 335–354. [CrossRef]

21.	Franck, U.; Odeh, S.; Wiedensohler, A.; Wehner, B.; Herbarth, O. The effect of particle size on cardiovascular disorders—The smaller the worse. *Sci. Total Environ.* **2011**, *409*, 4217–4221. [CrossRef]

22.	Kumar, P.; Morawska, L.; Birmili, W.; Paasonen, P.; Hu, M.; Kulmala, M.; Harrison, R.M.; Norford, L.; Britter, R. Ultrafine particles in cities. *Environ. Int.* **2014**, *66*, 1–10. [CrossRef]

23.	Buonanno, G.; Marks, G.B.; Morawska, L. Health effects of daily airborne particle dose in children: Direct association between personal dose and respiratory health effects. *Environ. Pollut.* **2013**, *180*, 246–250. [CrossRef] [PubMed]

24.	Giechaskiel, B.; Alföldy, B.; Drossinos, Y. A metric for health effects studies of diesel exhaust particles. *J. Aerosol Sci.* **2009**, *40*, 639–651. [CrossRef]

25.	Brown, D.; Wilson, M.; MacNee, W.; Stone, V.; Donaldson, K. Size-Dependent Proinflammatory Effects of Ultrafine Polystyrene Particles: A Role for Surface Area and Oxidative Stress in the Enhanced Activity of Ultrafines. *Toxicol. Appl. Pharmacol.* **2001**, *175*, 191–199. [CrossRef] [PubMed]

26. Hamoir, J.; Nemmar, A.; Halloy, D.; Wirth, D.; Vincke, G.; Vanderplasschen, A.; Nemery, B.; Gustin, P. Effect of polystyrene particles on lung microvascular permeability in isolated perfused rabbit lungs: Role of size and surface properties. *Toxicol. Appl. Pharmacol.* **2003**, *190*, 278–285. [CrossRef]

27. Strak, M.; Boogaard, H.; Meliefste, K.; Oldenwening, M.; Zuurbier, M.; Brunekreef, B.; Hoek, G. Respiratory health effects of ultrafine and fine particle exposure in cyclists. *Occup. Environ. Med.* **2009**, *67*, 118–124. [CrossRef] [PubMed]

28. Landkocz, Y.; LeDoux, F.; André, V.; Cazier, F.; Genevray, P.; Dewaele, D.; Martin, P.J.; Lepers, C.; Verdin, A.; Courcot, L.; et al. Fine and ultrafine atmospheric particulate matter at a multi-influenced urban site: Physicochemical characterization, mutagenicity and cytotoxicity. *Environ. Pollut.* **2017**, *221*, 130–140. [CrossRef] [PubMed]

29. Longhin, E.; Gualtieri, M.; Capasso, L.; Bengalli, R.; Mollerup, S.; Holme, J.A.; Øvrevik, J.; Casadei, S.; Di Benedetto, C.; Parenti, P.; et al. Physico-chemical properties and biological effects of diesel and biomass particles. *Environ. Pollut.* **2016**, *215*, 366–375. [CrossRef] [PubMed]

30. Sager, T.; Castranova, V. Surface area of particle administered versus mass in determining the pulmonary toxicity of ultrafine and fine carbon black: Comparison to ultrafine titanium dioxide. *Part. Fibre Toxicol.* **2009**, *6*, 15. [CrossRef]

31. Schmid, O.; Stöger, T. Surface area is the biologically most effective dose metric for acute nanoparticle toxicity in the lung. *J. Aerosol Sci.* **2016**, *99*, 133–143. [CrossRef]

32. Nygaard, U.C.; Samuelsen, M.; Aase, A.; Løvik, M. The Capacity of Particles to Increase Allergic Sensitization Is Predicted by Particle Number and Surface Area, Not by Particle Mass. *Toxicol. Sci.* **2004**, *82*, 515–524. [CrossRef]

33. Cauda, E.G.; Ku, B.K.; Miller, A.L.; Barone, T.L. Toward Developing a New Occupational Exposure Metric Approach for Characterization of Diesel Aerosols. *Aerosol Sci. Technol.* **2012**, *46*, 1370–1381. [CrossRef] [PubMed]

34. Oberdörster, G.; Oberdörster, E.; Oberdörster, J. Nanotoxicology: An Emerging Discipline Evolving from Studies of Ultrafine Particles. *Environ. Health Perspect.* **2005**, *113*, 823–839. [CrossRef]

35. Tran, C.L.; Buchanan, D.; Cullen, R.T.; Searl, A.; Jones, A.D.; Donaldson, K. Inhalation of poorly soluble particles. II. Influence Of particle surface area on inflammation and clearance. *Inhal. Toxicol.* **2000**, *12*, 1113–1126. [CrossRef] [PubMed]

36. Buonanno, G.; Stabile, L.; Morawska, L.; Giovinco, G.; Querol, X. Do air quality targets really represent safe limits for lung cancer risk? *Sci. Total Environ.* **2017**, *580*, 74–82. [CrossRef] [PubMed]

37. Sze-To, G.N.; Wu, C.L.; Chao, C.Y.H.; Wan, M.P.; Chan, T.C. Exposure and cancer risk toward cooking-generated ultrafine and coarse particles in Hong Kong homes. *HVAC&R Res.* **2012**, *18*, 204–216. [CrossRef]

38. Stabile, L.; Buonanno, G.; Ficco, G.; Scungio, M. Smokers' lung cancer risk related to the cigarette-generated mainstream particles. *J. Aerosol Sci.* **2017**, *107*, 41–54. [CrossRef]

39. Pacitto, A.; Stabile, L.; Viana, M.; Scungio, M.; Reche, C.; Querol, X.; Alastuey, A.; Rivas, I.; Álvarez-Pedrerol, M.; Sunyer, J.; et al. Particle-related exposure, dose and lung cancer risk of primary school children in two European countries. *Sci. Total Environ.* **2018**, *616–617*, 720–729. [CrossRef]

40. European Parliament and Council of the European Union. EU Directive 2008/50/EC of the European Parliament and of the Council of 21 May 2008 on ambient air quality and cleaner air for Europe, 2008. *Off. J. Eur. Union* **2008**, *L 152/1*, 1–44.

41. Buonanno, G.; Dell'Isola, M.; Stabile, L.; Viola, A. Uncertainty Budget of the SMPS–APS System in the Measurement of PM1, PM2.5, and PM10. *Aerosol Sci. Technol.* **2009**, *43*, 1130–1141. [CrossRef]

42. Buonanno, G.; Dell'Isola, M.; Stabile, L.; Viola, A. Critical aspects of the uncertainty budget in the gravimetric PM measurements. *Measurement* **2011**, *44*, 139–147. [CrossRef]

43. Moreno, T.; Pacitto, A.; Fernández, A.; Amato, F.; Marco, E.; Grimalt, J.; Buonanno, G.; Querol, X. Vehicle interior air quality conditions when travelling by taxi. *Environ. Res.* **2019**, *172*, 529–542. [CrossRef] [PubMed]

44. Kaur, S.; Nieuwenhuijsen, M.; Colvile, R. Pedestrian exposure to air pollution along a major road in Central London, UK. *Atmos. Environ.* **2005**, *39*, 7307–7320. [CrossRef]

45. Stabile, L.; Arpino, F.; Buonanno, G.; Russi, A.; Frattolillo, A. A simplified benchmark of ultrafine particle dispersion in idealized urban street canyons: A wind tunnel study. *Build. Environ.* **2015**, *93*, 186–198. [CrossRef]

46. Manigrasso, M.; Stabile, L.; Avino, P.; Buonanno, G. Influence of measurement frequency on the evaluation of short-term dose of sub-micrometric particles during indoor and outdoor generation events. *Atmos. Environ.* **2013**, *67*, 130–142. [CrossRef]

47. Scungio, M.; Arpino, F.; Cortellessa, G.; Buonanno, G. Detached eddy simulation of turbulent flow in isolated street canyons of different aspect ratios. *Atmos. Pollut. Res.* **2015**, *6*, 351–364. [CrossRef]

48. Kumar, P.; Ketzel, M.; Vardoulakis, S.; Pirjola, L.; Britter, R. Dynamics and dispersion modelling of nanoparticles from road traffic in the urban atmospheric environment—A review. *J. Aerosol Sci.* **2011**, *42*, 580–603. [CrossRef]

49. Scungio, M.; Buonanno, G.; Arpino, F.; Ficco, G. Influential parameters on ultrafine particle concentration downwind at waste-to-energy plants. *Waste Manag.* **2015**, *38*, 157–163. [CrossRef]

50. Neft, I.; Scungio, M.; Culver, N.; Singh, S. Simulations of Aerosol Filtration by Vegetation: Validation of Existing Models with Available Lab data and Application to Near-Roadway Scenario. *Aerosol Sci. Technol.* **2016**, *50*, 937–946. [CrossRef]

51. Buonanno, G.; Fuoco, F.C.; Stabile, L. Influential parameters on particle exposure of pedestrians in urban microenvironments. *Atmos. Environ.* **2011**, *45*, 1434–1443. [CrossRef]

52. Rizza, V.; Stabile, L.; Buonanno, G.; Morawska, L. Variability of airborne particle metrics in an urban area. *Environ. Pollut.* **2017**, *220*, 625–635. [CrossRef]

53. Pacitto, A.; Stabile, L.; Moreno, T.; Kumar, P.; Wierzbicka, A.; Morawska, L.; Buonanno, G. The influence of lifestyle on airborne particle surface area doses received by different Western populations. *Environ. Pollut.* **2018**, *232*, 113–122. [CrossRef] [PubMed]

54. Buonanno, G.; Stabile, L.; Morawska, L. Personal exposure to ultrafine particles: The influence of time-activity patterns. *Sci. Total Environ.* **2014**, *468*, 903–907. [CrossRef] [PubMed]

55. Buonanno, G.; Marini, S.; Morawska, L.; Fuoco, F.C. Individual dose and exposure of Italian children to ultrafine particles. *Sci. Total Environ.* **2012**, *438*, 271–277. [CrossRef] [PubMed]

56. Brent, R.L.; Weitzman, M. The Vulnerability, Sensitivity, and Resiliency of the Developing Embryo, Infant, Child, and Adolescent to the Effects of Environmental Chemicals, Drugs, and Physical Agents as Compared to the Adult: Preface. *Pediatrics* **2004**, *113*, 933–934.

57. Makri, A.; Stilianakis, N.I. Vulnerability to air pollution health effects. *Int. J. Hyg. Environ. Health* **2008**, *211*, 326–336. [CrossRef] [PubMed]

58. Ginsberg, G.; Foos, B.P.; Firestone, M.P. Review and Analysis of Inhalation Dosimetry Methods for Application to Children's Risk Assessment. *J. Toxicol. Environ. Health Part A* **2005**, *68*, 573–615. [CrossRef]

59. Heinrich, J.; Slama, R. Fine particles, a major threat to children. *Int. J. Hyg. Environ. Health* **2007**, *210*, 617–622. [CrossRef]

60. Bateson, T.F.; Schwartz, J. Children's Response to Air Pollutants. *J. Toxicol. Environ. Health Part A Curr. Issues* **2007**, *71*, 238–243. [CrossRef]

61. Selgrade, M.K.; Plopper, C.G.; Gilmour, M.I.; Conolly, R.B.; Foos, B.S.P. Assessing The Health Effects and Risks Associated with Children's Inhalation Exposures—Asthma and Allergy. *J. Toxicol. Environ. Health Part A* **2007**, *71*, 196–207. [CrossRef]

62. Buonanno, G.; Giovinco, G.; Morawska, L.; Stabile, L. Tracheobronchial and alveolar dose of submicrometer particles for different population age groups in Italy. *Atmos. Environ.* **2011**, *45*, 6216–6224. [CrossRef]

63. Buonanno, G.; Morawska, L.; Stabile, L.; Wang, L.; Giovinco, G. A comparison of submicrometer particle dose between Australian and Italian people. *Environ. Pollut.* **2012**, *169*, 183–189. [CrossRef] [PubMed]

64. The Organisation for Economic Co-Operation and Development. *How Long Do Students Spend in the Classroom?* OECD: Paris, France, 2012.

65. Stabile, L.; Dell'Isola, M.; Russi, A.; Massimo, A.; Buonanno, G. The effect of natural ventilation strategy on indoor air quality in schools. *Sci. Total Environ.* **2017**, *595*, 894–902. [CrossRef] [PubMed]

66. Stephens, B.; Siegel, J.A. Penetration of ambient submicron particles into single-family residences and associations with building characteristics. *Indoor Air* **2012**, *22*, 501–513. [CrossRef] [PubMed]

67. Zhu, Y.; Hinds, W.C.; Krudysz, M.; Kuhn, T.; Froines, J.; Sioutas, C. Penetration of freeway ultrafine particles into indoor environments. *J. Aerosol Sci.* **2005**, *36*, 303–322. [CrossRef]

68. Viana, M.; Díez, S.; Reche, C. Indoor and outdoor sources and infiltration processes of PM1 and black carbon in an urban environment. *Atmos. Environ.* **2011**, *45*, 6359–6367. [CrossRef]

69. Tippayawong, N.; Khuntong, P.; Nitatwichit, C.; Khunatorn, Y.; Tantakitti, C. Indoor/outdoor relationships of size-resolved particle concentrations in naturally ventilated school environments. *Build. Environ.* **2009**, *44*, 188–197. [CrossRef]

70. Health Effects Institute. HEI Perspectives 3. In *Understanding the Health Effects of Ambient Ultrafine Particles*; Health Effects Institute: Boston, MA, USA, 2013.

71. Pacitto, A.; Amato, F.; Moreno, T.; Pandolfi, M.; Fonseca, A.; Mazaheri, M.; Stabile, L.; Buonanno, G.; Querol, X. Effect of ventilation strategies and air purifiers on the children's exposure to airborne particles and gaseous pollutants in school gyms. *Sci. Total Environ.* **2020**, *712*, 135673. [CrossRef]

72. Salimi, F.; Mazaheri, M.; Clifford, S.; Crilley, L.R.; Laiman, R.; Morawska, L.; Clifford, S.J. Spatial Variation of Particle Number Concentration in School Microscale Environments and Its Impact on Exposure Assessment. *Environ. Sci. Technol.* **2013**, *47*, 5251–5258. [CrossRef]

73. Mazaheri, M.; Reche, C.; Rivas, I.; Crilley, L.R.; Alvarez-Pedrerol, M.; Viana, M.; Tobias, A.; Alastuey, A.; Sunyer, J.; Querol, X.; et al. Variability in exposure to ambient ultrafine particles in urban schools: Comparative assessment between Australia and Spain. *Environ. Int.* **2016**, *88*, 142–149. [CrossRef]

74. Rivas, I.; Viana, M.; Moreno, T.; Bouso, L.; Pandolfi, M.; Alvarez-Pedrerol, M.; Forns, J.; Alastuey, A.; Sunyer, J.; Querol, X. Outdoor infiltration and indoor contribution of UFP and BC, OC, secondary inorganic ions and metals in PM2.5 in schools. *Atmos. Environ.* **2015**, *106*, 129–138. [CrossRef]

75. Stabile, L.; Massimo, A.; Rizza, V.; D'Apuzzo, M.; Evangelisti, A.; Scungio, M.; Frattolillo, A.; Cortellessa, G.; Buonanno, G. A novel approach to evaluate the lung cancer risk of airborne particles emitted in a city. *Sci. Total Environ.* **2019**, *656*, 1032–1042. [CrossRef] [PubMed]

76. Buonanno, G.; Johnson, G.; Morawska, L.; Stabile, L. Volatility Characterization of Cooking-Generated Aerosol Particles. *Aerosol Sci. Technol.* **2011**, *45*, 1069–1077. [CrossRef]

77. Buonanno, G.; Morawska, L.; Stabile, L. Particle emission factors during cooking activities. *Atmos. Environ.* **2009**, *43*, 3235–3242. [CrossRef]

78. See, S.W.; Balasubramanian, R. Chemical characteristics of fine particles emitted from different gas cooking methods. *Atmos. Environ.* **2008**, *42*, 8852–8862. [CrossRef]

79. Wallace, L.A.; Ott, W.R.; Weschler, C.J. Ultrafine particles from electric appliances and cooking pans: Experiments suggesting desorption/nucleation of sorbed organics as the primary source. *Indoor Air* **2014**, *25*, 536–546. [CrossRef] [PubMed]

80. Rivas, I.; Kumar, P.; Hagen-Zanker, A. Exposure to air pollutants during commuting in London: Are there inequalities among different socio-economic groups? *Environ. Int.* **2017**, *101*, 143–157. [CrossRef]

81. Moreno, T.; Reche, C.; Rivas, I.; Minguillon, M.C.; Martins, V.; Vargas, C.; Buonanno, G.; Parga, J.; Pandolfi, M.; Brines, M.; et al. Urban air quality comparison for bus, tram, subway and pedestrian commutes in Barcelona. *Environ. Res.* **2015**, *142*, 495–510. [CrossRef]

82. Fierz, M.; Houle, C.; Steigmeier, P.; Burtscher, H. Design, Calibration, and Field Performance of a Miniature Diffusion Size Classifier. *Aerosol Sci. Technol.* **2011**, *45*, 1–10. [CrossRef]

83. Fernández, A.; Amato, F.; Moreno, N.; Pacitto, A.; Reche, C.; Marco, E.; Grimalt, J.O.; Querol, X.; Moreno, T. Chemistry and sources of $PM_{2.5}$ and volatile organic compounds breathed inside urban commuting and tourist buses. *Atmos. Environ.* **2020**, *223*, 117234. [CrossRef]

84. Stabile, L.; Trassierra, C.V.; Dell'Agli, G.; Buonanno, G. Ultrafine Particle Generation through Atomization Technique: The Influence of the Solution. *Aerosol Air Qual. Res.* **2013**, *13*, 1667–1677. [CrossRef]

85. Goel, A.; Kumar, P. A review of fundamental drivers governing the emissions, dispersion and exposure to vehicle-emitted nanoparticles at signalised traffic intersections. *Atmos. Environ.* **2014**, *97*, 316–331. [CrossRef]

86. Kumar, P.; Pirjola, L.; Ketzel, M.; Harrison, R.M. Nanoparticle emissions from 11 non-vehicle exhaust sources—A review. *Atmos. Environ.* **2013**, *67*, 252–277. [CrossRef]

87. Marra, J.; Voetz, M.; Kiesling, H.-J. Monitor for detecting and assessing exposure to airborne nanoparticles. *J. Nanopart. Res.* **2009**, *12*, 21–37. [CrossRef]

88. Kruskal, W.H.; Wallis, W.A. Errata: Use of Ranks in One-Criterion Variance Analysis. *J. Am. Stat. Assoc.* **1953**, *48*, 907. [CrossRef]

89. Buonanno, G.; Giovinco, G.; Morawska, L.; Stabile, L. Lung cancer risk of airborne particles for Italian population. *Environ. Res.* **2015**, *142*, 443–451. [CrossRef]

90. Stabile, L.; Buonanno, G.; Frattolillo, A.; Dell'Isola, M. The effect of the ventilation retrofit in a school on CO2, airborne particles, and energy consumptions. *Build. Environ.* **2019**, *156*, 1–11. [CrossRef]

91.   Buonanno, G.; Fuoco, F.; Morawska, L.; Stabile, L. Airborne particle concentrations at schools measured at different spatial scales. *Atmos. Environ.* **2013**, *67*, 38–45. [CrossRef]

92.   Stafoggia, M.; Schneider, A.; Cyrys, J.; Samoli, E.; Andersen, Z.J.; Bedada, G.B.; Bellander, T.; Cattani, G.; Eleftheriadis, K.; Faustini, A.; et al. Association Between Short-term Exposure to Ultrafine Particles and Mortality in Eight European Urban Areas. *Epidemiology* **2017**, *28*, 172–180. [CrossRef] [PubMed]

93.   Chen, X.; Zhang, L.-W.; Huang, J.-J.; Song, F.; Zhang, L.-P.; Qian, Z.-M.; Trevathan, E.; Mao, H.-J.; Han, B.; Vaughn, M.; et al. Long-term exposure to urban air pollution and lung cancer mortality: A 12-year cohort study in Northern China. *Sci. Total Environ.* **2016**, *571*, 855–861. [CrossRef] [PubMed]

9

An Accident Model with Considering Physical Processes for Indoor Environment Safety

Zhengguo Yang *, Yuto Lim and Yasuo Tan

School of Information Science, Japan Advanced Institute of Science and Technology 1-1 Asahidai, Nomi, Ishikawa 923-1292, Japan; ylim@jaist.ac.jp (Y.L.); ytan@jaist.ac.jp (Y.T.)
* Correspondence: yangzg@jaist.ac.jp

Abstract: Accident models provide a conceptual representation of accident causation. They have been applied to environments that have been exposed to poisonous or dangerous substances that are hazardous in nature. The home environment refers to the indoor space with respect to the physical processes the of indoor climate, e.g., temperature change, which are not hazardous in general. However, it can be hazardous when the physical process is in some states, e.g., a state of temperature that can cause heat stroke. If directly applying accident models in such a case, the physical processes are missing. To overcome this problem, this paper proposes an accident model by extending the state-of-the-art accident model, i.e., Systems-Theoretic Accident Model and Process (STAMP) with considering physical processes. Then, to identify causes of abnormal system behaviors that result in physical process anomalies, a hazard analysis technique called System-Theoretic Process Analysis (STPA) is tailored and applied to a smart home system for indoor temperature adjustment. The analytical results are documented by a proposed landscape genealogical layout documentation. A comparison with results by applying the original STPA was made, which demonstrates the effectiveness of the tailored STPA to apply in identifying causes in our case.

Keywords: STAMP; STPA; physical process; indoor environment safety; smart home systems

1. Introduction

Accident models provide a conceptual representation of accident causation [1]. Their state-of-the-art development is on the phase of systemic models [1–3], i.e., accident models based on system theory rather than reliability. They were specifically applied to understand accidents in industrial areas, e.g., deepwater well control [4], railway [5], and aviation [6]. Some others relate to places that have been exposed to poisonous or dangerous substances, e.g., oil transportation [7] and nuclear power plants [8]. These poisonous or dangerous substances are hazards in nature, which can directly cause harm when leaked or released in workplaces. The workplace is a strictly managed environment for work. Safety-critical systems are taken as preventative measures for leakage and release.

The home environment refers to the indoor space with respect to physical processes of indoor climate, e.g., temperature change, which is different from safety-critical environments in workplaces. Hereafter, we use physical process to represent the physical process of indoor climate. The home environment is not a hazardous place in general. However, it can become hazardous when the physical process transfers from a normal state (e.g., temperature for thermal comfort), then through some intermediate states (e.g., temperature for thermal discomfort), finally reaching a hazardous state (e.g., temperature for heat stroke). The home environment is a place for everyday living, and it is not as strictly managed as that in workplaces. Smart home systems are developed to maintain the home environment in desired states, not only as preventative measures.

In systemic accident models [1–3], accidents are the result of the violation of a set of constraints on the behaviors of the system components, i.e., management, humans, and technology. If directly applying a systemic accident model to the smart home system, the information of physical processes is missing. For example, a smart home system violated its constraint on adjusting the indoor temperature for thermal comfort, and resulted in high temperature for heat stroke. The physical process of temperature change from one of thermal comfort, to some intermediate states for discomfort, then to a high-temperature state that can cause, e.g., heat stroke, is missing. This process is important. First, it assists in understanding how system behaviors could result in accident through intermediate state(s). Second, we need the anomalies' information of the physical process and their causes to deploy reactions and precautionary measures. We can take advantage of hazard analysis techniques to identify the causes in the system to the anomalies. When an intermediate abnormal state or hazardous state of the physical process is detected, with considering the corresponding causes in the system, an effective precautionary or reaction measures can be selected. Generally, if something undesired is not considered in the very beginning of risk analysis, the causes in the following analysis cannot be identified [2,9]. Therefore, it is necessary to extend the systemic accident model by including the physical process.

A newly developing systemic accident model, i.e., Systems-Theoretic Accident Model and Process (STAMP) [2], is considered in this paper, as its underlying rationale has been widely acknowledged by comparing with other systemic accident models [10]. It is based on general system theory for understanding accident causality of sociotechnical systems. A brief introduction of it is presented in Section 3.1. We extend it by considering the following facets. First, the home environment is not inherently hazardous. It can ensure a comfortable life in some physical process states and cause harms in others. Therefore, we take the physical process into account in understanding accident formation. Second, the indoor environment is greatly affected by the behaviors of smart home systems. This is because physical processes are the result of smart home systems and the outdoor climate. However, in a limited period, e.g., days, the outdoor climate can be considered with no big changes. Third, the role of people in the home environment. The characteristics of workers in workplaces and occupants in the home environment are different.

In this paper, we extend the STAMP model with considering physical processes (hereafter denoted by STAMP-PP) to understand accident formation. Smart home systems interact with the home environment through its behaviors, e.g., warm up and cool down. Thus, the STAMP-PP connects the physical world through the behaviors of the systems. Under this consideration, accidents are the result of the violation of a set of constraints on the behaviors of the systems to cause abnormal changes in physical processes, and finally result in personal harm. The extended STAMP-PP model demonstrates accident formation with respect to system behaviors and physical processes. The system behaviors can be controlled either by the smart home system directly or by occupants indirectly.

The information related to physical processes is important, and the abnormal behaviors of systems under specific operation scenarios must be known to select the appropriate reactions and precautionary measures. To this end, hazard analysis techniques [9,11] that can assist in analyzing potential causes of accidents are required. We adopted a hazard analysis technique to identify causes of abnormal system behaviors under related operation scenarios, which can result in abnormal changes in physical processes. A new approach to hazard analysis, called System-Theoretic Process Analysis (STPA) [2,9], is based on the STAMP model, and it is tailored and applied to the smart home system [12] for adjusting the indoor temperature, to demonstrate how to identify causes to abnormal system behaviors that result in physical process anomalies. The STPA can be used to identify unsafe control actions of a controller and are also the reasons why unsafe control actions can happen under specific scenarios. As abnormal behaviors of smart home systems that result in intermediate states of physical processes are also considered, the STPA is then tailored also for identifying causes to these abnormal behaviors. Landscape Genealogical Layout Documentation (LGLD) is proposed for documenting the analytical results, and the relations among the results are clearly and straightforwardly represented by comparing with conventional ways of documentation, i.e., tables and lists. We compared the results with that of applying the original STPA, which demonstrate the effectiveness of the tailored STPA in identifying causes of abnormal system behaviors and the LGLD documentation in representing the relations among the results.

The contributions of this paper are as follows.

- We discussed the characteristics of the smart home in the viewpoint of occupants and the safety of the home environment.
- The concept of the Performers System, which emphasizes the behaviors performed by various home appliances, is proposed.
- We propose the STAMP-PP model for understanding accident formation, i.e., abnormal system behaviors that result in abnormal changes in physical processes and cause hazards.
- We tailored the STPA and applied it to a smart home system to identify inappropriate and unsafe control actions that cause abnormal system behaviors, which result in abnormal changes in physical processes, and hazards.
- An LGLD approach is proposed for documenting the STPA analytical results.

This paper extends a conference paper [13] that introduces the STAMP-PP model while considering physical processes. The application of the tailored STPA and the new way of documenting the results in this paper are novel.

The rest of this paper is organized as follows. Section 2 discusses some knowledge about smart home and home environment safety. Section 3 introduces the proposed accident model STAMP-PP. Then, the hazard analysis technique STPA is tailored and applied to a smart home system for indoor temperature adjustment in Section 4. In Section 5, a discussion is given. Section 6 introduces the related work. Finally, Section 7 concludes this paper and points out the future work.

2. Preliminaries

In this section, we discuss the characteristics of the smart home and home environment safety before introducing the STAMP-PP model and the application of the STPA.

2.1. Smart Home

A home is a place for people like individual or family members, etc. to live. It is the sum of the place where people live permanently and the social unit—family. Since the 20th century, with the introduction of electricity, informtion, and communication technologies, great changes have taken place in the home [14]. One representation of this change is the development of the concept of the smart home since the 1990s [15]. The primary objective of the smart home is to increase occupants' comfort and make daily life easier. An example of possible techniques that make a home smart is machine learning [16]. The smart home has certain characteristics [15,17], e.g., adaptability, connectivity, controllability, and computability. These are discussed from the viewpoint of technology. This section discusses the characteristics of the smart home from the viewpoint of occupants.

- Partial Automation: Although home life has been automated a lot more than ever before due to the development of electricity, electronics, and network technologies, there are still elements of our lifestyles that have been left unchanged in practice. For example, pots and pans are used for cooking with gas in everyday meals. Thus, contemporary homes are only partially automated in practice.
- Application Area: The home is a place where people live. Various off-the-shelf products are used to improve the quality of life. These are manufactured by different manufacturers for different purposes. Occupants are not professional in understanding their rationale, particularly that of high-tech products. Occupants only learn to use them through product instructions or other occupants.
- Complexity: The complexity of a home is owing to three aspects, i.e., variety of appliances and products; variety of occupants in terms of age, health condition, knowledge, gender, etc.;

and outdoor environment. Off-the-shelf appliances and products indoors have various purposes and are produced by different manufacturers. In smart homes, they are connected together by the smart home network to enable a variety of services [15,17]. Different from workers in workplaces which rely on skills, rules, and knowledge for a specific job [18], occupants are usually reliant on their own life experiences to lead their lives with respect to various indoor items. Outdoor climate, air quality, etc. can affect the indoor environment, which in turn impacts the working of indoor appliances or devices. All these add to the complexity of the smart home.

2.2. Home Environment Safety

If we refer to a dictionary, the definition of safety could be the condition of being protected from or unlikely to cause harm, injury or loss. In system safety [2], safety is taken as an emergent property of systems. It is defined as freedom from conditions that can cause death, injury, occupational illness, damage to or loss of equipment or property, or damage to the environment [19]. However, it cannot be freedom from the conditions in practice. Safety is, thus, the condition of risk that has been reduced to an acceptable level, e.g., as low as reasonably practicable [20]. Home environment safety could be understood as that the risk of indoor climate has been reduced to a level of no harm to the health of occupants.

As is known to all, indoor climate is affected by outdoor climate. Bad weather such as heatwaves has been occurring frequently in recent years due to global warming and weather anomalies around the world. For example, many places globally experienced intense heat in the summer of 2018. It was observed that the highest temperature record in Japan was broken and reached a new level of 41.1 °C (Japan Meteorological Agency [visited 2018.10.03] http://www.jma.go.jp/jma/index.html). Indoor climate can thus be endangered by outdoor climate anomalies.

The indoor climate is adjusted by dedicated home appliances, e.g., use of air-conditioner to adjust indoor temperature. Different home appliances are integrated via home networks that yield value-added integrated services [21,22]. In order to ensure thermal comfort, an indoor temperature adjustment service is an example of the integrated service, which can potentially adopt a window, curtain, and air-conditioning unit. Each involved home appliance may have safety instructions. However, due to the complexity of the smart home, they still could be used in scenarios that cause safety problems. For example, the heating mode of an air-conditioner was used when it should not be. Since the smart home is an application area, appliances inside it may be replaced from time to time. Once an appliance is introduced into the smart home, it also brings about risk. For example, the predefined integrated service may not be aware of the new item and cause safety problems when using it. Therefore, home environment safety depends on the proper use of the indoor climate adjustment service with respect to related home appliances if they were properly designed and manufactured.

To understand home environment safety, we also need to discuss how people relate to the smart home. This is mostly because home appliances are produced and operated by people. One group of people are professionals. These relate to the activities of design, manufacture, transport, installation, and disposal of appliances, home networks, and so on. One distinguishing characteristic is that they have expertise in a certain field. The other group is occupants who are non-professionals. They are the customers who use the various home appliances. Both groups of people can affect home environment safety in different ways. Professionals are responsible for designing, manufacturing, etc. safe systems and home appliances. Occupants care more about operational safety, since they are more error-prone in operations. We talk about occupants in this paper.

3. Accident Model

As the smart home is an application area, the STAMP-PP model is discussed with respect to system operations rather than system development. It aims to understand how accident formation relates to system behaviors in adjusting the home environment, i.e., how abnormal system behaviors cause indoor climate anomalies. Concrete information about indoor climate anomalies can be used for indoor climate anomaly detection [23], and abnormal system behaviors under operation scenarios can further be used in selecting reactions and precautionary measures when the corresponding indoor climate anomaly is detected.

The STAMP-PP model starts with system behaviors to describe accident formation. The behaviors can result in abnormal changes in physical processes, which can cause discomfort or harm. The connection between smart home systems and the home environment is the system behaviors. In this section, we first give a brief introduction of the STAMP model, then discuss in detail the proposed STAMP-PP model.

3.1. STAMP

In systems theory, systems are viewed as interrelated components kept in a state of dynamic equilibrium through feedback control loops. Figure 1 presents a standard contrl loop [2]. The STAMP model is based on system theory rather than the reliability that traditional accident models are grounded on. Safety, in STAMP, is an emergent property of systems. Accidents are the result of the lack of or inappropriate constraints imposed on the system design and operations. The STAMP model consists of three building blocks, i.e., safety constraints, a hierarchical safety control structure, and process models.

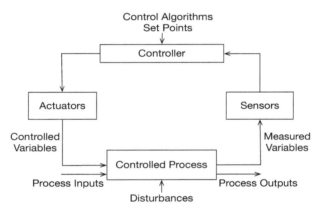

Figure 1. A standard control loop.

Safety constraints are a basic concept in the STAMP model. Losses occur only because safety constraints were not successfully enforced. Systems, in system theory, are viewed as hierarchical structures, where each level imposes constraints on the activity of the levels beneath it. Constraints are enforced by control actions of a higher-level system component (controller) to the lower-level one (controlled process).

The hierarchical safety control structure presents all stakeholders with their internal structures within the system under analysis, and the control actions and feedbacks that link the independent stakeholders and their internal components [6]. Control processes operate between levels of a system to control the processes at lower levels. The feedbacks provide information about how effectively the control actions ensure the constraints are enforced. The higher level uses the feedbacks to adapt future controls to more readily achieve its goals. An accident occurs when control processes provide inadequate control that violates safety constraints. Inadequate control comes from missing constraints, inadequate safety control commands, commands that were not executed correctly at a lower level, and inadequately communicated or processed feedback about constraint enforcement.

Process models are used by the controller to determine appropriate control actions. It is up to the type of the controller. For an automated controller, the process model is embedded in the control logic. For a human controller, the process model is the mental model. In both situations, it contains information of the required relationship among the system variables, the current system state, and the ways the process can change state. There are four conditions required to control a process, i.e., goal, action condition, observability condition, and model condition. Accidents related to component interaction can usually be explained in terms of an incorrect process model. The process model used by the controller does not match the controlled process that results in interaction accidents.

In the STAMP model, safety is achieved when the behaviors of components of a system appropriately ensured safety constraints. Accidents are the results of flawed processes involving interactions among people, societal and organizational structures, engineering activities, and physical system components that lead to violating the system safety constraints. The process leading up to an accident is described in terms of an adaptive feedback function that fails to maintain safety as system performance changes over time to meet a complex set of goals and values.

3.2. STAMP-PP

Since the STAMP-PP model focuses on the behaviors of systems to affect physical processes, we first define the concept of systems that emphasize behaviors, i.e., Performers System. The reason for choosing the word "performer" is to highlight that the behaviors are performed by the systems. Then, based on the Performers System, we can describe accident formation considering physical processes.

3.2.1. Performers System

The indoor environment is adjusted by indoor environment adjustment services with respect to various home appliances. To differentiate the home appliances with other indoor items, e.g., router or furniture, we define the concept of Performer.

Definition 1 (Performer). *A performer is a network-enabled home appliance that can adjust the indoor environment independently.*

There are two points to explain this definition. First, a Performer has networking capability so that it can be used by indoor environment adjustment services. Second, it has functions of adjusting the indoor environment, e.g., adjusting indoor temperature, or dehumidification.

There are two types of Performers based on the way it adjusts the indoor environment. One is direct adjustment, e.g., an air-conditioner which heats or cools indoor air directly; another is indirect adjustment, e.g., an electric window which adjusts, e.g., indoor temperature by introducing an air flow or solar radiation of the outdoor environment. In the latter case, the outdoor climate is passively used to adjust the indoor environment. Then, concept of Performer is used to define the Performers System.

Definition 2 (Performers System). *It is a system of all installed Performers in a house that are connected to the same home network.*

By connecting to the same home network, we can ensure that the Performers System can be used by the same indoor environment adjustment service. The Performers System has a goal prescribed by the indoor environment adjustment service. The goal is achieved by taking advantage of the functions of the Performers of the Performers System. Each Performer is taken as a subsystem of the Performers System. However, there is no need to have all related Performers working at the same time. Figure 2 shows an example of the Performers System. It consists of an air-conditioner, an electric window, and an electric curtain, which are taken as Performers. They connect to the same network and have the ability to adjust the indoor temperature. When adjusting the indoor temperature for thermal comfort (the goal), the indoor temperature adjustment service could use any combination of the Performers, but not necessarily all of them. The indoor temperature adjustment service is executed in the smart home system core. The Performers can of course be operated by occupants who are also the beneficiaries of the adjustment.

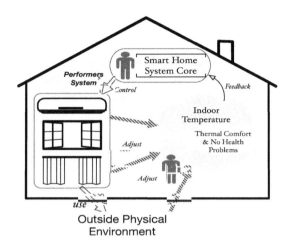

Figure 2. An example of the Performers System.

The Performers System can adjust physical processes with respect to various physical properties, e.g., temperature and humidity, by utilizing the functions the Performers provide. These physical processes may have different forms, for example, increasing or decreasing indoor temperature to the prescribed temperature level for thermal comfort.

3.2.2. Service

In this section, we discuss the behaviors of the Performers System, and when the behaviors can be taken as Services.

Definition 3 (Behavior of the Performers System). *The way the* Performers System *behaves to adjust the indoor environment.*

The Behaviors of the Performers System are the representation of the functions of related Performers. For example, the Performers System in Figure 2 has the ability to cool the temperature down (behavior), which could be achieved by setting a lower temperature level under the cool mode of the air-conditioner (function). The representation of the Behaviors is the physical process, e.g., temperature change.

Definition 4 (Service). *The* Behavior of the Performers System *exhibited in order to fulfill occupants' comfort requirement.*

One example of the Service can be the Performers System in Figure 2, which increases the indoor temperature to 22 °C for thermal comfort. Comfort means psychological and physical satisfaction with the state of the indoor environment, e.g., thermal comfort. It is the way to evaluate the Behaviors of the Performers System, and thus implicitly constrain the Behaviors. The comfort has different contents for different goals of the Performers System, e.g., thermal comfort; comfort in terms of humidity levels.

There are two ways to evaluate comfort. Let us take thermal comfort as an example. The first is based on the perception of occupants. If occupants feel uncomfortable, one can manually set a desired temperature level to the Performers System. It is easy and accurate but limited in some scenarios. For example, babies and elderly people may not be sensitive to temperature change due to their nervous system not being well developed or being degenerated. The second way is for comfort to be automatically evaluated by the smart home system core. This depends on various indices [24,25] for evaluating the physical environment for thermal comfort. For example, the PMV–PPD (Predicted Mean Vote and Predicted Percentage of Dissatisfied) index [25] is used to evaluate thermal comfort with respect to environmental factors and personal factors. PMV and PPD are short for predicted mean vote and predicted percentage of dissatisfaction, respectively, and are calculated based on Fomulas (1) and (2), where TS denotes the thermal sensation transfer coefficient and is determined by metabolic rate; MV means internal heat production in the human body and is determined by the metabolic rate and external work; and $HL1$, $HL2$, $HL3$, $HL4$, $HL5$, and $HL6$ represent heat losses through skin, sweating, latent respiration, dry respiration, radition and convection, respectively. Then combine with the ASHRAE (American Society of Heating, Regrigerating and Air-Conditioning Engineers) thermal sensation scale as shown in Table 1 to determine the comfortableness. The details are referred to in [25].

$$PMV = TS \times (WM - HL1 - HL2 - HL3 - HL4 - HL5 - HL6) \tag{1}$$

$$PPD = 100 - 95 \times e^{-0.03353 \times PMV^4 - 0.2179 \times PMV^2} \tag{2}$$

Table 1. ASHRAE thermal sensation scale.

Hot	Warm	Slightly Warm	Neutral	Slightly Cool	Cool	Cold
+3	+2	+1	0	−1	−2	−3

Environmental factors like humidity can be acquired through humidity sensors. Personal factors like metabolic rate can be roughly evaluated based on the occupant's activities, e.g., sedentary. Another way is to take advantage of wearable devices to measure personal information and send these data to the smart home system core to evaluate comfortableness. However, as far as the author is aware, cloth insulation cannot be evaluated through wearable devices for now. It can only be roughly evaluated through scenarios like in hot summer, where the value of cloth insulation is small, and an average value is assigned for the summer season.

Definition 5 (Service Manner). *The Service under a specific condition is called a Service Manner.*

The condition can be personal or environmental. For a Service, e.g., adjust the indoor temperature for thermal comfort by an air-conditioner, changes in conditions result in different Service content. For example, compare a sweating man in summer and a sedentary man in winter. The conditions in the former are high metabolic rate, high outdoor temperature, etc., and he needs for the temperature to be cooled down. For the latter, the conditions may be low metabolic rate, low outdoor temperature, etc., and he needs for the temperature to go up.

Definition 6 (Critical Service Manner). *The Service Manner in demand is called Critical Service Manner.*

This definition is given from the viewpoint of people who need the Service. In the example from Definition 5, the adjustment of indoor temperature to achieve a cool environment for thermal comfort is the Critical Service Manner for the sweating man; the adjustment of indoor temperature to achieve a warm environment for thermal comfort is the Critical Service Manner for the sedentary man.

3.2.3. Accident Formation

Accidents can be understood as the resilience [26] of the Performers System, i.e., adjusting indoor environment anomalies to maintain a normal performance has failed and resulted in undesired consequences. A Service may fail and result in uncomfortableness, then further evolve into hazards and cause harm to occupants. Accident formation is to describe how Services may fail and further evolve into a hazard to cause accidents. The causes on the system part, i.e., the Performers System, can be understood by the STAMP model. The relation between the Behaviors of the Performers System and the physical processes in accident formation is discussed in this section.

There are some considerations about physical processes from the viewpoint of engineering. Before a hazard is detected, physical process anomalies should be detected as early as possible so as to trigger precautionary measures. The information of hazards is also important for triggering reaction measures. All in all, accident formation, which is shown in Figure 3 focuses on the physical process anomalies resulting from abnormal Behaviors.

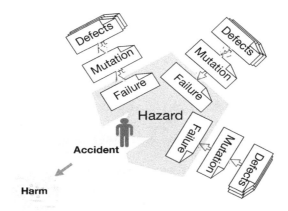

Figure 3. The accident causality model.

Next, let us discuss the terms and related rationales of the accident causality model in Figure 3.

Definition 7 (Defect). Defects *are the direct causes of a* Mutation.

A Mutation (Definition 8) relates to abnormal Behaviors of the Performers System. Abnormal Behaviors are due to unexpected control actions of the Performers System. Thus, Defects are inappropriate control actions (ICAs) and unsafe control actions (UCAs) of the Performers System to the abnormal Behaviors in adjusting the home environment. ICAs occur when Mutations result in Service Failures (Definition 9), and UCAs occur when Service Failures evolve into Hazards (Definition 10). For example, an ICA can be wrongly setting the heating mode of a Performer in hot summer, which will cause thermal discomfort. To differentiate from causes introduced in the STAMP model, Defects refer to more superficial reasons. As the home environment is an application area, the deeper causes for system development defects are not considered here. When physical process anomalies are detected, the home environment is expected to react to them immediately with respect to Defects. Therefore, the Defects are the direct causes and should be controllable, e.g., through reconfiguration.

As discussed at the beginning of Section 3, the STAMP-PP model relates to system operations. The occurrence of Defects is under the scenarios of system operations. The operations can be categorized into three types. One is operations by occupants. Another is by some controllers, e.g., Performers. The other is a mixture of by occupants and the controller. To react to a physical process anomaly efficiently, it is necessary to know the Defect with its corresponding operation scenario. To identify Defects with respect to the scenarios, we applied the hazard analysis technique STPA [2,9] in Section 4.

Definition 8 (Mutation). *A* Mutation *is the violation of constraint on the* Behavior *of the* Performers System.

Abnormal changes in physical processes are the representation of the abnormal Behavior of the Performers System. Therefore, the constraint on the Behavior of the Performers System is the prescription of changes in physical processes that satisfy the comfort purpose. Thus, the Mutation is the change of physical processes under adjustment unacceptably deviating from the prescribed curve(s), which results in uncomfortableness. For example, the amplitude of temperature fluctuation should not exceed a threshold value for thermal comfort [23]. The Mutation in this case is the amplitude of temperature fluctuation exceeding the threshold value.

The Mutation is a state of physical processes between the state that brings about comfort and the state that results in a hazard. This concept is important not only for the understanding of accident formation, but also because the information can trigger precautionary measures. First, a Mutation indicates the current indoor environment adjustment is inefficient. Second, by combining the *Mutation* with Defects under certain operation scenarios, one can select appropriate precautionary measures to restore the state of physical processes that bring about comfort.

Definition 9 (Service Failure). *The* Behavior of the Performers System *exhibits failed to fulfill occupants' comfort requirement.*

The concept of Mutation is related to physical processes, while Service Failure refers to both physical processes and the perception of occupants. The occurrence of Service Failure is when occupants perceive the uncomfortableness brought about by the Mutation.

There are two ways to determine whether a Service has failed. One is directly determined by the perception of occupants. Another one resorts to various indices [24,25], by which the smart home system core can conclude whether the home environment satisfies the comfort requirement with occupants on the scene.

Definition 10 (Hazard). *It is an indoor environment state that will cause harm to occupants.*

A Hazard will harm the health condition of occupants who are on the scene. One or more Service Failures will form or further evolve into a hazardous situation. For example, a Service Failure for thermal discomfort evolves into a hazard if the indoor temperature reaches a level that can cause heat stroke if occupants are present.

Taking thermal related hazards as an example, the evaluation of a hazard depends on heat stress indices [27] and cold stress indices [28]. These indices are sophisticated techniques to represent thermal sensations to hot and cold conditions. Heat stress [27] is defined as the net heat load to which an occupant is exposed from the combined contributions of metabolic heat, environmental factors, and clothing that results in an increase in heat storage in the body. Cold stress [28] is the climatic condition under which the body heat exchange is just equal to or too large for heat balance at the expense of significant heat and sometimes heat debt. Similarly to the PMV–PPD index, they also have a complex relations with environmental and personal factors, which can be measured in practice.

Definition 11 (Accident). *It is an unintentional event where a* Hazard *results in harm of occupants.*

It involves both the home environment and occupants. The Hazard has harmed occupants. The Accident can be detected by evaluating the Hazard and the health condition of the occupants. The latter can be measured by taking advantage of wearable devices.

Definition 12 (Harm). *Death, physical injury or damage to the health of occupants.*

It is the consequence of the Accident. It varies with respect to Hazards and the health conditions of occupants. For example, it may cause heat illnesses or even death to elderly people due to heat exposure [29]; it may also affect sleep and the circadian rhythm that can cause cardiac autonomic response during sleep due to cold exposure [30].

4. Application of STPA

To identify Defects under operation scenarios, the hazard analysis technique STPA is adopted. In this section, we first introduce the STPA steps, then discuss the way to tailor it and a new way of documenting the analytical results, and finally illustrate the application results and compare them with the application of the original STPA.

4.1. STPA

The Systems-Theoretic Process Analysis (STPA) [2,9] is a new hazard analysis technique based on the STAMP model. The goal is to identify causes that lead to hazards and result in losses so they can be eliminated or controlled. The STPA has three steps, the latter two of which are taken as the main steps.

The first step is to establish the system engineering foundation. Three things should be done in this step. The first is to define the interested accident and its related system hazards. They should be specific and concise. Then, the system level safety requirements and design constraints need to be specified to prevent system hazards from occurring. The last is to define the safety control structure that takes the system level safety requirements and design constraints as inputs.

The second step is to identify potentially unsafe control actions (UCAs). Every controller of systems usually has one or more control actions. System hazards thus result from inadequate control or enforcement of the safety constraints. Four taxonomies are provided to look for the UCAs:

1. A control action required for safety is not provided or not followed;
2. An unsafe control action is provide;
3. A potentially safe control action is provided too early or too late (at the wrong time or in the wrong sequence);
4. A control action required for safety is stopped too soon or applied too long.

Then, we translate the identified UCAs into safety requirement and designed constraints on system component behaviors.

The third step determines how each UCA identified in step two could occur. For each UCA, examine the parts of the control loop as shown in Figure 4 to see if they can cause the UCA under some scenarios. Then, design controls and mitigation measures if they do not already exist or evaluate existing measures.

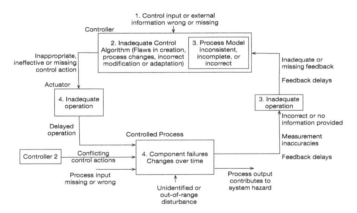

Figure 4. A classification of controls flaws leading to system hazards.

4.2. Tailor

The understanding of accident formation within the indoor environment has extended the STAMP model, i.e., the STAMP-PP model, by also considering physical processes. As disscussed in Section 3.2.3, abnormal Behaviors result in Service Failures, and abnormal Behaviors are due to inappropriate control actions of the Performers System. Thus, it is important to know what these inappropriate control actions are. Therefore, the STPA need to be tailored for this main purpose.

In the first step of STPA, in addition to the definition of accident, system hazard, and safety requirement, information about the Service Failure and requirements of no failure, i.e., the reliability of the Performers System to deliver Services, should also be provided. Design constraints are not required as this work is not for implementing a safe system.

When a Service Failure occurs, which means the corresponding control action is inappropriate, then precautionary measure(s) should be provided. Thus, in the second step of STPA, ICAs should be identified with regard to the taxonomies provided in step two of STPA. Namely, given a state of physical processes, under a specific operation scenario, we need to consider whether a control action with respect to the taxonomies will cause a Service Failure. Precautionary measures are determined by considering the context information which consists of the ICAs, operation scenarios, and the Service Failure. The process of the determination heavily depends on the expertise of concerned areas, which is directed by the reliability requirements. The precautionary measures are also control actions. Ineffective precautionary measures will result in the occurrence of a Hazard (then, reaction measures are required). For each precautionary measure, UCAs are identified by considering the taxonomies provided in step two of STPA under the conditions of operation scenarios and of a state of physical processes. Safety requirements to the UCAs also need to be identified to guide the selection of reaction measures.

The third step of STPA is not necessary. This is because the analysis in our case is to identify Defects under operation scenarios, which can be utilized in selecting appropriate precautionary and reaction measures, but not in designing and manufacturing a system.

STPA adopted tables and lists for documenting analytical results [2,9]. After applying the tailored STPA, the results, i.e., control actions, ICAs, and their related reliability requirements and operation scenarios; and precautionary measures, UCAs, and their related safety requirements and operation scenarios, could be documented by that used by the STPA. However, the relations among them are not clearly represented. In this paper, we propose a Landscape Genealogical Layout Documentation (denoted as LGLD) for documenting the results, which is illustrated in Figure 5. The ancestor is a control action. The first generation illustrates ICAs and their related reliability requirements and operation scenarios. The second generation represents precautionary measures. The third generation represents UCAs and their related safety requirements and operation scenarios. These results can be numbered for better reference, e.g., ICA-m for an ICA, which means the inappropriate control action m. This way of documentation also implies the analysis direction, i.e., from control actions to ICAs, then to precautionary measures, and finally to UCAs.

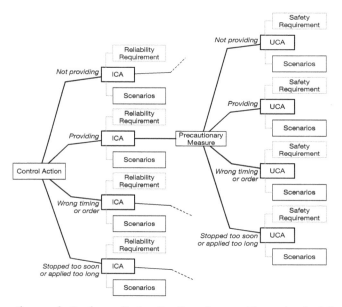

Figure 5. Documenting the analytical results by the Landscape Genealogical Layout Documentation (LGLD) approach.

For each control action, the taxomomies provided in step two of STPA to list the ICAs must be considered. Every ICA is attached with the reliability requirements and operation scenarios. Then, each ICA is connected with a precautionary measure. The precautionary measure is to prevent its connected ICA from changing the home environment from a Service Failure to a Hazard. For each precautionary measure, it is important to list the UCAs attached with safety requirements and operation scenarios.

4.3. Results

As discussed in Section 2.2, weather anomalies affect indoor environment, e.g., the heatwave of summer 2018. Thus, we consider the example of the high-temperature results in heat stroke [31,32]. Heat stroke is clinically diagnosed as a severe elevation in body temperature (a core body temperature of 40°C or higher) that occurs in the presence of central nervous system dysfunction and a history of environmental heat exposure or vigorous physical exertion. It can be classified into nonexertional (classic) heat stroke and exertional heat stroke. The former occurs in very young or older people, or those with chronic illness when the environmental temperature is high. The latter happens to young fit people and involves prolonged excessive activities like sports. This paper focuses on the former case.

Heat stroke can be assessed by heat stress indices, among which the most widely accepted and used one is the wet bulb globe temperature (WBGT) [33]. The Ministry of the Environment of Japan (Ministry of the Environment, Japan: http://www.wbgt.env.go.jp/en/) has recommended a criterion for thermal conditions based on the WBGT as shown in Table 2.

Table 2. Recommended criteria for thermal conditions.

WBGT (°C)	Threat Level
~21	Almost Safe
21~25	Caution
25~28	Warning
28~31	Severe Warning
31~	Danger

We applied the tailored STPA to a Performers System that is for indoor temperature adjustment for thermal comfort. The structure of the system consists of four parts, i.e., home, home gateway, service intermediary, and service provider [12], as shown in Figure 6. The home is the place that occupants live in, which is equipped with different kinds of items, e.g., home appliances, to meet everyday living requirements. The home gateway is the gateway of the networked Performers to the outside world of the home. The indoor temperature adjustment service is executed here to issue commands for the control of Performers. The service intermediary aggregates various services from different service providers and maintains them locally. It can also respond to service subscriptions from home gateways. The service provider designs, publishes, and updates concrete services to the service intermediary for future use. We applied STPA to investigate the Behaviors of the Performers System to elicit possible Defects under related operation scenarios. Thus, we focus on executing the indoor temperature adjustment service in the home gateway to adjust the indoor temperature. The home gateway, service intermediary, and service provider can be taken as the smart home system core.

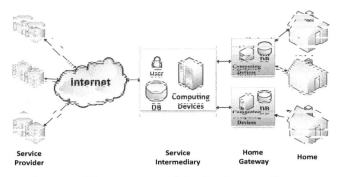

Figure 6. The structure of the Performers System.

In order to apply STPA, an assisting tool, i.e., STAMP Workbench (the STAMP Workbench is an open-source, free, easy-to-use tool for people who are interested in system safety analysis by using the STAMP/STPA. It was developed by the IT Knowledge Center of Information-Technology Promotion Agency, Japan. https://www.ipa.go.jp/english/sec/complex_systems/stamp.html) is adopted for the analysis. The STAMP Workbench is claimed to include features such as concentration on thinking and help analysis and is not just an editing tool, or guide analysis procedure, but an unlimited and intuitive operation. Analytical results can be exported into Excel files and images.

In the first step of STPA, we prepared some concepts for further analysis as shown in Table 3. For demonstration, Service Failure is defined as when the indoor WBGT temperature is adjusted within (25,28] °C, and the Hazard is when the indoor WBGT temperature is adjusted over 28 °C. The reliability requirement corresponds to the Service Failure, which represents the requirement to ensure a Service will not fail.

Table 3. Preparation for the tailored System-Theoretic Process Analysis (STPA).

Accident	Physical harm of occupants due to heat stroke
Service Failure	Indoor WBGT temperature is within (25,28] °C
Reliability Requirement	Indoor WBGT temperature should be adjusted bellow 25 °C
Hazard	Indoor WBGT temperature is over 28 °C
Safety Requirement	Indoor WBGT temperature should be adjusted bellow 28 °C

Figure 7 illustrates the safety control structure for indoor temperature adjustment. The Home Gateway is the controller, which is responsible for executing the indoor temperature adjustment service. The controlled process is the Home Environment. Performers are taken as actuators. We consider an air-conditioner and a window here as Performers. Empirically, for energy saving, these two cannot work at the same time. The control actions are listed on the arrow from the Home Gateway to the Performers. The control action "set to X °C" means to set Performers to adjust the indoor temperature to X °C. The feedback to the Home Gateway is the indoor temperature. Indoor temperature can be adjusted by the indoor temperature adjustment service that is executed in the Home Gateway, or by occupants to issue commands, i.e., control inputs to the Home Gateway.

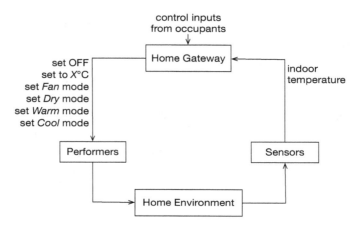

Figure 7. The safety control structure for indoor temperature adjustment.

Next, based on the results of the first step, we identifies ICAs and UCAs and elicited their related requirements and operation scenarios. Part of the results are shown in Figures 8 and 9, which adopted the LGLD approach introduced in Section 4.2 for the documentation. They illustrate the results of analyzing the control actions "set OFF" and "set to X °C". "N/A" denotes the taxonomy is not applicable to the corresponding control action. In Figure 8, the second "set OFF" can be considered as a reconfiguration compared with the first one. The precautionary measure "set Cool mode" can be deployed for the two ICAs that relate to the "set OFF". One reason could be that a different configuration has a higher possibility to restore the physical process to a comfortable state, as the "set

OFF" has caused the ICA. For the second operation scenario, not-providing ICA, it is because people are not sensitive to temperature change and thus did not issue the "set OFF" command manually.

For the results as shown in Figure 9, X satisfies $X < 25$ °C. There are two reasons for the not-providing ICA and UCA to say that the Performers System is working in the Fan mode. The first is if it works in the Warm mode, which in this case is inappropriate or hazardous, and we cannot make a solemn vow to conclude that not providing "set to X °C" is inappropriate or hazardous. Second, the other working modes, i.e., Dry and Cool, have a cooling effect based on our experience, which may not be an ICA or UCA even when "set to X °C" is not provided. One more thing that needs to be explained is that providing "set to X °C" at the "wrong time" is neither inappropriate nor hazardous. If it is not provided in time, the home environment would experience Service Failure or Hazard for some time. However, providing "set to X °C" at a different time should not be inappropriate or hazardous. Conversely, since "set to X °C" is provided, the home environment could be restored to a safe level, even though later than expected. In this case, providing "set to X °C" can be thought of as a precautionary or reaction measure, rather than an inappropriate or hazardous control action.

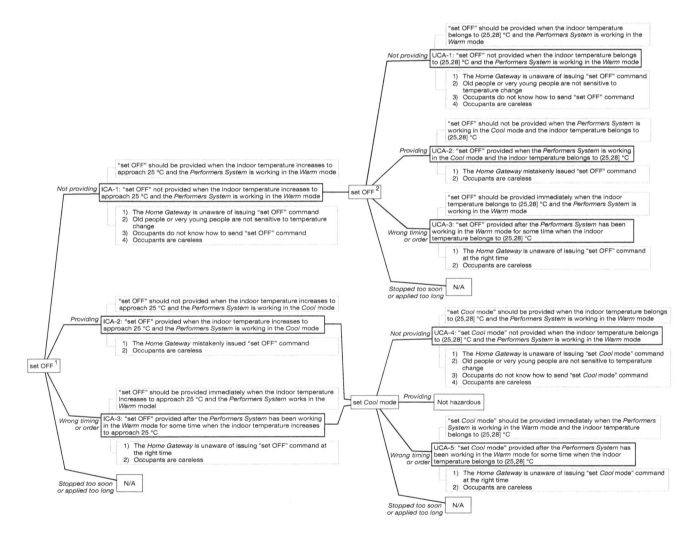

Figure 8. The analysis results for the control action "set OFF".

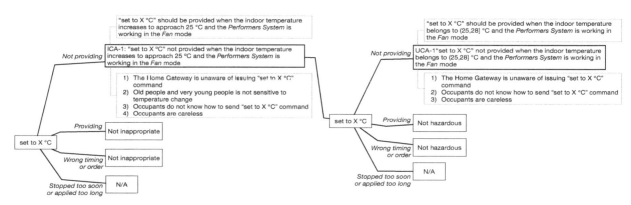

Figure 9. The analysis results for the control action "set to X °C".

4.4. Comparsion of Results

Originally, the STPA is the only hazard analysis technique was based on the STAMP model [2]. Thus, in this section, let us compare the results presented in Section 4.3 with those by adopting the original STPA. Since the third step of STPA was not taken into account, the comparison only considers the results that derived from the first two steps of STPA. For comparability, the temperature issue discussed in Section 4.3 is still part of our focus when adopting the original STPA.

As discussed, the goal of STPA is to identify causes that lead to hazards and result in losses, so they can be eliminated or controlled. The causes to be identified are UCAs, and flaws in the control loop (as shown in Figure 4) under some scenarios which are different from our case. The elimination or control usually resorts to designing and implementing a safe system, while in our case, the aim is to select appropriate precautionary and reaction measures which can restore a safe Service delivery.

The system engineering foundation is given first. The prepared definitions are illustrated in Table 4. The design constraint is system level constraint and is expected to further decompose into constraints that can be assigned to system components as the analysis evolves. Compared with what is shown in Table 3, Service Failures and reliability requirements to the system are gone, which indicates ICAs will not be identified afterward. This is because ICAs are supposed to result in Service Failure, and reliability requirements of ICAs can be taken as the decomposition of the system level reliability requirement. The safety control structure as shown in Figure 7 can also be used here.

Table 4. Preparation for the STPA analysis.

Accident	Physical harm of occupants due to heat stroke
Hazard	Indoor WBGT temperature is over 28 °C
Safety Requirement	Indoor WBGT temperature should be adjusted bellow 28 °C
Design constraint	The Performers System is capable of adjusting the indoor WBGT temperature bellow 28 °C

Next, the UCAs identified in step two of STPA are shown in Tables 5 and 6 for control actions "set OFF" and "set to X °C", respectively. Then, for each UCA, safety requirements and design constraints can be derived. For example, the safety requirement for UCA-1 in Figure 5 could be:

- "set OFF" should be provided when the indoor temperature belongs to (25,28] °C and the Performers System is working in the Warm mode.

The design constraints for UCA-1 could be:

- The Performers System should be accurately aware of the indoor temperature change;
- "set OFF" should be provided when needed.

Table 5. UCAs for the control action "set OFF".

	Hazard: Indoor WBGT temperature is over 28 °C			
Control Action	Not Providing	Providing	Wrong Timing or Order	Stopped Too Soon or Applied Too Long
set OFF	UCA-1: "set OFF" not provided when the indoor temperature belongs to (25,28] °C and the Performers System is working in the Warm mode	UCA-2: "set OFF" provided when the Performers System is working in the Cool mode and the indoor temperature belongs to (25,28] °C	UCA-3: "set OFF" provided after the Performers System has been working in the Warm mode for some time when the indoor temperature belongs to (25,28] °C	N/A

Table 6. UCAs for the control action "set to X°C".

	Hazard: Indoor WBGT Temperature Is over 28 °C			
Control Action	Not Providing	Providing	Wrong Timing or Order	Stopped Too Soon or Applied Too Long
set to X °C	UCA-4: "set to X °C" not provided when the indoor temperature belongs to (25,28] °C and the Performers System is working in the Fan mode	Not hazardous	Not hazardous	N/A

There are some differences in the comparison with the results derived in step two of STPA. The ICAs, operation scenarios for ICAs, reliability requirements, precautionary measures, and operation scenarios for UCAs cannot be obtained by adopting the original STPA. However, the design constraints can be derived. The UCAs identified by adopting the original and tailored STPA are equivalent. The results documented by tables and lists are separated. It is a trivial problem when checking the relations between the results that were documented by the conventional approach. The LGLD approach integrated the results and overcame this problem. We discuss the advantages of the tailored STPA at the end of Section 5.

5. Discussion

As one of our everyday living experiences, physical processes could be in hazardous states that cause harm to occupants [29,30]. The occurrence of hazardous states can be transferred from a normal state of a physical process that brings about comfortableness, through some intermediate states for uncomfortableness. The normal state is maintained by the smart home system, which takes advantage of home appliances or the outdoor climate. When the behaviors of the smart home system deviate from expectations, the above transformation will occur. Thus, this living experience validates the proposed STAMP-PP model.

To validate the terms defined in the STAMP-PP model, let us consider the eight abnormal indoor temperature situations that were discussed in our previous work [23]. Mutations can be undesired fluctuation, constantly cooler/warmer than expectation, undesired duration of the temperature that results in discomfort, and Service Failures when combining the Mutations with the feeling of occupants on the scene. Hazards can be unbearable hot/cold and undesired duration in hot/cold situations.

Mutations and Hazards provide detectable evidence of physical process anomalies. The information of Service Failure combines with ICAs under related operation scenarios—e.g., as depicted in Figures 8 and 9, one can predict whether Hazards would happen before providing precautionary measures, because Service Failures and the ICAs under related operation scenarios can be the context, in which a Hazard could occur if time goes on. This context information can be acquired from various sensors (e.g., temperature sensor), system states, etc. For example, in Figure 9, we can know whether the temperature is increasing to approach 25 °C through temperature sensors. Then, precautionary measures can be selected under the direction of reliability requirements. If a Hazard is unfortunately detected, reaction measures have to be selected based on the UCAs under operation scenarios and the Hazard, which are guided by safety requirements.

The purpose of precautionary measures is to restore Service delivery. This is achieved by reconfiguration of the Performers System. The reconfiguration has two forms. One is to reset the current working Performer(s), e.g., the precautionary measure "set OFF" (the second one) as shown in Figure 8, which is achieved by resetting the OFF command to the same Performer. Another is to reconfigure ready-to-use Performers to achieve the goal, e.g., the "set Cool mode" in Figure 8, which could be reconfigured to a standby Performer to the Cool mode. The purpose of reactions is to restore a safe home environment state, which could be a Service Failure (that needs further interference) or the process of normal Service delivery. It is also achieved by reconfiguring the Performers System, which has the same contents as those introduced for precautionary measures. The reactions should have another content that precautionary measures do not have, i.e., a warning mechanism. The warning mechanism is triggered when a Hazard is detected that implies precautionary measures have failed. In the very beginning, the warning signals will be sent to occupants who could leave the scene or do something else to ensure safety. If the reconfiguration in the reaction stage cannot restore the system to a safe situation, the warning mechanism will inform an emergency department, e.g., a hospital, through networks.

Accident models can be roughly classified into three categories [1,2], i.e., sequential models, epidemiological models, and systemic models. Sequential models describe accidents as the result of time-ordered sequences of discrete events. Epidemiological models view accidents as a combination of latent and active failures within a system, which is analogous to the spreading of a disease. Latent conditions, e.g., management practices or organizational culture, can lie dormant within a system for a long time, which can finally create conditions at a local level to result in active failures. The STAMP model is based on system theory and so is the STAMP-PP model. It describes physical process anomalies that cause uncomfortableness and health problems as a result of abnormal Behaviors of the Performers System. Furthermore, to better understand the causal relation between the Performers System and physical processes and better represent the time order of physical process evolvement, the terms in the STAMP-PP model can be taken as events.

Hazard analysis techniques are devised based on accident models. The STPA was invented based on the STAMP model to identify causes that conventional approaches overlooked [2,9]. We found it is an efficient tool for assisting in the identification of Defects, i.e., ICAs and UCAs (e.g., the results shown in Figures 8 and 9). However, hazard analysis heavily depends on expertise in a specific area. Accident models and hazard analysis techniques are to assist in understanding accidents and guiding for causal analysis. The STAMP-PP model offers a way to understand physical process anomalies with respect to the behaviors of smart home systems. To apply the tailored STPA to that in our case (some results of which are shown in Figures 8 and 9), one may need to possess knowledge in at least the fields of computer science, software engineering, computer networks, and even physics. This makes the authors feel that safety research in different areas is like different research fields.

In the first step of STPA, the accidents and system hazards provided mostly depend on the interest of an organization or the government [2]. For example, in this paper, Accident is defined as physical harm of occupants due to heat stroke, as illustrated in Table 3. It can also, however, encompass other symptoms aside from heat stroke, e.g., severe cold. Concrete accidents are different due to the variety of environments, e.g., workplaces and the home environment. It is also determined by budget, severity of occurrence, and frequency of occurrence [2].

In step two of STPA, originally, the taxonamies were provided to identify UCAs with respect to control actions. Physical process anomalies (not Hazards) are the result of abnormal Behaviors of the Performers System. The behaviors are the representations of the functions which are achieved by the control actions issued by the Performers Sytem. For example, stopping the Performers System from working is achieved by the control action "set OFF" as shown in Figure 8. Thus, the taxonomies are applicable for identifying ICAs. This ensures that all possible ICAs can be effectively identified.

The states of physical process change from a state of comfort to one of discomfort, then to one of hazard. When a state of discomfort is detected, precautionary measures are adopted to prevent the physical process from transferring into a state of hazard. Precautionary measures refer to control actions achieved by reconfigurations. Thus, the taxonomies are also applicable to the precautionary measures. For example, the control action "set OFF" in Figure 8 can both be the original control action and the precautionary measure. The same applies to the control action "set to X °C" as shown in Figure 9.

The purpose of adopting the hazard analysis technique is to identify Defects, so as to select appropriate precautionary and reaction measures for adjusting home environment anomalies to maintain a normal performance. By checking the comparison presented in Section 4.4, we found the tailored STPA can satisfy this purpose, while the original STPA cannot. First, ICAs under operation scenarios and the reliability requirements of the ICAs can be identified by the tailored STPA. This is due to Service Failure, and the system level reliability requirement is provided in the first step of STPA. Naturally, precautionary measures are not necessarily identified in the original STPA, because precautionary measures are selected for ICAs. Second, even though UCAs can be identified by both the original and tailored STPA, the operation scenarios have different contents. For the tailored STPA, the operation scenarios refer to occupants or controllers or both as discussed when introducing the concept of Defect. For the original STPA, the operation scenarios refer to the control flaws as shown in Figure 4, which are identified in step three of STPA.

As discussed in Section 4.2, the STPA takes advantage of tables and lists to document the analytical results (see the results shown in Tables 5 and 6). We proposed the LGLD approach to document the analytical results. The advantage of comparing tables and lists is that it can clearly represent the relations among the results in a straightforward way. For example, in Figures 8 and 9, the relations among control actions, ICAs, precautionary measures, and UCAs are clear. Further, it is clear to see the (reliability or safety) requirements attached to each ICA and UCA, and the related operation scenarios under which the ICA and UCA can occur. This kind of relations is not explicitly represented using tables and lists such as the ones presented in Section 4.4. To build such documentation, one can build along the way of analysis, because the analysis starts from control actions to identify ICAs under related operation scenarios, and reliability requirements, then to determine precautionary measures, and finally to identify UCAs under related operation scenarios, and safety requirements.

When applying the STAMP-PP model to understand accident formation, the Performers System has to fully control the home environment, or its behaviors may not be the "only" reason to cause Service Failure, then Hazard. This can be the limitation of the proposed STAMP-PP model.

6. Related Work

This section includes two parts. The first is about accident models based on system theory. The second discusses safety-related research in the smart home environment.

6.1. Accident Models

The systems approach is considered as the dominant paradigm in safety research [34]. It views accidents as the unexpected interactions among system components, i.e., technical, social, and human elements. Among various systemic models, there are three most cited models [10], that is, STAMP [2], Functional Resonance Analysis Method (FRAM) [3], and Accimap [35,36].

The concept of STAMP has been briefly introduced in Section 3.1. According to [10], over half of the reviewed papers were STAMP-related, which indicates a pervasive acknowledgement of its underlying rationale. Its application has attracted researchers from a broad field. The authors of [6] adopted it to investigate aircraft rapid decompression events. The authors of [4] applied it in the analysis of deepwater well control safety. In the field of railway, the authors of [5] investigated railway accidents and accident spreading by taking the China Jiaoji railway accident as the example. It has also been applied to the field related to poisonous or dangerous substances. The authors of [7] adopted it in analyzing the China Donghuang oil transportation pipeline leakage and explosion accident. The authors of [8] applied it to the Fukushima Daiichi nuclear disaster and to promote safety of nuclear power plants. Smart home systems are generally not considered as safety-critical systems. However, as weather anomalies, e.g., heatwaves, occur regularly due to global warming, smart home systems for indoor environment adjustment, in this context, can be taken as safety-critical. Thus, the STAMP model can be adopted for understanding accident formation in the home environment.

FRAM was developed to act as both an accident analysis and risk assessment tool [10]. The FRAM model graphically describes systems as interrelated subsystems and functions that will exhibit varying degrees of performance varation. Accidents result from the fact that the emergent variation produced from the performance variability of any system component to "resonate" with that of the rest of the elements is too high to control. It has been discussed that the FRAM and STAMP approaches focus better on qualitative modeling and description of systemic behavior and accidents [37]. FRAM also has applications in different fields, e.g., the authors of [38] applied it to railway traffic supervision to investigate interdisciplinary safety analysis of complex sociotechnological systems. The authors of [39] extended FRAM by including a framework with steps to support hazard analysis. Some efforts have been tried to quantify it, e.g., the authors of [40] developed a semiquantitative FRAM based on a Monte Carlo simulation.

The Accimap method is a graphical representation of a particular accident scenario that relates to systemwide failures, decisions, and actions [3,36]. Accimap is a generic approach and does not use taxonomies (that is different than that of the STPA [2,9]) of failures across the different levels of considered [41]. The Accimap produces less reliable accident analysis results compared to STAMP [42].

The selection of accident analysis techniques depends on the system characteristics, i.e., manageability and coupling [1]. The systemic approaches are usually adopted by systems with low manageability and tight coupling. Systemic approaches related to complex sociotechnical systems have their own strengths. For one such system, it is better to adopt multiple approaches which supplement each other, even though STAMP is considered much more effective and reliable in understanding accidents and hazard analysis [10,41,42].

6.2. Smart Home Safety

In the past, safety research inside the home environment used to be based on events or chains-of-events. With the emergence of the so-called smart homes, safety research inside the home environment also has new forms. Some of it refers to monitoring the home environment. With the purpose of detecting safety problems of indoor climate abnormal variations, the authors of [12] proposed a CPS (cyberphysical system) home safety architecture to support an event-based detection. The authors of [43] presented a method that maps the real home connection to a virtual home environment, together with related policies to ensure remote monitoring, to ensure home safety. Elderly safety in the smart home environment was achieved by analyzing and inferring locations, time slots, and periods of stay of elderly people [44]. Robot techniques were also employed, e.g., the authors of [45] developed a robot which can, for example, sense gas leakage and shut off the gas valve. Others focus on a specific part of the home. The authors of [46] proposed risk analysis and assessment when cooking to prevent potential risks. This is because the kitchen is also prone to safety problems like gas leakage and fire accidents. Electricity is also an important risk factor. The authors of [47] adopted an alert circuit with a voltage level indicator to prevent the smart solar home system from being overloaded and damaged. With cloud computing techniques becoming pervasive in implementing smart home systems, risks like cloud service unavailability have also been introduced. To overcome this, the authors of [48] discussed home resilience in the presence of possible unavailability and proposed RES-Hub, i.e., a standalone hub to ensure the continuity of required functionalities.

Most of the studies like those discussed above focus on implementing systems to deal with home safety problems. If not properly designed and implemented, the system itself can be a risk factor. Thus, requirement elicitation becomes critical. Conventional safety-related techniques are applied to safety-critical areas, e.g., aviation [6]. Our work employed these techniques to the smart home systems.

7. Conclusions and Future Work

We extended the accident model STAMP by considering physical processes in the home environment. As the home environment is adjusted by behaviors of the smart home system, we first proposed the concept of Performers System that emphasizes the behaviors performed by various home appliances. Then, based on the Behavior of the Performers System, we proposed accident formation with respect to physical processes going from normal state to some intermediate states that result in uncomfortableness, and finally to states that cause harm. In order to identify the Defects, i.e., ICAs and UCAs that result in abnormal system behaviors, the hazard analysis technique STPA was tailored and applied to the smart home system for indoor temperature adjustment. After comparison with the results derived by adopting the original STPA, we found that the tailored STPA is an efficient tool to assist in identifying the Defects. The analytical results of applying STPA were used to adopt tables and lists for documentation, but the relations among the results were not straight-forward and unclear. We then proposed the LGLD approach, whose advantages are demonstrated by the comparison of results.

For future work, we aim to map the context information in which physical process anomalies occurred in the cyber world, such that precautionary measures and reactions can be effectively selected in real time. This brings about the problem of how to parameterize the STPA analytical results, i.e., ICAs, operation scenarios, and reliability requirements; and UCAs, operation scenarios, and safety requirements.

Author Contributions: Conceptualization, Z.Y., Y.L., and Y.T.; methodology, Z.Y. and Y.T.; validation, Z.Y. and Y.T.; writing—original draft preparation, Z.Y.; writing—review and editing, Z.Y. and Y.T.; supervision, Y.T.

Acknowledgments: The authors would like to thank Toshiaki Aoki who gave constructive comments.

References

1. Underwood, P.; Waterson, P. *Accident Analysis Models and Methods: Guidance for Safety Professionals*; Loughborough University: Loughborough, UK, 2013; p. 28.
2. Leveson, N.G. *Engineering a Safer World: Systems Thinking Applied to Safety*; The MIT Press: Cambridge, MA, USA; London, UK, 2011.
3. Hollnagel, E. *FRAM: The Functional Resonance Analysis Method: Modelling Complex Socio-technical Systems*, 1st ed.; CRC Press: Boca Raton, FL, USA; London, UK; New York, NY, USA, 2012.
4. Meng, X.; Chen, G.; Shi, J.; Zhu, G.; Zhu, Y. STAMP-based analysis of deepwater well control safety. *J. Loss Prev. Process Ind.* **2018**, *55*, 41–52. [CrossRef]
5. Ouyang, M.; Hong, L.; Yu, M.H.; Fei, Q. STAMP-based analysis on the railway accident and accident spreading: Taking the China—Jiaoji railway accident for example. *Saf. Sci.* **2010**, *48*, 544–555. [CrossRef]
6. Allison, C.K.; Revell, K.M.; Sears, R.; Stanton, N.A. Systems Theoretic Accident Model and Process (STAMP) safety modelling applied to an aircraft rapid decompression event. *Saf. Sci.* **2017**, *98*, 159–166. [CrossRef]
7. Gong, Y.; Li, Y. STAMP-based causal analysis of China-Donghuang oil transportation pipeline leakage and explosion accident. *J. Loss Prev. Process Ind.* **2018**, *56*, 402–413. [CrossRef]
8. Daisuke, U. STAMP Applied to Fukushima Daiichi Nuclear Disaster and the Safety of Nuclear Power Plants in Japan. Master's Thesis, School of Engineering, Massachusetts Institute of Technology, Cambridge, MA, USA, 2016.
9. Leveson, N.G.; Thomas, J.P. *STPA Handbook*; Massachusetts Institute of Technology: Cambridge, MA, USA, 2018.
10. Underwood, P.; Waterson, P. A critical review of the STAMP, FRAM and Accimap systemic accident analysis models. In *Advances in Human Aspects of Road and Rail Transportation*; CRC Press: Boca Raton, FL, USA, 2012; pp. 385–394.

11. Ericson, C.A., II. *Hazard Analysis Techniques for System Safety*; John Wiley and Sons, Inc.: Hoboken, NJ, USA, 2005; Chapter 3, pp. 31–54.
12. Yang, Z.; Lim, A.O.; Tan, Y. Event-based home safety problem detection under the CPS home safety architecture. In Proceedings of the 2013 IEEE 2nd Global Conference on Consumer Electronics (GCCE), Tokyo, Japan, 1–4 October 2013; pp. 491–495. [CrossRef]
13. Yang, Z.; Lim, Y.; Tan, Y. A risk model for indoor environment safety. In Proceedings of the 2017 IEEE 6th Global Conference on Consumer Electronics (GCCE), Nagoya, Japan, 24–27 October 2017; pp. 1–5. [CrossRef]
14. Harper, R. *Inside the Smart House*; Springer: Berlin/Heidelberg, Germany, 2003.
15. Lobaccaro, G.; Carlucci, S.; Löfström, E. A Review of Systems and Technologies for Smart Homes and Smart Grids. *Energies* **2016**, *9*, 1–33. [CrossRef]
16. Djenouri, D.; Laidi, R.; Djenouri, Y.; Balasingham, I. Machine Learning for Smart Building Applications: Review and Taxonomy. *ACM Comput. Surv.* **2019**, *52*, 24:1–24:36. [CrossRef]
17. Toschi, G.M.; Campos, L.B.; Cugnasca, C.E. Home automation networks: A survey. *Comput. Stand. Interfaces* **2017**, *50*, 42–54. [CrossRef]
18. Rasmussen, J. Skills, rules, and knowledge; signals, signs, and symbols, and other distinctions in human performance models. *IEEE Trans. Syst. Man Cybern.* **1983**, *SMC-13*, 257–266. [CrossRef]
19. MIL-STD-882E. *Department of Defense Standard Practice: System Safety*; Standard, Department of Defense: Washington, DC, USA, 2012.
20. Defence Standard 00-56. *Safety Management Requirements for Defence Systems, Part 1, Requirements*; Standard, Ministry of Defence: London, UK, 2007.
21. Chemishkian, S. Building smart services for smart home. In Proceedings of the 2002 IEEE 4th International Workshop on Networked Appliances (Cat. No.02EX525), Gaithersburg, MD, USA, 15–16 January 2002; pp. 215–224. [CrossRef]
22. Nakamura, M.; Tanaka, A.; Igaki, H.; Tamada, H.; Matsumoto, K. Constructing Home Network Systems and Integrated Services Using Legacy Home Appliances and Web Services. *Int. J. Web Serv. Res. (IJWSR)* **2008**, *5*, 82–98. [CrossRef]
23. Yang, Z.; Aoki, T.; Tan, Y. Multiple Conformance to Hybrid Automata for Checking Smart House Temperature Change. In Proceedings of the 2018 IEEE/ACM 22nd International Symposium on Distributed Simulation and Real Time Applications (DS-RT), Madrid, Spain, 15–17 October 2018; pp. 1–10. [CrossRef]
24. Anderson, G.B.; Bell, M.L.; Peng, R.D. Methods to Calculate the Heat Index as an Exposure Metric in Environmental Health Research. *Environ. Health Perspect.* **2013**, *121*, 1111–1119. [CrossRef]
25. ANSI/ASHRAE Standard 55-2010. *Thermal Environmental Conditions for Human Occupancy*; Standard, American Society of Heating, Refrigerating and Air-Conditioning Engineers, Inc.: Atlanta, GA, USA, 2010.
26. Woods, D.D.; Hollnagel, E.; Leveson, N. *Resilience Engineering: Concepts and Precepts*, 1st ed.; CRC Press: Boca Raton, FL, USA, 2006.
27. Jacklitsch, B.; Williams, W.J.; Musolin, K.; Coca, A.; Kim, J.H.; Turner, N. *NIOSH Criteria for a Recommended Standard: Occupational Exposure to Heat and Hot Environments*; Recommended Standard 2016-106; U.S. Department of Health and Human Services, Centers for Disease Control and Prevention, National Institute for Occupational Safety and Health: Cincinnati, OH, USA, 2016.
28. ISO 11079. *Ergonomics of the Thermal Environment—Determination and Interpretation of Cold Stress When Using Required Clothing Insulation (IREQ) and Local Cooling Effects*; Standard, International Organization for Standardization: Geneva, Switzerland, 2007.
29. van Loenhout, J.; le Grand, A.; Duijm, F.; Greven, F.; Vink, N.; Hoek, G.; Zuurbier, M. The effect of high indoor temperatures on self-perceived health of elderly persons. *Environ. Res.* **2016**, *146*, 27–34. [CrossRef] [PubMed]
30. Okamoto-Mizuno, K.; Mizuno, K. Effects of thermal environment on sleep and circadian rhythm. *J. Physiol. Anthropol.* **2012**, *31*, 14. [CrossRef] [PubMed]
31. Leon, L.R.; Bouchama, A. Heat Stroke. In *Comprehensive Physiology*; American Cancer Society: Atlanta, GA, USA, 2015; pp. 611–647, doi:10.1002/cphy.c140017. [CrossRef]
32. Gaudio, F.G.; Grissom, C.K. Cooling Methods in Heat Stroke. *J. Emerg. Med.* **2016**, *50*, 607–616. [CrossRef] [PubMed]

33. Budd, G.M. Wet-bulb globe temperature (WBGT)—Its history and its limitations. *J. Sci. Med. Sport* **2008**, *11*, 20–32. [CrossRef] [PubMed]

34. Salmon, P.; Williamson, A.; Lenné, M.; Mitsopoulos-Rubens, E.; Rudin-Brown, C. Systems-Based Accident Analysis in the Led Outdoor Activity Domain: Application and Evaluation of a Risk Management Framework. *Ergonomics* **2010**, *53*, 927–39. [CrossRef] [PubMed]

35. Svedung, I.; Rasmussen, J. Graphic representation of accident scenarios: Mapping system structure and the causation of accidents. *Saf. Sci.* **2002**, *40*, 397–417. [CrossRef]

36. Rasmussen, J. Risk management in a dynamic society: A modelling problem. *Saf. Sci.* **1997**, *27*, 183–213. [CrossRef]

37. Bjerga, T.; Aven, T.; Zio, E. Uncertainty treatment in risk analysis of complex systems: The cases of STAMP and FRAM. *Reliab. Eng. Syst. Saf.* **2016**, *156*, 203–209. [CrossRef]

38. Belmonte, F.; Schön, W.; Heurley, L.; Capel, R. Interdisciplinary safety analysis of complex socio-technological systems based on the functional resonance accident model: An application to railway traffic supervision. *Reliab. Eng. Syst. Saf.* **2011**, *96*, 237–249. [CrossRef]

39. Tian, J.; Wu, J.; Yang, Q.; Zhao, T. FRAMA: A safety assessment approach based on Functional Resonance Analysis Method. *Saf. Sci.* **2016**, *85*, 41–52. [CrossRef]

40. Patriarca, R.; Gravio, G.D.; Costantino, F. A Monte Carlo evolution of the Functional Resonance Analysis Method (FRAM) to assess performance variability in complex systems. *Saf. Sci.* **2017**, *91*, 49–60. [CrossRef]

41. Salmon, P.M.; Cornelissen, M.; Trotter, M.J. Systems-based accident analysis methods: A comparison of Accimap, HFACS, and STAMP. *Saf. Sci.* **2012**, *50*, 1158–1170. [CrossRef]

42. Filho, A.P.G.; Jun, G.T.; Waterson, P. Four studies, two methods, one accident—An examination of the reliability and validity of Accimap and STAMP for accident analysis. *Saf. Sci.* **2019**, *113*, 310–317. [CrossRef]

43. Yang, L.; Yang, S.H.; Yao, F. Safety and Security of Remote Monitoring and Control of intelligent Home Environments. In Proceedings of the 2006 IEEE International Conference on Systems, Man and Cybernetics, Taipei, Taiwan, 8–11 October 2006; Volume 2, pp. 1149–1153. [CrossRef]

44. Kim, S.C.; Jeong, Y.S.; Park, S.O. RFID-based Indoor Location Tracking to Ensure the Safety of the Elderly in Smart Home Environments. *Pers. Ubiquitous Comput.* **2013**, *17*, 1699–1707. [CrossRef]

45. Lee, K.H.; Seo, C.J. Development of user-friendly intelligent home robot focused on safety and security. In Proceedings of the ICCAS 2010, Gyeonggi-do, Korea, 27–30 October 2010; pp. 389–392. [CrossRef]

46. Yared, R.; Abdulrazak, B.; Tessier, T.; Mabilleau, P. Cooking Risk Analysis to Enhance Safety of Elderly People in Smart Kitchen. In *Proceedings of the 8th ACM International Conference on PErvasive Technologies Related to Assistive Environments*; ACM: New York, NY, USA, 2015; pp. 12:1–12:4. [CrossRef]

47. Hasan, T.; Nayan, M.F.; Iqbal, M.A.; Islam, M. Smart Solar Home System with Safety Device Low Voltage Alert. In Proceedings of the 2012 UKSim 14th International Conference on Computer Modelling and Simulation, Cambridge, UK, 28–30 March 2012; pp. 201–204. [CrossRef]

48. Doan, T.T.; Safavi-Naini, R.; Li, S.; Avizheh, S.; K., M.V.; Fong, P.W.L. Towards a Resilient Smart Home. In Proceedings of the 2018 Workshop on IoT Security and Privacy, Budapest, Hungary, 20–20 August 2018; ACM: New York, NY, USA, 2018; pp. 15–21. [CrossRef]

10

The Impact of Indoor Malodor: Historical Perspective, Modern Challenges, Negative Effects and Approaches for Mitigation

Pamela Dalton [1],*, Anna-Sara Claeson [2] and Steve Horenziak [3]

[1]   Monell Chemical Senses Center, Philadelphia, PA 19104, USA
[2]   Department of Psychology, Umeå University, 90187 Umeå, Sweden; anna-sara.claeson@umu.se
[3]   The Procter & Gamble Company, Cincinnati, OH 45202, USA; horenziak.sa@pg.com
*   Correspondence: dalton@monell.org.

Abstract: Malodors, odors perceived to be unpleasant or offensive, may elicit negative symptoms via the olfactory system's connections to cognitive and behavioral systems at levels below the known thresholds for direct adverse events. Publications on harm caused by indoor malodor are fragmented across disciplines and have not been comprehensively summarized to date. This review examines the potential negative effects of indoor malodor on human behavior, performance and health, including individual factors that may govern such responses and identifies gaps in existing research. Reported findings show that indoor malodor may have negative psychological, physical, social, and economic effects. However, further research is needed to understand whether the adverse effects are elicited via an individual's experience or expectations or through a direct effect on human physiology and well-being. Conversely, mitigating indoor malodor has been reported to have benefits on performance and subjective responses in workers. Eliminating the source of malodor is often not achievable, particularly in low-income communities. Therefore, affordable approaches to mitigate indoor malodor such as air fresheners may hold promise. However, further investigations are needed into the effectiveness of such measures on improving health outcomes such as cognition, mood, and stress levels and their overall impact on indoor air quality.

Keywords: malodor; indoor air; human olfaction; volatile organic comound (VOC); microbial volatile organic compound (MVOC); VOC; MVOC; health effects; smell; malodor mitigation; air fresheners; fragrance

1. Introduction

The sense of smell is a fundamental means of navigating the sensory world and orienting ourselves to ecologically and socially appropriate behavior. Evolutionarily, this chemical sense was the original means by which the earliest organisms achieved adaptive regulation of action and can be considered the origin of behavior [1]. For humans today, volatile molecules can travel for long distances and thus can provide important information about people, places, food and things that cannot otherwise be immediately detected by other sensory systems. Beyond its informational content, odor can attract, intrigue, impress and entice, as well as repel, offend, disgust, or evoke pity. Malodors are odors perceived to be unpleasant or offensive and, while not necessarily occurring at the known thresholds for direct adverse events, may elicit negative symptoms via the olfactory system's connections to other cognitive and behavioral systems.

Malodors are sometimes depicted as an inconvenience or annoyance of relatively minor importance to human perception and experience [1]. An understanding of malodors as a merely "aesthetic" issue, however, ignores their potential for negative impact on human health and social relations [2]. Malodors propagate a variety of psychological, social and economic disturbances, many of which are preventable. As defined at the International Health Conference, "health is a state of complete physical, mental, and social well-being and not merely the absence of disease or infirmity" [3]. Although crafted in 1946, this definition of health has remained in use by organizations such as the World Health Organization. Combating the sources and mitigating the impacts of malodors therefore represents an important public health undertaking.

Throughout history, people have used perfumes, incense, herbs and other means at their disposal to rid their indoor environments of the malodors that occur in the course of human life and industry. However, within a social structure where people cannot always remove the sources of malodors or move themselves to avoid odors, or where odors are the result of industries that sustain food and energy supplies, frequent and intense malodors that are left unchecked can contribute to larger challenges. Malodors can directly affect physical health if the malodorous chemical represents an irritant or harmful airborne substance and occurs at a high enough concentration to exceed observable adverse effect levels. Additionally, malodors may act indirectly, as a mediator of mood, performance and health symptoms, effects which are the focus of this review [4]. Negative effects of low-level chemical exposures (e.g., malodor exposure) are further discussed in Section 5. As the World Sanitation Foundation notes, malodors that result from poor sanitation can compound sanitation issues in under-resourced and developing areas [5]. In rural India, malodors resulting from poor sanitation in pit latrines can indirectly result in open defecation and thus spark a variety of new community-wide health hazards, including compounded malodor issues [6–8].

While eliminating the source of indoor malodor can be a direct mode of intervening in odorous environments, it is often not achievable with the resources at hand. Even in today's urbanizing societies, where malodors now concentrate indoors and in private spaces, people, especially those in low-income communities, may not have the resources to remove the sources of malodors or to relocate their residence. In this review, we examine and discuss the current state of understanding on the role of indoor malodors for impacting human behavior, performance and health, including the individual factors that may govern such responses and identify research priorities to address the data gaps where they exist. Malodors have been reported to have a number of negative psychological, physical, social and economic consequences, as will be discussed herein. Conversely, removal of malodor by increased ventilation or filtering has been reported to increase performance and subjective responses in workers (e.g., ratings of air freshness and air quality), highlighting the potential benefits of mitigating malodors in indoor spaces [9,10]. However, such interventions may not be feasible in many indoor environments. Malodor is an important part of indoor air quality, and accessible and affordable malodor solutions such as air fresheners should be studied to determine if similar benefits are observed.

2. A Historical Perspective

Throughout history and across different parts of the world, malodors have varied in intensity and cultural impact as has the use of perfumes to mitigate malodors. Influenced by the Egyptians, the ancient Romans used perfumes intensely and even applied them to domestic animals to mitigate malodors. In the 4th century, the use of perfumes and pleasant aromas was condemned as an indulgence and idolatry by the Christian church [11]. Partially as a result of this policy, the European cities of the medieval and renaissance ages are known to be among the most foul-smelling environments in human history. Without proper sanitation infrastructures, their close quarters and high population density led to high concentrations of malodors [12]. Rotten food, excrement and slaughtered animal remains frequently littered the streets [11]. To mitigate indoor malodors, Medieval Europeans scattered herbs throughout their homes, sewed aromatic leaves into pillows, or polished wood with myrrh. To perfume themselves, they sprinkled rose water on their clothes or wore pomanders.

Out of this environment emerged the beginnings of modern commercial perfumes. Perfumes with essential oils were made for royalty by Italian chemists in the 14th century. With Caterina de'

Medici's marriage to Henry II in the mid-16th century, Renaissance Italy's perfumes traveled to France where they continued to flourish centuries later. Throughout these periods, fragrances were used to mitigate the negative impact caused by malodors and functioned as a social symbol of higher class [12]. Perspectives on odors have changed significantly since the Renaissance, though people today still seek out means of combating malodors and asserting control over unpleasant smells in their lives, often through the use of pleasantly scented products like air fresheners.

3. Modern Indoor Malodor Challenges Associated with Urbanization

As malodors in public spaces have generally decreased with post-industrialism sanitation improvements, the domestic household has become a prominent site for exposure to malodors. While malodors experienced in historical periods were concentrated in shared areas, contemporary experiences with malodors are frequently experienced in personal spaces [13]. Contemporary building and insulation techniques used in modern homes can allow malodors to concentrate within the household [14].

Household odors are a combination of external odors that enter the home and odors produced within the home. External odors that invade the home include emissions from industry and pollution. Odors produced within the home arise from aggregate effects of low concentrations of volatile organic compounds (VOC) caused by cooking, pet, and human body odors, and the use of personal and household cleaning products, among others. They can also arise from microbial volatile organic compounds (MVOC) formed by the metabolic processes of fungi and bacteria present on building materials [15]. Over time, these VOC will become absorbed by the porous surfaces in homes such as carpets, soft furnishings, curtains, wall paper and even the grout between tiles. The combination of the bouquet of VOC present in households imparts each home with its own unique smell [16].

Exposure to certain mixtures of MVOC and VOC has been shown to increase reports of poor air quality within indoor spaces [17]. However, it should be noted that VOC as a class of compounds are not inherently toxic or malodorous. With respect to establishing toxicity, one must measure the levels and refer to the known threshold for adverse effect for each specific VOC. Many indoor VOC are perceived to have a pleasant smell and can have positive associations, including the wide variety of VOC that are released during such activities as baking bread or cooking. Additionally, there is a certain amount of subjectivity in an individual's response to specific VOC, as one person may report a positive reaction to a certain VOC based on pleasant memories associated with that VOC while others may report it as a malodor.

The perception of VOC also differs with respect to concentration and context. For instance, the substance skatole (3-methylindole) is present in flowers and essential oils and is frequently used in fine fragrances at low concentrations. However, at higher concentrations, it is perceived as having a distinct fecal odor, pointing to the importance of concentration with respect to malodor perception [18]. The context in which a VOC is interpreted is also critical to how it is perceived. Participants who were told that isovaleric acid (a cheesy-smelling fatty acid) was a body odor rated it as far more unpleasant than participants who were told it was a food odor [19]. Finally, genetic variation across the population has resulted in a 'highly personalized inventory of functional olfactory receptors' that not only determine what any individual can smell but how pleasant or unpleasant an odorant is perceived to be [20–22].

In today's urban societies, people can spend nearly 90% of their time in indoor environments [23,24]. As such, there has been investigation into whether the experience of indoor malodors should be regarded as a health issue or a merely aesthetic one [2,25,26]. This line of inquiry has encompassed field studies of industries that emit high quantities of malodorous compounds in residential areas, surveys of workplace productivity in specific chemical environments, psychological laboratory tests of odor exposure and case studies of heightened olfactory sensitivity, in addition to genetic and neurophysiological studies of olfaction and related biological systems. These types of studies have yielded important insights into understanding of the diverse effects of odors on health and social interactions.

4. The Human Perception of Malodors

Perception of a malodor occurs when a molecule activates receptor cells linked to one of several cranial nerves associated with chemoreception. The olfactory nerves of the nasal epithelium are the most significant in odor perception and transmit information from the nasal cavity to the olfactory bulb, which in turn transmits olfactory information to other areas of the brain. In addition, the trigeminal nerve transmits information about pungency from the mouth and eyes as well as the nose. The chorda tympani nerve, glossopharyngeal nerve or vagus nerve may additionally be activated if the compound enters via the mouth [27].

Pungency and odor perception have been determined to be separate chemical senses, as anosmics, who lack the ability to smell, can still sense chemicals through their pungency effects in the nose, mouth and elsewhere [28]. While unpleasant olfactory sensations define malodors, at sufficiently high concentrations these sensations can be further accompanied by unpleasant pungency sensations.

The chemical senses of odor and pungency perception vary in several significant ways. For one, the threshold detection for pungency is generally several orders of magnitude higher in concentration than what is required to perceive the odor; people most often perceive a smell before it becomes so strong as to sting their eyes [29]. Though people may adapt to a constant odor in a matter of minutes or hours, adaptation to the perception of pungency occurs over longer periods [30].

Detection thresholds for malodors vary dramatically depending on the specific chemical in question, with thresholds generally declining with the carbon chain length of the compound [28,29]. Humans can detect common indoor malodors like hexyl acetate at concentrations as low as 2.9 parts per billion [30]. Malodors from sulfur compounds like isoamyl mercaptan can be detected at concentrations as low as 0.77 parts per trillion (ppt) [31], and MVOC can be detected as low as 0.2 ppt (i.e., from 2-Isopropyl-3-methoxy-pyrazine) [15]. From an evolutionary point of view, this ability to detect extremely low concentrations of chemical compounds in the air affords identification of various sources of danger, such as spoiled food or harmful chemicals.

Like all senses, olfaction is a product of biological evolution whose features are linked to survival and adaptation [32]. To this end, major connections have been identified between the olfactory system and cognitive processes, such as associative learning [33] and emotional memory [34–36], as well as "fight or flight" response [37]. "Top-down" cognitive functions, such as risk and danger perception, can also influence "bottom-up" information from the odor stimulus by allocating greater attentional resources to malodors, thus increasing their negative impact [38].

While the perception of malodors across individuals follows the same physiological pathways, the intensity of and response to the perception can vary. Although there can be differences across individuals in their sensitivity to specific malodors, the more important drivers of an individual's hedonic response may be due to their expectations, past experiences and the context in which an odor is experienced. For instance, the smell of smoke around a campfire can evoke a positive scent experience. However, the smell of smoke within a home will evoke an entirely different response, as the smoke is a signal of danger in this context.

5. The Negative Effects of Malodors

Utilizing the search term "malodor harm" or variations thereof (e.g., synonyms of "malodor" and "harm") via publicly available databases such as ScienceDirect (https://www.sciencedirect.com/) revealed scientific research across an array of disciplines documenting various negative effects associated with malodors. These negative effects were observed to cluster into six categories as identified by the authors of this review (Figure 1).

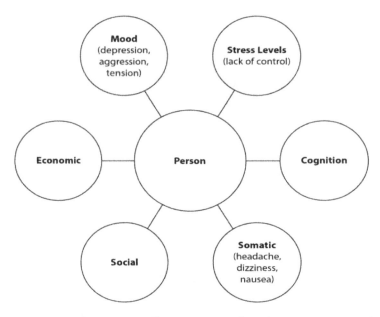

Figure 1. Reported negative effects associated with exposure to malodors.

Studies evaluating the negative effects of malodors have been conducted in both the field (observational) and the laboratory (experimental) employing a variety of dependent measures (Table 1).

Table 1. Measurement of malodor effects conducted in the field via observations and in the laboratory (Lab) via experimental methods.

Effect	Assessment Approaches Utilized	References
Mood	Profile of mood states (POMS) Mood scales Motivation on tasks	Field [39–42] Lab [33,43]
Stress levels	Heart rate Blood Pressure Salivary cortisol, alpha-amylase Anxiety/stress scales	Field [26,29,41]
Cognition	Cognitive tasks—simple and complex Creative problem solving	Lab [43–45]
Somatic	Symptom reports Pulmonary function Airway inflammation	Field [46–50] Lab [51]
Social	Self-reported behaviors/evaluations Pro-social behaviors (helping, friendliness)	Field [52,53] Lab [53–55]
Economic	Property valuation (homes, cars) Consumer choice behaviors (hotels, B&Bs)	Field [56–59]

Several studies have reported the negative effect of malodors on mood. Self-reported feelings of depression [39,40], fatigue [39,40], confusion [39,41], aggression [40,60,61], and tension [39,40] have all been positively correlated with malodor exposure, whereas subjective well-being [40] has been negatively correlated with such exposures. Even when no malodor is present, expectations of malodor exposure may cause negative effects on mood [43].

Malodors may cause individuals to feel a lack of control over their environment, adversely affecting stress levels. Indoor household malodors of external origin that are consistent and uncontrollable may produce feelings of helplessness [39,62]. Perceived control has also been shown to affect tolerance of a given malodor [44]. Individual coping style, however, may also affect odor annoyance and symptom prevalence. Studies have suggested that those who have "palliative" or avoidance coping styles generally report less annoyance and symptoms than those with "instrumental" or problem-oriented coping styles [52]. Stress about the perceived toxicological effects of malodors may further allow odors to act as a trigger for other symptoms and behaviors [48–50].

Malodors have been shown to have detrimental effects on cognition. Rotton [44] has shown that exposure to malodor does not affect simple cognitive tasks, but that it has a detrimental effect on more complex tasks, such as proofreading. Cognitive deficits resulting from malodor exposure may be due to their negative effect on focus [45].

Malodors have been shown to elicit somatic symptoms. Somatic symptoms that have been reported with malodor exposure include vomiting, nausea, dizziness, headache, loss of appetite, sleep disorders and irritation of eyes, throat and nose [2,63]. Malodors can also cause somatic symptoms via "odor-worry" and stress [46]. Asthmatics, for example, may experience exacerbation of symptoms from non-irritating odors that are perceived as harmful [51]. Others may experience the stress effects of malodors because of "environmental worry" [46], an association with the odor as socially taboo or by perceiving possible property devaluation resulting from the odor [39]. Exposure to certain malodors has also been shown to affect the immune system, an effect probably mediated by perceived stress [47].

Social relations are also threatened by indoor malodors. Habituation to the odors of one's background can make people acutely aware of intrusive malodors, which may cause a variety of problems in social settings. Subjective ratings of odor unpleasantness have been shown to correlate with perception of socially undesirable traits [54]. Odor perception may also integrate with higher-order visual processes, such as facial perception. Unpleasant facial expressions paired with malodors have been shown to increase people's ratings of odor intensity and decrease their respiratory amplitude [53]. Judgments of interpersonal attraction are also influenced by the presence of malodor [64]. Indoor malodors can also reduce social interactions by causing inhabitants to experience shame or embarrassment about the malodor, even when they are external in origin [55]. It is reasonable to assume that this occurs with household odors as well.

Malodors can have economic effects. Unlike other "less visible" forms of pollution, malodors are readily identified and capitalized into property values [56]. Some industries, such as tanneries, paint factories, pulp mills and livestock operations, are regulated by legislated minimum "setback distances" that facilities must maintain from surrounding properties. Setbacks are used to minimize the economic effects of pollution, including malodors. Not only do properties surrounding these facilities decrease in value, but if the facilities' setback distance from surrounding properties is either unenforced or inaccurately determined [4,57], then malodors emitted from the facility can result in net economic loss, despite efficiency gains made by the offending firm [58]. Business and home owners alike can also suffer economic consequences. Malodors can affect car sales [59], worker productivity at call centers [9] and consumer satisfaction within the hospitality industry [65].

It is important to note that these negative effects are not necessarily independent measures and that individual effects can often act to compound economic ones. Levy and Yagil [66], for example, suggest that low Air Quality Index scores within stock exchanges may affect mood and risk aversion, thus resulting in lower stock returns. Fist, Black and Brunner [10] note that improvements to indoor environmental quality has a strong effect on workplace productivity and health, estimating a potential annual gain of $20 billion from improvements in office buildings in the United States.

6. Contemporary Approaches and Benefits of Mitigating Malodors

Efforts to protect the public from the adverse effects of outdoor malodors take the form of regulations in many jurisdictions. Regulations include concentration or exposure limits if the odors are produced by specific target pollutants, nuisance or annoyance laws and property setbacks. The odor impact criteria established by such regulations are most commonly based on field olfactometry-based concentration measurements, although instrumental concentration measurements or air dispersion models are also used [67].

In order to comply with odor regulations and to promote good relations with neighboring households, some facilities may install technology to control odors at the source. Control strategies include oxidation, adsorption, chemical reaction, chemical scrubbing, biofiltration/bioremediation and other methods [68].

To help households control indoor odors, product manufacturers often employ some of the same technical strategies used industrially to control odors. These include air filters and filter media, oxidizers, absorbents/adsorbents, surface and air sprays, and a variety of volatile ingredient diffusers. Air filtration to remove odors may be achieved with filters attached to heating, ventilation and air conditioning (HVAC) units or stand-alone filtering units. Such filters may utilize activated carbon or zeolite adsorbents, photocatalytic oxidants such as metal oxides (for example, patents US 8911670, US 8038935), or odor-reactive chemistry such as metallic salts or amine polymers (for example, patent US 4892719).

Household consumer goods products designed to eliminate household and automobile odors include air fresheners, pump and spray aerosols and diffusers. Such products may contain technologies designed to: capture or alter the molecular structure of VOC responsible for the underlying malodor; prevent perception of the malodorous VOC (MOVOC) by the olfactory system; and/or mask MOVOC via fragranced ingredients. Technologies used in air freshening sprays and diffusers that are designed to capture or alter specific types of malodor molecules are summarized in (Table 2). Spray products may utilize one or a variety of such technologies to eliminate the molecular source(s) and/or perception of MOVOC.

Table 2. Malodor classification and patented technologies that can be used in air fresheners to mitigate common indoor malodors.

Malodor	Common Molecular Components	Eliminated By	Reference
Smoke (Tobacco)	Cyclic compounds (e.g., methyl pyrole, pyridine)	Volatility reduction via complexation with cyclodextrin	[69–71]
Greasy Cooking Odors	Aldehydes (octanal, nonanal)	Capture via reaction with polyamine polymer	[72]
Body Odor	Acidic compounds (isovaleric acid); Thiols (methanthiol)	Salt formation by pH neutralization	[73]
Kitchen Odors	Amines (butylamine, trimethylamine)	Salt formation by pH neutralization; Reaction with carbonyl compounds	[74]
	Sulfur compounds (dipropyl sulfide)	Reaction with carbonyl compounds	[75]
	Amines; Fatty acids	Salt formation by pH neutralization; Reaction with carbonyl compounds	[76]
	2-Penethylfuran, Thiazoles, and Thiols (2-ethyl-1-hexanethiol)	Volatility reduction via complexation with cyclodextrin	[77]
	Sulfites; Amines (trimethylamine); Acid Compounds (acetic acid)	Salt formation by pH neutralization; Reaction with carbonyl compounds	[76]
	Fatty Acids; Amines; Thiols	Salt formation by pH neutralization; Reaction with carbonyl compounds	[78]
Bathroom Odors	Skatole; Morpholines; Acids (thioglycolic acid)	Olfactory receptor antagonists; Reaction with carbonyl compounds	[79]
	Bacterial VOC	Reaction with carbonyl compounds; Antimicrobial agents	[80]
Mold & Mildew	Fungal VOC	Reaction with carbonyl compounds + Antimicrobial	[80]
Pet Odor	Acidic Compounds; Sulfur Compounds; Amines	Salt formation by pH neutralization; Reaction with carbonyl compounds; Olfactory receptor antagonists	[74,79]

One such technology approach to mitigate indoor malodors is the use of cyclodextrin (CD) or cyclodextrin derivatives to trap MOVOC by complexation. Cyclodextrin has a macro-ring structure consisting of glucopyranose units. It is produced naturally by bacteria including *Bacillus macerana* and *Bacillus circulans*, and is made industrially in bioreactors utilizing engineered glucosyl transferase enzymes [81]. The cavity of the cyclodextrin ring is apolar, so that less polar MOVOC readily displace water and become "trapped" upon interaction with aqueous cyclodextrin. Once complexed, the volatility of the MOVOC is significantly reduced and the malodors remain trapped in the CD cavity as long as the complex stays dry [69,82]. Patent documents indicate that consumer product companies make use of cyclodextrin technology in spray air freshening and other consumer products (for example, patents US 5760475, US 6077318, US 6248135 and US 6451065).

Spray air freshening products may also utilize pH buffers to neutralize acid or basic odors and convert them to non-volatile salts. Acidic odors include short chain fatty acids such as isovaleric acid, heptanoic acid and basic odors include amines such as ammonia, butyl amine, and trimethyl amine. Both types of neutralizable MOVOC are constituents of household odors such as food waste odors, and human odors [16]. Buffer systems used in spray air freshening products to neutralize odors may include for example, citrate or carbonate buffers. Neutralization of acid and amine odors to pH in the range of 5–8 by pH buffering converts these MOVOC to non-volatile salts, reducing or eliminating the odor [83].

Enzyme inhibitors may be used in consumer products, including air freshening products, to prevent production of odorous metabolites (patent US 9200269). For example, urease inhibitors and β-glucuronidase inhibitors have been described to prevent the formation MOVOC from urine by microorganisms on fibrous consumer products [84].

Unsaturated aldehydes are MOVOC components of household odors that are derived from the oxidation of skin oils or from oxidation of lipids during cooking [16,85,86]. Amine-functional polymers are known to bind with and capture aldehydes (including formaldehyde) through the formation of imine bonds [87,88], and have been used in air freshening products to bind with odors (patent US 9273427).

Additional technologies used in air freshening sprays include anti-microbial agents such as quaternary ammonium compounds to eliminate odor-causing microbes, which use salts of transition metals, particularly zinc and copper to complex with odors (for example, patents US 5783544, US 6503413), and oxidizing agents such as chloramine (patent US 6743420) that eliminate MOVOC through both antimicrobial and oxidative mechanisms.

Diffusion-type products, like heated or unheated fragrance diffusers, may typically contain reactive materials such as carbonyl compounds designed to react with nucleophilic or electrophilic malodorous molecules such as amines (for example, patents US 8992889, US 5795566, US 7998403) to form covalently bonded, non-odorous products.

Technologies designed to prevent the olfactory perception of malodors based on mechanisms such as olfactory receptor antagonism have also been explored by consumer product companies (for example, patents EP 2812316, US 9526680, US 9254248). Such approaches target specific olfactory receptors known to be activated by malodorous agonists with antagonistic agents to block activation of these receptors by the malodor.

Consumer products designed to control malodor often, but not always, contain fragrance in addition to the technologies described above, or may contain fragrance without additional technology. Fragrance can mitigate malodors by masking their smell. The mechanism by which fragrances mask malodors is not well understood and may be achieved through a combination of signal interference (such as by receptor antagonism as discussed above) and through top-down processing effects (e.g., blending malodor with other odors to create perception of a new, non-malodorous aroma). Additionally, as noted above, some fragrances may contain reactive materials, such as carbonyl compounds designed to react with nucleophilic or electrophilic malodorous molecules. Pleasant fragrances have also been shown to have beneficial effects by increasing positive emotions, decreasing negative mood states and reducing indices of stress [89,90]. It is postulated that fragrances exert these effects through emotional learning, conscious perception and belief/expectation [91].

While seeking solutions to mitigate malodors, consumers and regulators must balance the economic, environmental and health costs of indoor malodors against the benefits delivered by the odor mitigating approaches. Companies that manufacture odor control technologies that emit fragrance or odor mitigating molecules into the air follow safety assessment paradigms that are widely recognized to ensure consumer safety when used according to label instructions. These assessments are aligned with the process outlined by EU Scientific Committee on Consumer Safety and are based on an understanding of both the inherent hazards of any materials in a product formulation as well as the level of exposure to those materials based on usage scenarios including extreme consumer usage [76,92]. In addition, the Research Institute for Fragrance Materials (RIFM) has published extensive industry guidance for conducting safety assessments of fragrance ingredients [93]. A 2007 US Environmental Protection Agency (EPA) review found that 0.23% of reported air freshener exposures involved an adverse reaction and that the number of reported exposure incidents for air fresheners was relatively small when compared to the reported exposure incidents for other product categories [94].

Household consumer products designed to eliminate odors are widely used by consumers in the United States, with 75.9% of US households purchasing an air care product as of March 2019, according to Nielsen HomeScan panel data [95]. Air care products are broadly available at retail outlets at relatively low cost. Buying rates of air care products are highest in households with annual incomes less than $20,000 [96]. This may be due in part because lower-income households are disproportionately affected by environmental odors, odors arising from crowded conditions, and by economic limitations on their ability to deal with odor sources, such as those associated with sub-standard housing.

Despite the potential negative effects of malodors and the widespread use of consumer products designed to eliminate odors, the health and quality-of-life benefits of the use of such consumer products has not been widely studied. A review of published literature on the health impacts of using air cleaning devices was recently completed by Kelly and Fussell [97]. The studies reviewed focused mostly on indoor air cleaning devices that reduced particle and VOC concentrations by using filters, adsorbents, oxidative technologies or combinations thereof. The studies generally showed no or low levels of improvement in the health outcomes measured for households with good ambient air quality and modest improvements for households with very poor ambient air quality. However, none of these studies, and few other published studies, have specifically examined the impact of indoor malodor reduction on health outcomes such as cognition, mood, and stress levels, among others.

It may be inferred that eliminating the perception of malodor can reduce psychological effects of malodors, such as the feeling of a lack of control [41,44,47]. While studies have shown that people with problem-oriented coping strategies experience more stress and stress-related symptoms due to malodor exposure [52], air care and cleaning technologies offer a solution that allows people with problem-oriented coping styles to directly address the problems caused by malodors. Air care products designed to eliminate malodors can provide a more widely affordable solution compared to more costly alternatives such as home filtration systems, especially for low-income households who are economically unable to purchase such systems, replace malodorous household structures or items, or relocate away from substandard housing or industrial sources of malodor.

7. Conclusions

While there are complex issues at play in the distribution and effects of malodors (e.g., pollution concentrated near low-income communities, lack of access to proper sanitation), malodors are a general fact of daily human life. Indoor malodors are particularly challenging for people in developing countries or in low-income communities [2], which may lack the financial resources or opportunities to directly change the sources or living situations that harbor malodors. In these scenarios, malodors ultimately contribute to broad issues of structural inequality [41,98].

Viewing malodors as a merely "aesthetic" issue ignores their potential for negative impact on human health, previously defined as a "state of complete physical, mental and social well-being and not merely the absence of disease or infirmity" [3]. Review of the current literature includes several studies from a diverse range of disciplines reporting negative psychological, physical, social and economic consequences of indoor malodor. Conversely, removal of malodor has been reported to increase performance and subjective responses in workers [10], highlighting the potential benefits of mitigating malodors in indoor spaces. However, there are several gaps in the current research. Specifically, there is a lack of understanding regarding the mechanisms by which malodors can elicit any adverse effects, whether through an individual's experience or expectations that provide the interpretative context in which a VOC is experienced or through a more direct effect on human physiology and well-being. Thus, well-controlled studies examining the emotional, behavioral and performance-related outcomes induced by exposure to malodors are needed, as are studies that include a formal examination of the individual variables (i.e., personality, gender, age, culture) that may influence the magnitude and direction of malodor effects.

Eliminating the source of malodor can be a direct mode of intervening in odorous indoor environments, though it is often not achievable with the resources at hand, particularly in low-income communities. Therefore, easily accessible and affordable approaches to eliminate malodors such as air fresheners with odor eliminating technologies (Table 2) may hold promise for reducing some of the negative effects of indoor malodor. However, we found relatively few investigations into the effectiveness of such measures on improving health outcomes such as cognition, mood, and stress levels, among others. Therefore, further study is recommended on the impact of air fresheners and other odor mitigating products on health outcomes via malodor elimination and/or emission of pleasant fragrances, as well as their impact on measures of overall indoor air quality.

Author Contributions: The authors (P.D., A.-S.C., S.H.) contributed equally to the work by providing substantial contributions to the conception and design of the work and the acquisition, analysis, and interpretation of data; drafting the work and revising it critically for important intellectual content; and agreement to be accountable for all aspects of the work in ensuring that questions related to the accuracy or integrity of any part of the work are appropriately investigated and resolved. All authors have read and agreed to the published version of the manuscript.

Acknowledgments: The authors would like to thank Eric Moorhead of Spectrum Science Communications and Mary Begovic Johnson of The Procter & Gamble Company for their helpful reviews and comments during the writing of this manuscript.

Conflicts of Interest: Pamela Dalton is a consultant/grantee or speaker for the following companies: American Chemistry Council, Ajinomoto Co., Inc., Altria Group, Campbell Soup Company, Church & Dwight, The Coca-Cola Company, Diageo, plc, Diana Ingredients, Estee Lauder Inc., Firmenich Incorporated, Fragrance Creators Association, Givaudan SA, GlaxoSmithKline, Intelligent Sensor Technology, Inc., Japan Tobacco Inc., Johnson & Johnson Consumer Products, Kao Corporation, Kellogg, Kerry, Mars, McCormick & Company, Inc., Mead Johnson Nutritionals, Mondel ez International, PepsiCo, Inc., Pfizer, Inc., Procter & Gamble, Reckitt Benckiser Group, Roquette, Royal DSM, Sensonics International, Suntory Holdings Ltd., Symrise, Takasago International Corporation, Tate & Lyle, Unilever Research & Development, Wm. Wrigley Jr. Company, Young Living Essential Oils and Zensho Holdings Co. Ltd. Dr. Dalton received an honorarium from the HCPA for the preparation of this manuscript. Anna-Sara Claeson has no conflicts of interest. Dr. Claeson received an honorarium from the HCPA for the preparation of this manuscript. Steve Horenziak is an employee of The Procter & Gamble Company. The funder (HCPA) had a role in the design of the review, the compilation of published work on the subject and the decision to publish the results. The funder had no role in the analyses or interpretation of the reported data. The authors had full editorial control over the content.

References

1.   Semin, G.R.; de Groot, J.H.B. The chemical bases of human sociality. *Trends Cogn. Sci.* **2013**, *17*, 427–429. [CrossRef] [PubMed]

2.   Wing, S.; Horton, R.A.; Marshall, S.W.; Thu, K.; Tajik, M.; Schinasi, L.; Schiffman, S.S. Air pollution and odor in communities near industrial swine operations. *Environ. Health Perspect.* **2008**, *116*, 1362–1368. [CrossRef] [PubMed]

3.   Card, A.J. Moving beyond the WHO definition of health: A new perspective for an aging world and the emerging era of value-based care: Redefining health. *World Med. Health Policy* **2017**, *91*, 127–137. [CrossRef]

4. Hooiveld, M.; van Dijk, C.; van der Sman-de Beer, F.; Smit, L.A.; Vogelaar, M.; Wouters, I.M.; Heederik, D.J.; Yzermans, C.J. Odour annoyance in the neighbourhood of livestock farming—Perceived health and health care seeking behaviour. *Ann. Agric. Environ. Med.* **2015**, *22*, 55–61. [CrossRef]

5. Gates, B. A Perfume that Smells Like Poop? Gates Notes 2016. Available online: https://www.gatesnotes. com/Development/Smells-of-Success (accessed on 26 November 2019).

6. Nakagiri, A.; Niwagaba, C.B.; Nyenje, P.M.; Kulabako, R.N.; Tumuhairwe, J.B.; Kansiime, F. Are pit latrines in urban areas of Sub-Saharan Africa performing? A Review of usage, filling, insects and odour nuisances. *BMC Public Health* **2016**, *16*, 120. [CrossRef]

7. Tobias, R.; O'Keefe, M.; Künzle, R.; Gebauer, H.; Gründl, H.; Morgenroth, E.; Pronk, W.; Larsen, T.A. Early testing of new sanitation technology for urban slums: The case of the Blue Diversion Toilet. *Sci. Total Environ.* **2017**, *576*, 264–272. [CrossRef]

8. Seleman, A.; Bhat, M.G. Multi-criteria assessment of sanitation technologies in rural Tanzania: Implications for program implementation, health and socio-economic improvements. *Technol. Soc.* **2016**, *46*, 70–79. [CrossRef]

9. Wargocki, P.; Wyon, D.P.; Fanger, P.O. The performance and subjective responses of call-center operators with new and used supply air filters at two outdoor air supply rates. *Indoor Air* **2004**, *14*, 7–16. [CrossRef]

10. Fist, W.J.; Black, D.; Brunner, G. Benefits and costs of improved IEQ in U.S. offices. *Indoor Air* **2011**, *21*, 357–367.

11. Classen, C.; Howes, D.; Synnott, A. Following the scent from the Middle Ages to modernity. In *Aroma: The Cultural History of Smell*; Routledge: London, UK, 1994; pp. 51–55.

12. Diggs, B. Making Good scents: Fragrance in the Middle Ages and Renaissance. *Renaiss. Mag.* **2009**, *65*, 36–40.

13. MacPhee, M. Deodorized culture: Anthropology of smell in America. *Ariz. Anthropol.* **1992**, *8*, 89–102.

14. Howieson, S.G.; Sharpe, T.; Farren, P. Building tight- ventilating right? How are new air tightness standards affecting indoor air quality in dwellings? *Build Serv. Eng. Res. Technol.* **2013**, *35*, 475–487. [CrossRef]

15. Korpi, A.; Järnberg, J.; Pasanen, A.-L. Microbial volatile organic compounds. *Crit. Rev. Toxicol.* **2009**, *39*, 139–193. [CrossRef] [PubMed]

16. Hammond, C.J. Chemical composition of household odors: An overview. *Flavour Frag. J.* **2013**, *28*, 251–261. [CrossRef]

17. Claeson, A.-S.; Nordin, S.; Sunesson, A.-L. Effects on perceived air quality and symptoms of exposure to microbially produced metabolites and compounds emitted from damp building materials. *Indoor Air* **2009**, *19*, 102–112. [CrossRef] [PubMed]

18. Barden, T.C. Indoles: Industrial, agricultural and over-the-counter uses. In *Heterocyclic Scaffolds II: Topics in Heterocyclic Chemistry*; Gribble, G., Ed.; Springer: Berlin/Heidelberg, Germany, 2010; Volume 26, pp. 31–46.

19. de Araujo, I.E.; Rolls, E.T.; Velazco, M.I.; Margo, C.; Cayeux, I. Cognitive modulation of olfactory processing. *Neuron* **2005**, *46*, 671–679. [CrossRef]

20. Olender, T.; Waszak, S.M.; Viavant, M.; Khen, M.; Ben-Asher, E.; Reyes, A.; Nativ, N.; Wysocki, C.J.; Ge, D.; Lancet, D. Personal receptor repertoires: Olfaction as a model. *BMC Genom.* **2012**, *13*, 414. [CrossRef]

21. Trimmer, C.; Keller, A.; Murphy, N.R.; Snyder, L.L.; Willer, J.R.; Nagoi, M.H.; Katsanism, N.; Vosshall, L.B.; Matsunami, H.; Mainland, J.D. Genetic variation across the human olfactory receptor repertoire alters odor perception. *Proc. Natl. Acad. Sci. USA* **2019**, *116*, 9475–9480. [CrossRef]

22. Keller, A. Different noses for different mice and men. *BMC Biol.* **2012**, *10*, 75. [CrossRef]

23. U.S. Environmental Protection Agency. *Report to Congress on Indoor Air Quality: Volume II—Assessment and Control of Indoor Air Pollution*; EPA/400/1-89/001C; U.S. Environmental Protection Agency: Washington, DC, USA, 1989. Available online: https://nepis.epa.gov/Exe/ZyNET.exe/9100LMBU.TXT? ZyActionD=ZyDocument&Client=EPA&Index=1986+Thru+1990&Docs=&Query=&Time=&EndTime= &SearchMethod=1&TocRestrict=n&Toc=&TocEntry=&QField=&QFieldYear=&QFieldMonth= &QFieldDay=&IntQFieldOp=0&ExtQFieldOp=0&XmlQuery=&File=D%3A%5Czyfiles%5CIndex% 20Data%5C86thru90%5CTxt%5C00000022%5C9100LMBU.txt&User=ANONYMOUS&Password= anonymous&SortMethod=h%7C-&MaximumDocuments=1&FuzzyDegree=0&ImageQuality=r75g8/ r75g8/x150y150g16/i425&Display=hpfr&DefSeekPage=x&SearchBack=ZyActionL&Back=ZyActionS& BackDesc=Results%20page&MaximumPages=1&ZyEntry=1&SeekPage=x&ZyPURL (accessed on 3 December 2019).

24. Klepeis, N.E.; Nelson, W.C.; Ott, W.R.; Robinson, J.P.; Tsang, A.M.; Switzer, P.; Behar, J.V.; Hern, S.C.; Engelmann, W.H. The National Human Activity Pattern Survey (NHAPS): A resource for assessing exposure to environmental pollutants. *J. Expo. Anal. Environ. Epidemiol.* **2001**, *11*, 231–252. [CrossRef]

25. McGinley, M.A.; McGinley, C.M. The "gray line" between odor nuisance and health effects. In Proceedings of the Air and Waste Management Association 92nd Annual Meeting, St. Louis, MO, USA, 20–24 June 1999.

26. Shusterman, D. Critical Review: The health significance of environmental odor pollution. *Arch. Environ. Health* **1992**, *47*, 76–87. [CrossRef] [PubMed]

27. Silver, W.L. Neural and Pharmacological Basis for Nasal Irritation. *Ann. N. Y. Acad. Sci.* **1992**, *64*, 152–163. [CrossRef] [PubMed]

28. Cometto-Muñiz, J.E.; Cain, W.S. Sensory irritation: Relation to indoor air pollution. *Ann. N. Y. Acad. Sci.* **1992**, *641*, 137–151. [CrossRef] [PubMed]

29. Schiffman, S.S. Livestock odors: Implications for human health and well-being. *J. Anim. Sci.* **1998**, *76*, 1343–1355. [CrossRef] [PubMed]

30. Cometto-Muñiz, J.E.; Cain, W.S.; Abrahamm, M.H.; Gil-Lostes, J. Concentration-detection functions for the odor of homologous n-acetate esters. *Physiol. Behav.* **2008**, *95*, 658–667. [CrossRef]

31. Nagata, Y. Measurement of odor threshold by triangle odor bag method. *Bull. Jpn. Environ. Sanit Cent* **1990**, *17*, 77–89.

32. Stevenson, R.J. An initial evaluation of the functions of human olfaction. *Chem. Senses* **2010**, *35*, 3–20. [CrossRef]

33. Herz, R.S.; Schankler, C.; Beland, S. Olfaction, emotion and associative learning: Effects on motivated behavior. *Motiv. Emot.* **2004**, *28*, 363–383. [CrossRef]

34. Kay, L.M.; Freeman, W.J. Bidirectional processing in the olfactory-limbic axis during olfactory behavior. *Behav. Neurosci.* **1998**, *112*, 541–553. [CrossRef]

35. Herz, R.S.; Eliassen, J.; Beland, S.; Souza, T. Neuroimaging evidence for the emotional potency of odor-evoked memory. *Neuropsychologia* **2004**, *42*, 371–378. [CrossRef]

36. Willander, J.; Larsson, M. Smell your way back to childhood: Autobiographical odor memory. *Psychon. Bull. Rev.* **2006**, *13*, 240–244. [CrossRef] [PubMed]

37. Wisman, A.; Shrira, I. The smell of death: Evidence that putrescine elicits threat management mechanisms. *Front. Psychol.* **2015**, *6*, 1274. [CrossRef] [PubMed]

38. Dalton, P. Odor perception and beliefs about risk. *Chem. Senses* **1996**, *4*, 447–458. [CrossRef] [PubMed]

39. Schiffman, S.S.; Sattely Miller, E.A.; Suggs, M.S.; Graham, B.G. The effect of environmental odors emanating from commercial swine operations on the mood of nearby residents. *Brain Res. Bull.* **1995**, *37*, 369–375. [CrossRef]

40. Eltarkawe, M.; Miller, S. The impact of industrial odors on the subjective well-being of communities in Colorado. *Int. J. Environ. Res. Public Health* **2018**, *15*, 1091. [CrossRef]

41. Horton, R.A.; Wing, S.; Marshall, S.W.; Brownley, K.A. Malodor as a trigger of stress and negative mood in neighbors of industrial hog operations. *Am. J. Public Health* **2009**, *99*, S610–S615. [CrossRef]

42. Cavalini, P.M.; Koeter-Kemmerling, L.G.; Pulles, M.P.J. Coping with odour annoyance and odour concentrations: Three field studies. *J. Environ. Psychol.* **1991**, *11*, 123–142. [CrossRef]

43. Knasko, S.C. Ambient odor's effect on creativity, mood, and perceived health. *Chem. Senses* **1992**, *17*, 27–35. [CrossRef]

44. Rotton, J. Affective and cognitive consequences of malodorous pollution. *Basic Appl. Soc. Psychol.* **1983**, *4*, 171–191. [CrossRef]

45. Nordin, S.; Aldrin, L.; Claeson, A.-S.; Andersson, L. Effects of negative affectivity and odor valence on chemosensory and symptom perception and perceived ability to focus on a cognitive task. *Perception* **2017**, *46*, 431–446. [CrossRef]

46. Shusterman, D.; Lipscomb, J.; Neutra, R.; Satin, K. Symptom Prevalence and odor-worry interaction near hazardous waste sites. *Environ. Health Perspect.* **1991**, *94*, 25–30.

47. Avery, R.C.; Wing, S.; Marshall, S.W.; Schiffman, S.S. Odor from industrial hog farming operations and mucosal immune function in neighbors. *Arch. Environ. Health* **2004**, *59*, 101–108. [CrossRef] [PubMed]

48. Shusterman, D. Odor-associated health complaints: Competing explanatory models. *Chem. Senses* **2001**, *26*, 339–343. [CrossRef]

49. Claeson, A.-S.; Lidén, E.; Nordin, M.; Nordin, S. The role of perceived pollution and health risk perception in annoyance and health symptoms: A population-based study of odorous air pollution. *Int. Arch. Occup. Environ. Health* **2013**, *86*, 367–374. [CrossRef] [PubMed]

50. Tjalvin, G.; Lygre, S.H.L.; Hollund, B.E.; Moen, B.E.; Bråtveit, M. Health complaints after a malodorous chemical explosion: A longitudinal study. *Occup. Med.* **2015**, *65*, 202–209. [CrossRef] [PubMed]

51. Jaén, C.; Dalton, P. Asthma and odors: The role of risk perception in asthma exacerbation. *J. Psychosom. Res.* **2014**, *77*, 302–308. [CrossRef]

52. Steinheider, B.; Winneke, G. Industrial odours as environmental stressors: Exposure-annoyance associations and their modification by coping, age and perceived health. *J. Environ. Psychol.* **1993**, *13*, 353–363. [CrossRef]

53. Cook, S.; Kokmotou, K.; Soto, V.; Fallon, N.; Tyson-Carr, J.; Thomas, A.; Giesbrecht, T.; Field, M.; Stancak, A. Pleasant and unpleasant odour-face combinations influence face and odour perception: An event-related potential study. *Behav. Brain Res.* **2017**, *333*, 304–313. [CrossRef]

54. McBurney, D.H.; Levine, J.M.; Cavanaugh, P.H. Psychophysical and social ratings of human body odor. *Personal. Soc. Psychol. Bull.* **1976**, *3*, 135–138. [CrossRef]

55. Tajik, M.; Muhammad, N.; Lowman, A.; Thu, K.; Wing, S.; Grant, G. Impact of odor from industrial hog operations on daily living activities. *New Solut.* **2008**, *18*, 193–205. [CrossRef]

56. Anstine, J. Property Values in a Low Populated Area when Dual Noxious Facilities are Present. *Growth Chang.* **2003**, *34*, 345–358. [CrossRef]

57. Cameron, T.A. Directional heterogeneity in distance profiles in hedonic property value models. *J. Environ. Econ. Manag.* **2006**, *51*, 26–45. [CrossRef]

58. Bazen, E.F.; Fleming, R.A. An economic evaluation of livestock odor regulation distances. *J. Environ. Qual.* **2004**, *33*, 1997–2006. [CrossRef] [PubMed]

59. Matt, G.E.; Romero, R.; Ma, D.S.; Quintana, P.J.; Hovell, M.F.; Donohue, M.; Messer, K.; Salem, S.; Aguilar, M.; Boland, J.; et al. Tobacco use and asking prices of used cars: Prevalence, costs, and new opportunities for changing smoking behavior. *Tob. Induc. Dis.* **2008**, *4*, 2–10. [CrossRef]

60. Jones, J.W.; Bogat, G.A. Air pollution and human aggression. *Psychol. Rep.* **1978**, *43*, 721–722. [CrossRef]

61. Rotton, J.; Frey, J.; Barry, T.; Milligan, M.; Fitzpatrick, M. The air pollution experience and physical aggression. *J. Appl. Soc. Psychol.* **1979**, *9*, 397–412. [CrossRef]

62. Evans, G.W.; Jacobs, S.V. Air pollution and human behavior. *J. Soc. Issues* **1981**, *37*, 95–125. [CrossRef]

63. Steinheider, B. Environmental odours and somatic complaints. *Zent. Hyg. Umweltmed.* **1999**, *202*, 101–119. [CrossRef]

64. Rotton, J.; Barry, T.; Frey, J.; Soler, E. Air pollution and interpersonal attraction. *J. Appl. Soc. Psychol.* **1978**, *8*, 57–71. [CrossRef]

65. Ren, L.; Qiu, H.; Wang, P.; Lin, P.M.C. Exploring customer experience with budget hotels: Dimensionality and satisfaction. *Int. J. Hosp. Manag.* **2016**, *52*, 13–23. [CrossRef]

66. Levy, T.; Yagil, J. Air pollution and stock returns in the US. *J. Econ. Psychol.* **2011**, *32*, 374–383. [CrossRef]

67. Brancher, M.; Griffiths, K.D.; Franco, D.; de Melo Lisboa, H. A review of odour impact criteria in selected countries around the world. *Chemosphere* **2017**, *168*, 1531–1570. [CrossRef] [PubMed]

68. Rafson, H. (Ed.) *Odor and VOC Control Handbook*; McGraw-Hill Professional: New York, NY, USA, 1998.

69. Hedges, A.R. Industrial applications of cyclodextrins. *Chem. Rev.* **1998**, *98*, 2035–2044. [CrossRef] [PubMed]

70. Behan, J.M.; Goodall, J.A.; Perring, K.D.; Piddock, C.C.; Provan, A.F. Anti-Smoke Perfumes and Compositions. U.S. Patent 5,676,163, 30 November 1993.

71. Pilosof, D.; Cappel, J.P.; Geis, P.A.; McCarty, M.L.; Trinh, T.; Zwerdling, S.S. Fabric Treating Composition Containing Beta-Cyclodextrin and Essentially Free of Perfume. U.S. Patent 5,534,165, 12 August 1994.

72. Flachsmann, F.; Gautschi, M.; Sgaramella, R.P.; McGee, T. Dimethylcyclohexyl Derivatives as Malodor Neutralizers. U.S. Patent Application 20100209378, 7 September 2007.

73. Nguyen, P.N.; Bhaveshkumar, S. Aerosol Odor Eliminating Compositions Containing Alkylene Glycol(s). U.S. Patent Application 20110311460, 18 June 2010.

74. Joulain, D.; Racine, P. Deodorant Compositions Containing at Least Two Aldehydes and the Deodorant Products Containing Them. U.S. Patent 5,795,566, 29 May 1989.

75. Joulain, D.; Maire, F.; Racine, P. Agent Neutralizing Bad Smells from Excretions and Excrements of Animals. U.S. Patent 4,840,792, 29 May 1986.

76. Woo, R.A.; Readnour, C.M.; Olchovy, J.J.; Malanyaon, M.N.; Liu, Z.; Jackson, R.J. Malodor Control Composition Having a Mixture of Volatile Aldehydes and Methods Thereof. U.S. Patent Application 20110150814, 17 December 2009.

77. Woo, R.A.; Trinh, T.; Cobb, D.S.; Schneiderman, E.; Wolff, A.M.; Rosenbalm, E.L.; Ward, T.E.; Chung, A.H.; Reece, S. Uncomplexed Cyclodextrin Compositions for Odor Control. U.S. Patent 5,942,217, 9 June 1997.

78. Parekhm, P.P.; Nicoll, S.P.; Ramsammy, V.; Colt, K.K.; Betz, A.; Deshpande, V.M.; VanKippersluis, W.H.; Boden, R.M. Synergistically-Effective Composition of Zinc Ricinoleate and One or More Substituted Monocyclic Organic Compounds and Use Thereof for Preventing and/or Suppressing Malodors. U.S. Patent Application No. 20050106192, 13 November 2003.

79. Aussant, E.J.; Bassereau, M.B.; Fraser, S.B.; Warr, F.J. Use of Fragrance Compositions for the Prevention of the Development of Indole Base Malodours from Fecal and Urine Based Soils. U.S. Patent No. 20080032912, 4 August 2006.

80. Levorse, A.T.; Monteleone, M.G.; Tabert, M.H. Novel Malodor Counteractant. U.S. Patent Application 20130149269, 12 December 2011.

81. Shejtli, J. Introduction and general overview of cyclodextrin chemistry. *Chem. Rev.* **1998**, *98*, 1743–1753. [CrossRef] [PubMed]

82. Ho, T.M.; Howes, T.; Bhandari, B.R. Encapsulation of gases in powder solid matrices and their applications. *Powder Technol.* **2014**, *259*, 87–108. [CrossRef]

83. Qamaruz-Zaman, N.; Kun, Y.; Rosli, R.-N. Preliminary observation on the effect of baking soda volume on controlling odour from discarded organic waste. *Waste Manag.* **2015**, *35*, 187–190. [CrossRef] [PubMed]

84. Dutkiewicz, J.K. Cellulosic materials for odor and pH control. In *Medical and Healthcare Textiles, Woodhead Publishing Series in Textiles*; Woodhead Publishing Limited: Cambridge, UK, 2010; pp. 140–147.

85. Duan, Y.; Zheng, F.; Chen, H. Analysis of volatiles in Dezhou braised chicken by comprehensive two-dimensional gas chromatography/high resolution-time of flight mass spectrometry. *Food Sci. Technol.* **2015**, *60*, 1235–1242. [CrossRef]

86. Ara, K.; Hama, M.; Akiba, S.; Koike, K.; Okisaka, K.; Hagura, T.; Kamiya, T.; Tomita, F. Foot odor due to microbial metabolism and its control. *Can. J. Microbiol.* **2006**, *52*, 357–364. [CrossRef]

87. Gesser, H.D.; Fu, S. Removal of aldehyde and acidic pollutants from indoor air. *Environ. Sic. Technol.* **1990**, *24*, 495–497. [CrossRef]

88. Swasy, M.I.; Campbell, M.L.; Brummel, B.R.; Guerra, F.D.; Attia, M.F.; Smith, G.D.; Alexis, F. Whitehead, D.C. Poly(amine) modified kaolinite clay for VOC capture. *Chemosphere* **2018**, *213*, 19–24. [CrossRef]

89. Herz, R.S. The role of odor-evoked memory in psychological and physiological health. *Brain Sci.* **2016**, *6*, 22. [CrossRef] [PubMed]

90. Matsunaga, M.; Bai, Y.; Yamakawa, K.; Toyama, A.; Kashiwagi, M.; Fukuda, K.; Oshida, A.; Sanada, K.; Fukuyama, S.; Shinoda, S.; et al. Brain–immune interaction accompanying odor-evoked autobiographic memory. *PLoS ONE* **2013**, *8*, e72523. [CrossRef] [PubMed]

91. Herz, R.S. Aromatherapy facts and fictions: A scientific analysis of olfactory effects on mood, physiology and behavior. *Int. J. Neurol.* **2009**, *119*, 263–290. [CrossRef] [PubMed]

92. European Commission Scientific Committee on Consumer Safety. The SCCS Notes of Guidance for the Testing of Cosmetic Ingredients and Their Safety Evaluation, 10th Revision. SCCS/1602/18: Adopted at the SCCS Plenary Meeting 24–25 October 2018. Available online: https://ec.europa.eu/health/sites/health/files/scientific_committees/consumer_safety/docs/sccs_o_224.pdf (accessed on 3 December 2019).

93. Api, A.M.; Belsito, D.; Bruze, M.; Cadby, P.; Calow, P.; Dagli, M.L.; Dekant, W.; Ellis, G.; Fryer, A.D.; Fukuyama, M.; et al. Criteria for the Research Institute for Fragrance Materials, Inc. (RIFM) safety evaluation process for fragrance ingredients. *Food Chem. Toxicol.* **2015**, *82*, S1–S19. [CrossRef]

94. Notice of Receipt. Notice of Receipt of TSCA Section 21 Petition. *Fed. Regist.* **2007**, *72*, 60016–60018. Available online: https://www.govinfo.gov/content/pkg/FR-2007-12-21/pdf/07-6176.pdf (accessed on 14 January 2020).

95. Johnson, M.B.; Kingston, R.; Utell, M.J.; Wells, J.R.; Singal, M.; Troy, W.R.; Horenziak, S.; Dalton, P.; Ahmed, F.K.; Herz, R.S.; et al. Exploring the science, safety, and benefits of air care products: Perspectives from the inaugural air care summit. *Inhal. Toxicol.* **2019**, *31*, 12–24. [CrossRef]

96. Nielsen Holdings Plc. Data Retrieved through a Paid Subscription on March 2019. For More Information about the Nielsen Homescan Database Visit. Available online: https://catalog.data.gov/dataset/nielsen-homescan (accessed on 13 January 2020).

97. Kelly, F.J.; Fussell, J.C. Improving indoor air quality, health and performance within environments where people live, travel, learn and work. *Atmos. Environ.* **2019**, *200*, 90–109. [CrossRef]

98. Kole de Peralta, K. Mal Olor and colonial Latin American history: Smellscapes in Lima, Peru, 1535–1614. *Hisp. Am. Hist. Rev.* **2019**, *99*, 1–30. [CrossRef]

11

Combined Model for IAQ Assessment: Part 1—Morphology of the Model and Selection of Substantial Air Quality Impact Sub-Models

Michał Piasecki * and Krystyna Barbara Kostyrko

Department of Thermal Physic, Acoustic and Environment, Building Research Institute, Filtrowa 1,
00-611 Warsaw, Poland; k.kostyrko@itb.pl
* Correspondence: m.piasecki@itb.pl.

Abstract: Indoor air quality (IAQ) is one of the most important elements affecting a building user's comfort and satisfaction. Currently, many methods of assessing the quality of indoor air have been described in the literature. In the authors' opinion, the methods presented have not been collected, systematized, and organized into one multi-component model. The application purpose of the assessment is extremely important when choosing IAQ model. This article provides the state-of-the-art overview on IAQ methodology and attempts to systematize approach. Sub-models of the processes that impact indoor air quality, which can be distinguished as components of the IAQ model, are selected and presented based on sensory satisfaction functions. Subcomponents of three potential IAQ models were classified according to their application potential: IAQ quality index, IAQ comfort index, and an overall health and comfort index. The authors provide a method for using the combined IAQ index to determine the indoor environmental quality index, IEQ. In addition, the article presents a method for adjusting the weights of particular subcomponents and a practical case study which provides IAQ and IEQ model implementation for a large office building assessment (with a BREEAM rating of excellent).

Keywords: indoor environment quality; IEQ; PPD; IAQ; TVOC; BREEAM assessment; occupant satisfaction

1. Introduction

1.1. State-of-the-Art Indoor Air Quality Measurement Systems

Approximately 30 years ago, people began to realize that buildings not only provide them with a sense of security, but can also significantly affect their health and well-being. This is particularly important due to the fact that people spend an increasing amount of time in closed indoor environment. Air quality and ventilation approaches were initially based on the users' dissatisfaction with the scent of the human body and, as such, the understanding of indoor air quality (IAQ) had serious limitations. Large quantities of pollutants and their sources clearly influence the indoor comfort of building inhabitants, as well as their health. In 1998, Fanger [1] presented an approach to the quantitative determination of perceived IAQ based on the level of dissatisfaction of residents caused by bad odors and irritants, smoke, and other sources of pollution. This approach provided two new measures of IAQ: the olf, which quantifies the pollution generated from a strong source of human bio-pollutant in the range of the impact of emitted odors on perceived air quality, and the decipol, measuring the perceived air quality in an indoor space with a source of pollution of one olf at a ventilation rate of 10 l/s. The number of emitted olfs per floor unit in different types of buildings and the amounts of pollutants from tobacco smoking (in olfs) can then be determined. Consideration of only the odour of the human body, without taking into account the influence of pollutants from various other sources (for example emissions from construction products), was very limited. At this stage, the determination of IAQ did not take into account the significant differences among contaminants and did not distinguish their specific impacts on health or comfort. The study of emissions undetectable by the senses (such as carbon monoxide and other pollutants that affect health at concentrations below their odor threshold), together with health-effects thresholds, has become particularly important. The range of air pollutants that should be considered as IAQ components is very difficult to determine, because the composition of pollutants constantly changes due to the fact of their dynamic nature, secondary reactions, sorption processes, and other physical and chemical phenomena occurring in indoor environments, thus it cannot be inherently determined, as illustrated in Figure 1.

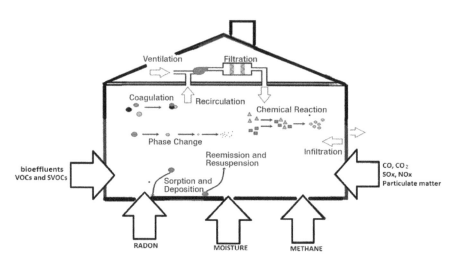

Figure 1. Overview of the physical and chemical processes of potential pollution sources in the indoor environment of a building. (VOCs—Volatile Organic Compounds, SVOCs—semi Volatile Organic Compounds).

Smoking alone emits more than 7000 different compounds, many of which are harmful [2] for humans and animals and may transport biological pollutants that can act as allergens. People and household animals emit gases which are unpleasant, transfer pathogens, and cause diseases. These examples show that there are many paths for penetration of and exposure to the sources of pollution in indoor environments.

In connection with the growing need to determine levels of indoor air pollution, new centers performing tests and new methods have been created considering the ability to analyze an increasing number of harmful substances. Measurement of pollutant concentrations in the air is generally a task performed by experts mainly in accredited laboratories and the results are published in scientific journals, technical reports, and, eventually, in guidelines, e.g., those of American Society of Heating Refrigerating and Air Conditioning Engineers ASHRAE [3]. The presence and concentrations of pollutants are often detected and measured without careful consideration of the significance of these measurements, and the pollutants measured may not be the most widespread or the most harmful. Some emissions are incorrectly grouped together; for example, more than one million volatile organic compounds (VOCs) are known and their toxicities are generally unknown, but they are often reported as a single value and referred to as the total VOCs (TVOCs) component. Frequently, carbon dioxide is used as an indicator of IAQ, although it does not have such a negative effect on the health of residents in the concentrations in which it is usually found in buildings. In our opinion, CO_2 is rather a marker of human bioeffluents. Examples of different understandings of the set of typical pollutants in an indoor environment are shown in Table 1 [4]. This table provides recommended values from the results of the European project HealthVent [5], which aimed to develop health-based ventilation guidelines. Table 1 also includes recommendations provided by the Word Health Organization (WHO) on the acceptable levels of pollutant concentrations [6,7], as well as recommendations from other organizations, such as China's IAQ standard values [8]. Different approaches to the IAQ issue mean that the exposure limits assumed in the various source materials differ.

Table 1. Acceptable levels of pollutant concentrations occurring in indoor air, according to World Health Organization (WHO) recommendations supported by the results of the EU research program HealthVent, values recommended by the standard EN 16798-1:2019 (replacing EN 15251:2007), values recommended by the Chinese indoor air quality (IAQ) standard GB/T 18883-2002 and for acceptable levels of pollution for the certification of office buildings in Hong Kong.

Pollutant	WHO Guidelines for IAQ with Updates [6,7]	HealthVent Project [5,9]	EN 16798-1:2019 [10]	IAQ Standards for China [8,11,12]	IAQ Certification Hong Kong [13]
CO_2			<500 ppm beyond outdoor level	485 ppm	<1000 ppmv 1800 mg/m³ (8 h)
CO	100 mg/m³ (15'); 35 mg/m³ (1 h); 10 mg/m³ (8 h); 7 mg/m³ (24 h)	19 mg/m³ (8 h)	100 mg/m³ (15'); 35 mg/m³ (1 h); 7 mg/m³ (24 h)	1 mg/m³	10 mg/m³ 7000 µg/m³ (8 h)
Formaldehyde HCHO	0.1 mg/m³ (30')	0.03 mg/m³ (30')	0.1 mg/m³ (30')	10 µg/m³	<0.1 mg/m³ (8 h)
Benzene	>0.17 mg/m³	<outdoor concentration	No safe level can be determined	0.11 mg/m³	17 µg/m³
NO_2	40 µg/m³ (1 year); 200 µg/m³ (1 h)	40 µg/m³ (1 week)	40 µg/m³ (1 year); 200 µg/m³ (1 h)	10 µg/m³	150 µg/m³ (8 h)
SO_2	20 µg/m³ (24 h)		20 µg/m³ (24 h)	20 µg/m³ (24 h)	
Naphthalene	0.01 mg/m³; 0.02 (1 year)	0.01 mg/m³ (1 year)	0.01 mg/m³ (1 year)		0.01 mg/m³ (8 h)
Trichloroethene	>2.3 µg/m³				230 µg/m³ (8 h)
Tetrachloroethene	0.25 mg/m³ (1 year)	0.25 mg/m³ (1 year)	0.25 mg/m³ (1 year)		0.25 mg/m³ (8 h)
Respirable particulate matter $PM_{2.5}$	10.0 µg/m³ (1 year)		10 µg/m³ (1 year)	15 µg/m³ (1 year); 35 µg/m³ (24 h)	100 µg/m³ (8 h)
PAH *	>0.012 ng/m³		No safe level can be determined		1.2 ng/m³ (8 h)
TVOC **			1000 µg/m³	600 µg/m³	600 µg/m³ (8 h)

* PAH Para-Aminohippuric Acid—cyclic aromatic hydrocarbons; ** TVOC Total Volatile Organic Compounds; Reference [13] provides additional TVOC certification tests for new office buildings by determining (at the ppbv level) the content of carbon tetrachloride, chloroform, 1,2- and 1,4-dichlorobenzene, ethylbenzene, toluene, and o-, m- and, p-xylene.

Theoretical work on a combined IAQ model allowing aggregation of the results of the assessment of components affecting humans [14] is not yet well recognized in the literature. However, studies on IAQ indicators, which aim to provide a quantitative description of indoor air pollution, have been conducted since the nineties. In 2003, a significant study by Sekhar et al. [15] was published related to the standard indoor pollutant index (IPSI), the disease symptom index in the building symptom index (BSI), and to the often-cited works by Moschandres and Sofuoglu [16,17] on the indoor environmental index (IEI), indoor air pollution index (IAPI), and the indoor pollutant standard index (IPSI). The IAPI characterizes air pollution in an office with a single number: the index. The index value ranges between zero (lowest pollution level, i.e., best indoor air quality) and 10 (highest pollution level i.e., worst indoor air quality). The IAPI is a composite index; sub-indices ed are aggregated using the arithmetic mean in conjunction with a tree-structured calculation scheme. This scheme gives rise to some reservations, because at the top of the tree-structured calculation scheme is the IEI (calculated as the arithmetic mean of the IAPI and the IDI (indoor air discomfort index), and the combination of IAQ sensation and thermal conditions does not appear until later.

While considering the indicators for the quantitative description of pollution, the proposal of the IEA Working Group named "Defining the Metrics of IAQ" should also be mentioned. This group prepared, in 2017, the document entitled "In the Search of Indices to Evaluate the Indoor Air Quality of Low-Energy Residential Buildings" [18]. The group made the following assumption for the categorization of various indicators: there should be one index per individual pollutant and a dimensionless coefficient should be specified to evaluate the IAQ, provided that the current (observed) concentrations of a given pollutant c_j are related to the ELVs (exposure limit values) concentration $c_{j,ELV}$.

$$IAQ_{index}(j) = I_j = \frac{c_j}{c_{j,ELV}} \qquad (1)$$

The index is calculated for each individual pollutant [18], which is specific only for this exact pollutant. The report showed that aggregation can be performed by addition, by taking the maximum value or by other methods, in an attempt to define metrics that can be used to evaluate IAQ. The assumption was that the reference value usually refers to health risks (accounting for chronic or acute effects), but other metrics can also be used, (e.g., odor or irritation threshold). There are two important properties to be considered when aggregating sub-indices: ambiguity and eclipsing. As a result of the analysis, the authors concluded "that there are problems with model aggregation methods. In the aggregation model $I_{agg} = I_1 + I_2$, ambiguity creates a false alarm and in the aggregation model $I_{agg} = 1/2(I_1 + I_2)$, eclipsing underestimates the effect" [18]. Therefore, the discussion remains open [18]. The report also showed how there are large spreads of concentrations of individual pollutants (up to seven rows), even in the group of pollutants for which sub-indices were built. It determined the difficulties of building a weighted scheme based on the simplest percentage adjustment of the concentration shares and, thus, the share of the mass of pollutants to be removed by ventilation.

The current state of knowledge does not provide information authorizing the omission of certain pollutants. Hence, taking into account the lack of data on the characteristics of each chemical compound and consideration of the "removal efficiency" [19] requires us to abandon thinking about the adjustment of many individual pollutants, and to focus only on the creation of a model based on the representative and target components. In this state of knowledge, there are hopeful studies and proposals with a grey combined $\sum IAQ_{index}$ model and the grey clustering model for IAQ indicators proposed by Zhu and Li in 2017 [20] is particularly interesting, especially when the relationships between system factors and the system's IAQ behavior and the interrelationships among the factors are uncertain. At first, all specific indoor air pollutants and related parameters should be measured. However, this is a very complex and time-consuming process. On the basis of the characteristics and correlations of the pollutants, the indoor air quality can be characterized by representative indicators. Studies [20] have pointed out that respirable particulates, CO_2 and TVOCs, were the three most representative and independent environmental parameters which can be used as an evaluation index of indoor air quality in office buildings. Since each indicator represents a class of pollutants with similar sources and dissemination characteristics, this index group avoids unreliability due to the fact that these indicators are "too small" because of critical concentration depression. A data pretreatment method must be used in the calculation procedure, reflecting the differences in concentration levels among different pollutants, but also expressing their influence on the comfort and health of the indoor occupants. Moreover, the measured pollutant concentrations can be used to predict the probable levels of other parameters, and good agreement was found between the predictions and measured values.

1.2. The Research Questions

The main research question contained in the paper concerned whether it was possible with the current state of knowledge to create and use in practice an IAQ model that was based on a unified and coherent approach for input indoor air parameters (such as pollutant concentrations, odor levels, and moisture content) and provided one output parameter (we proposed occupant satisfaction, IAQ_{index} (in %)). The authors looked for physical equations for the IAQ_{index}'s subcomponents and dependencies for their predicted occupant satisfaction functions with a pollutant concentration c_j ($PD = f(c_j)$ in %) which could be used as a model for subcomponents.

This paper's intention was to provide an IAQ model with a step-by-step process which can be used to determine the value of the overall indoor environmental quality index (in %) including another three components: thermal comfort, acoustic comfort, and lighting quality. The innovative approach and added value of this article is in the use of the proposed IAQ model in practice and the relatively simple calculation of the overall IEQ value (with an uncertainty estimation) using the actual results of measurements in the Building Research Establishment Environmental Assessment Method BREEAM certified case study office building. The authors also provided occupant satisfaction functions for CO_2, TVOCs, and formaldehyde HCHO in two variants: with experimental %PD values taken from the literature and for these pollutants' %PD values converted from an Air Quality Index system (see Section 2.2.).

2. Methods

2.1. Research Content and Strategy

The proposed IAQ model is presented in Sections 2.2–2.7. The model is later used to analyze the case study of an office building described in Section 2.9. Figure 2 presents the subsequent research steps from theory to practical application. Section 2.8 shows the method for determining indoor environmental quality IEQ_{index} where IAQ index is a subcomponent/part of the IEQ_{index} model. In order to determine the IAQ and IEQ, physical measurements of the indoor environment in the building were conducted using the experimental approach provided in Section 2.10. Based on these physical indoor measurements the IAQ and IEQ indexes (number of occupants satisfied with the indoor air and overall indoor quality, respectively) were assessed (see Section 3) and discussed (see Section 4).

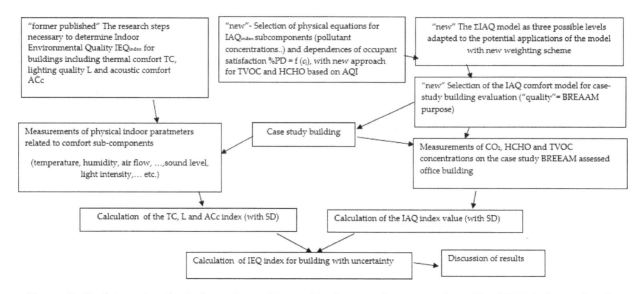

Figure 2. To determine the indoor air quality and indoor environmental quality (IEQ) indexes for the case study. (TC—thermal comfort, L—light quality, ACc—acoustic comfort).

2.2. The IAQ Model Proposal—Basic Assumptions

In the IAQ model construction process (our proposal), the commonly accepted approach is to transform individual concentrations of pollutants into subcomponents before they are aggregated into a single index (occupant satisfaction in %). However, summation of sub-indices can lead to situations in which all are under individual health thresholds, but the final indicator shows when the threshold has been exceeded. Conversely, the averaging of partial sub-indices can lead to an overall indicator showing an acceptable IAQ, even though one or more partial indicators are larger than their individual thresholds. One solution is to use the maximum value of all sub-indices to create the final form of the $\sum IAQ_{index}$. Taking these issues into consideration, the authors created the $\sum IAQ$ model with three complication levels adapted to the purposes of potential applications of the model, as presented in Figure 3;

i.   Certification of a building, e.g., via the BREEAM system using (three sub-indices), called "quality";

ii.   Design, including perceptible contaminants affecting comfort and using the IAQ index when calculating the IEQ (five sub-indices), called "comfort";

iii.   Complex design, with the $\sum IAQ_{index}$ representing both comfort and health (seven sub-indices) called "comfort/health".

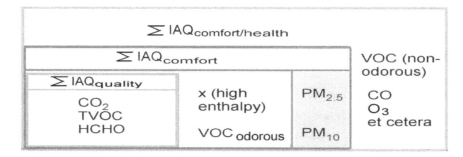

Figure 3. $\sum IAQ$ model has three possible levels (i.e., similar to a Russian doll structure) adapted to the potential applications of the model.

The simplest one, "quality", is an inner part of the $\sum IAQ_{comfort}$ model and can be used separately for simple applications with the main purpose of supporting a green building certification, e.g., via the

BREEAM system using three components, i.e., CO_2, HCHO, and TVOC. This model is used later on the case study of a BREEAM building.

Figure 1 shows the processes influencing the morphology of the $\sum$IAQ model and Figure 3 provides a list of the pollutants for which three IAQ submodels were built, containing human-perceived contaminants ($IAQ_{quality}$ and $IAQ_{comfort}$ models), but also the $IAQ_{comfort/health}$ model for both perceptible and imperceptible pollutants, i.e., those that are not perceptible by humans but affect health and require additional energy for intensive ventilation for health reasons. There are potential sub-indices, such as $IAQ(VOC_{non-odorous})$ or $IAQ(CO)$ [5,7]. Dust pollutants may have their sub-index both in the comfort model (if reliable curves of human sensory perception of PM concentrations are known) or in the $IAQ_{comfort/health}$ model if their health impact is considered to be the dominant feature. Considering the types of pollutants harmful to health assigned to the sub-indices of IAQ, we only consider the most important air pollutants (i.e., target emissions) that were given in the WHO guide in 2010 [5,7]. Submodels of processes that impact on air quality in indoor environments, which can be distinguished as components of the IAQ model, were based on sensory satisfaction functions (index of occupant dissatisfaction (PD) with the level of air pollution). Subcomponents of the three potential IAQ models were classified according to their future potential applications: in the assessment of environmental quality index IEQ (models $IAQ_{quality}$ and $IAQ_{comfort}$) or in the design of ventilation taking into account all possible harmful-to-health pollutants (model $IAQ_{comfort/health}$). In our opinion, such systematization creates order and has a practical dimension as presented later on in the case study. The following are the target pollutant groups:

i.   In the air quality model $\sum(IAQ)_{quality}$, the IAQ index subcomponents were assigned to the selected three pollutants. The submodels for the IAQ were CO_2, TVOC, and formaldehyde HCHO, as recommended by References [5,10,21,22];

ii. In the $\sum(IAQ)_{comfort}$ model, thepreviously provided simplified $IAQ_{quality}$ subcomponents for the three main pollutants were extended with a set of selected compounds $VOC_{odorous}$, related to the collection of IAQ sub-indices ($VOC_{odorous}$) with an unknown cardinality, increased appropriately for the number of dominant pollutants. In addition, we provided a conditional deluge of two more components: (1) calculated using the enthalpy of hot and humid air (high enthalpy $h > 55$ kJ/kg [23]), the percentage of persons dissatisfied with respiratory cooling with humid air at relatively high temperatures and (2) the percentage of persons dissatisfied with indoor pollution with respect to dust pollution (PM_{10} and $PM_{2.5}$), measured via panel tests. The introduction of a dust-pollution subcomponent to the IAQ model may be debatable, because some experimenters [24,25] underline the unique results of sensory tests of discomfort from dust, and the influence of "emissions" of respiratory dust particles on satisfaction is still under-researched. Considering the above, we expected two variants of the comfort model: with PD (PM_{10}, $PM_{2.5}$) or without this factor;

iii. In the overall $\sum$IAQ model, comfort $(IAQ)_{comfort}$ and health risk $(IAQ)_{health}$ indicators were used, and, hence, this model was called $(IAQ)_{comfort/health}$. Models for subcomponents of IAQ not perceived by humans but influencing health, can be borrowed from the index set in the AQI (air quality index) system [26–28], which was adapted to assess the quality of indoor air based on, and in accordance with, the concepts of the air quality assessment system used globally by the American EPA.

Values of AQI indices published on active EPA websites using the air quality index system were introduced for application in US federal regulations in 1999 [28]. Currently, the AQI system for outdoor air includes the following pollutants: ozone, particulate pollutants (PM_{10} and $PM_{2.5}$), carbon monoxide CO, sulfur dioxide SO_2, and nitrogen dioxide NO_2. To convert a specific air pollutant concentration to an AQI, the EPA developed a tool called the AQI Calculator, which is an open resource [29]. This system (referring to the index from 2004 [16] and the indoor pollutant standard index (IPSI)) was further developed, and the proposed IAQI for indoor air presented by Wang et al. [30] in 2008 and a newer proposal [27] from 2017 for a similar but narrower set of indices, also for indoor air, were both modeled on it. The AQI and IAQI indicators showed an increase in the level of impact on human health with increasing concentrations of air pollution. There are some detected difficulties here, since "AQI is a piecewise linear function of the pollutant concentration" [27]. The calculated values of the AQI [31] or IAQI [30] indices, over the entire 0–500 scale calculated from the measured concentrations of selected contaminants or in the part of the scale corresponding to the IAQ rating, ranged from "good" to "unhealthy", and can be converted to $PD\%$ for use in the model equation, $(IAQ)_{comfort/health}$. Concentrations will be significant when the uncertainties of scale conversions are estimated. Authors believe that their way of converting the AQI scale to $PD\%$ (which is similar to the method of conversion of the IEQ components' ordinal scales from the OFFICAIR EC project [32] to $PD\%$ scale; for example, the occupant percentage dissatisfied with noise [33,34] should be accepted in light of the expected results of a metrological analysis of the reliability of the combined $\sum IAQ$ model.

Subcomponent models (physical functions, $PD\%$) of the IAQ model for all individual air pollutants are presented later in this section.

2.3. $\sum IAQ$ Model Weighting Scheme Considering Air Pollution Ventilating

To obtain a comprehensive picture of IAQ in a building, it is necessary to measure the number of pollutants with different individual concentrations. There are methods that weigh sub-indices [21] but the problem is finding an effective weighting scheme and understanding how to adjust them in the overall model of all the pollutants in $\sum IAQ$. For this reason, we proposed an adjustment method for the weights. In our opinion, provided in detail in References [33,34], and also according to Reference [22], the best weighting scheme, which would lead to a credibly aggregated model of IAQ composed of many extractable components (sub-indices), would be a system based on concentration values (the "excess masses" of pollutants to be regarded as loadings for the ventilation system). Therefore, the we aimed to determine the individual pollutants assigned to the IAQ model, their concentrations, c_j, as the inputs of the IAQ submodels, and their "excess concentrations" originating from emissions or determined within indoor environments. Thus, it was possible to determine directly the energy requirements for ventilation purposes and the required minimum global ventilation rate. Determining the input concentration value, c_j, for each IAQ sub-index enables the determination of the total mass of pollutants in the air, which is the basis for determining the air change rate $N_{1,\dots 7}$ (overall air change rate), assuming that the model includes all significant IAQ pollutants.

Currently, according to References [35–37], the most common assumption made is that pollution from VOC_j compounds arises only from emissions due to the presence of construction or finishing materials (for $j = 1, 2, \dots n$) (it can be assumed that the source i of an emission is the entire indoor environment and then $i = 1$) from the zero state. The physical model for determining the ventilation rate in indoor environments polluted with VOC-type pollutants from building materials is given by EN 16798-1:2019 [10], assuming that design parameters for indoor air quality are derived using limit values for substance concentrations. In accordance with ECA Report Number 11 [36], the design ventilation rate required to dilute an individual substance emitted from building materials is calculated as:

$$Q_h = \frac{G_h}{C_{h,j} - C_{h,0}} \times \frac{1}{\varepsilon_v} \tag{2}$$

where Q_h is the ventilation rate required for dilution in m^3 per second, G_h is the emission rate of the substance in micrograms per second, $C_{h,j}$ is the guideline value of the substance in micrograms per m^3, $C_{h,o}$ is the concentration of the substance in the supply air in micrograms per m^3, and ε_v is the ventilation effectiveness.

In fact, in a building with active "indoor chemistry" (see Figure 1), the use of this formula seems to be increasing. Taking into account the dynamic nature of the processes of generating various pollutants, the approach to IAQ and its components should be changed and subcomponents should be treated as pollution load processes, increasing in number not only due to the emission processes but also due to the generation of bio-pollution, water evaporation, and even dust infiltration from outside. It is also possible to set steady-state (initial) concentrations of pollutants and to determine the expected time courses of removal of these pollutants by means of ventilation (curves $\sum c_j = f(\tau)$), at constant values of air change rate per hour ACH (h^{-1}). Such ventilation rate calculations for CO_2 were developed in 1997 by Persily [38] and similar ones were provided in 2017 by Gyot [39] at the Berkeley National Laboratory. These calculations are not very accurate, as shown in the general demonstration graph in Figure 4.

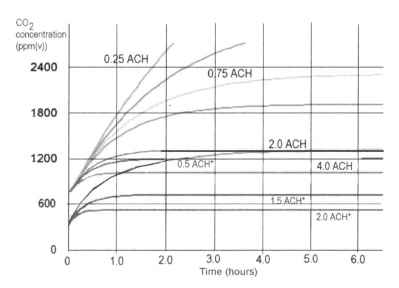

Figure 4. Of CO_2 above outdoors levels with two people in the contaminated building (ACH—air change rate) and in a typical clean office room (ACH*).

Less accurate time-dependent curves of the total minimum ventilation rate ACH needed for "contaminant exhaustion" can be determined using programs [40] based on generic engineering equations for the sum of pollutants $\sum c_j$. The generic equation for pollution concentration (the ratio of the amount of polluting product to the amount of fluid in the space (such as air in a room) can be calculated from the following equation:

$$c = q/(n \cdot V \cdot (1 - e^{-Nt}))$$ (3)

where c is the pollution concentration in the space (or in the room) with perfect mixing (m^3/m^3) or (kg/kg), q is the amount of pollution added to the space (m^3/h) (kg/h), N is the air change rate per hour (h^{-1}), V is the volume or mass of the space (m^3) or (kg), e is the number 2.72, and t is time (h). If the initial concentration (at $t = 0$) in the space and the concentration in the supply fluid is zero, after some time the concentration in the room will stabilize. The ventilation rate graph for an amount of pollution q = 1 and a volume of space V = 1, shows the values of $\sum c_j$, similar to Figure 4. In order to obtain more exact values of the VOC concentrations remaining in the room for the air change rate function, it is possible to use the published dependencies or for the assumed volumes V of ventilated spaces with determinate concentrations, as they can be determined experimentally. A simplified method for determining the ventilation rate N from a simple formula for the time, t, course of a trial ventilation was provided by the Japanese researchers Noguchi et al. [41] in 2016. A description of this method is worth reading. Based on the temporal changes of the TVOC concentration measured using a PID_{TVOC} meter [42], the air change rate N or the ventilation rate F was estimated using the following method. Assuming perfect mixing of the air in the room and a constant TVOC emission rate E, the concentration change of TVOC in the room can be expressed by the following equation:

$$C(t) = C_j + \left(\frac{E}{F}\right)\left(1 - e^{-\frac{F}{V}t}\right) = C_j + \left(\frac{E}{F}\right)\left(1 - e^{-Nt}\right)$$ (4)

Finally, Equation (5) can be expressed with one unknown parameter, the air exchange rate N as:

$$\log\left(\frac{C_{st} - C_j}{C_{st} - C(t)}\right) = Nt \tag{5}$$

The initial concentration c_j can be determined from the experimental results. After a long time, when the exponential term in Equation (5) can be assumed to be zero, the concentration $C(t)$ becomes constant. The steady-state concentration C_{st} can be determined from the temporal change in the experimental results where the concentration levels off.

The rule that the IAQ model should include a weighting scheme, referring to the variation in the share of pollutants in the IAQ, has been noted in References [35,37]. According to the first proposal, the weighting system is based on the differentiation of coefficients R_j, which are the ratios of the real concentration values (or mass of pollutants) to the values of reference concentrations, representing the so-called relative masses of non-eliminated pollutants. This can also be represented by the desirable reduction in the level of pollution by means of ventilation and, thus, also by the energy requirements. For one emission source, the proposed system with a coefficient R_j for a given pollutant C_j has the formula:

$$R_j = \frac{y_j}{I_j}, \quad j = 1, 2, \ldots m \tag{6}$$

where I_j is the ratio of the gas-phase concentration to the reference concentration value. For example, for the LCI (Lowest Concentration of Interest) value for the jth compound emitted from the building material, the factor R_j is dimensionless, since y_j is the gas-phase concentration for the jth compound in $\mu g \cdot m^{-3}$, I_j is the lowest concentration of interest (LCI) [41] for the jth compound in $\mu g \cdot m^{-3}$ and m is the number of all elected compounds. The weight coefficients W_j are used as the weights of the equations in the IAQ index, according to:

$$W_j = \frac{R_j}{\sum\limits_{j=1}^{m} R_j} \quad j = 1, 2, \ldots \ldots m \tag{7}$$

The authors of Reference [35] justified adjusting the coefficients where $\sum R_j \leq 1$, but they did not explain the physical meaning of this condition. We believe that further discussion should include the issue of whether the "relative mass" of contamination expressed by Equation (6) has a proper place in the weighting scheme for the equations. The dimensionless quantity (7) does not have a sound physical meaning [37]. In our opinion, this type of calculation method is debatable.

One should strive to cover all the sub-indices of the combined IAQ_{ij} model with a weighting scheme that would give VOCs a share in the total energy requirements for ventilation. The term "relative mass" should correspond to weights rationally proportional to the energy expenditure for ventilation of individual pollutants (IAQ sub-indices). Therefore, our future work will focus on introducing weights for all expressions in the overall IAQ model equation. However, since this is currently not possible without an adjustment method adapted to weight determination for very small concentrations, it was decided to present our model as an interim solution. We proposed the use of weights based on the "excess concentration" values within the pollutant categories only with similar and comparable orders of concentration values, for example the $VOC_{odorous}$ and $VOC_{non-odorous}$ categories. The rules of adjustment with boundary conditions will be provided and justified in the follow-up article to this report.

2.4. $\sum$IAQ Model Scheme Morphology

According to the new proposal, for models with air quality sub-indices $IAQ(P_j)$ (developed with the standard EN 16798-1:2019 as a reference for IEQ model creation [22,33] and with assumptions described in Reference [34]), in the case where indoor air has many pollutants, $P_1, \ldots j$, the combined ΣIAQ_{index} equation is:

$$\Sigma IAQ_{index} = W_{P1} \cdot IAQ(P_1)_{index} + W_{P2} IAQ(P_2)_{index}, \ldots, W_{Pj} IAQ(P_j)_{index} \qquad (8)$$

where the $W_{P1}, \ldots {Pj}$ weighting system for IAQ components is created on the basis of the arithmetic mean and the concept of "excess concentration" is introduced only for groups of pollutants with similar concentration values. There is a difference in concentration Δc_j between the observed concentration of pollutant c_j and the reference concentration c_{ref} (c_{ELV} or c_{LCI}), which is below the current concentration in contaminated rooms. Thus, the excess concentration is:

$$\Delta c_j = c_j - c_{ref} \qquad (9)$$

The weights $W_{1, \ldots j}$ for all three IAQ models are determined on the basis of arithmetic means or by adjusting all the values of Δc_j in a given model using Equation (10):

$$W_j(IAQ_{comfort/health}) = \frac{\Delta c_j}{\sum\limits_{j=1...7} \Delta c_{1...7}} \qquad (10)$$

where the sum of the adjusted weights W_j of all ventilated pollutants described with sub-indices should be unity. The weight values for a given IAQ model, (e.g., $IAQ_{comfort}$) may be different, but the sum of the sub-index weights must be ≤1.0.

The values of the reference concentrations are the concentration levels that are acceptable or recommended as limit values for various pollutants P_j. In the case of the $\Sigma IAQ_{quality}$ submodel as part of the IEQ_{index} model, weights should be used for the $VOC_{odorous}$ (HCHO and TVOC) reference threshold concentrations of odors. The weights in the weighting system should be adjusted to unity according to the Equation (11):

$$W_{HCHO} = \frac{\Delta c_j}{\sum\limits_{j=2,3} \Delta c_{2,3}} = \frac{(c_j - c_{ref})}{(c_{TVOC} - c_{ref}) + (c_{HCHO} - c_{ref})} \qquad (11)$$

There are, however, non-typical cases in which the scales have different values. This is the case for formaldehyde, the concentration of which is many times lower in the building than the threshold level c_{th}. According to the WHO [7], the admissible value c_{ref} is also higher than the concentration in the building. In this case, the authors recommend taking the reference value as zero. Then, the weight W_{HCHO} described by Equation (11) (for two pollutants), would not be negative ($c_{HCHO} - c_{ref}$). The ASHRAE Guideline 10 (2011) [3] recommends that the IEQ model (and appropriate weights, W_i) should contain synergy effects of environmental parameters included in the subcomponents and their sensory perceptions.

Figure 5 shows the extended IAQ_{index} model with its sub-indices treated as components of the IEQ_{index}, but also with sub-indices of the $IAQ_{comfort/health}$ type, i.e., pollutants that do not belong to the IEQ model but are important to health and the energy balance of a building with a mechanical ventilation system. The experimental dependencies of the percentage of persons dissatisfied, $\%PD$, and the values of the concentrations of pollutants, c_j, sensed in indoor air in the appropriate ranges are of fundamental significance in the sub-indices relevant to the IEQ model [43].

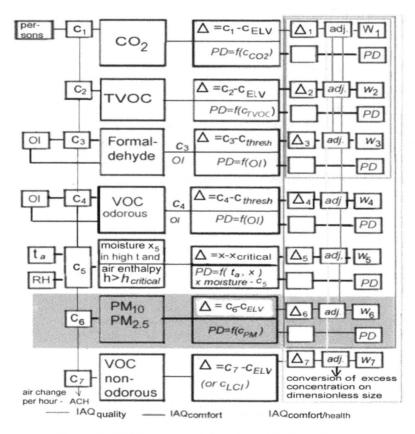

Figure 5. ΣIAQ_{index} model with weighting scheme.

From the dependencies, expressed as the curves for $PD(CO_2)$ or $PD(VOC_{odorous})$, the equations of the models are derived Equations (12)–(14):

$$\Sigma IAQ_{quality} = W_1 \cdot IAQ(CO_2) + W_2 \cdot IAQ(TVOC) + W_3 \cdot IAQ(HCHO) \tag{12}$$

$$\Sigma IAQ_{comfort} = W_1 \cdot IAQ(CO_2) + W_2 \cdot IAQ(TVOC) + W_3 \cdot IAQ(HCHO) \\ + W_4 \cdot IAQ(VOC_{odorous}) + W_5 \cdot IAQ(h) \tag{13}$$

$$\Sigma IAQ_{comfort/health} = W_1 \cdot IAQ(CO_2) + W_2 \cdot IAQ(TVOC) + W_3 \cdot IAQ(HCHO) + W_4 \cdot IAQ(VOC_{odorous}) \\ + W_5 \cdot IAQ(h) + W_6 \cdot IAQ(PM_{2.5}, PM_{10}) + W_{7a} \cdot IAQ(VOC_{non\text{-}odorous}) + W_{7b} \cdot IAQ(CO) + W_{7c} \cdot IAQ(NO_2) \tag{14}$$

The scheme of the $\Sigma IAQ_{comfort/health}$ model consists of seven (or more) components or IAQ submodels and these are models for the various types of pollutants: $IAQ(CO_2)$, $IAQ(TVOC)$, $IAQ(HCHO)$, $IAQ(VOC_{odorous})$, $IAQ(h)$, $IAQ(PM_{2.5}, PM_{10})$, and the selected $IAQ(VOC_{non\text{-}odorous})$. The $IAQ(VOC_{odorous})$ and $IAQ(VOC_{non\text{-}odorous})$ models should be multiplied, depending on the number of dominant VOC pollutants, and, hence, the $\Sigma IAQ_{comfort/health}$ model will, in practice, have more than seven components.

The inputs of each IAQ submodel are unit concentrations in air of a given pollutant, c_j (in the case of $IAQ(h)$. This is the moisture content x in well-known units "g of water vapor (g_w) per kg of dry air (kg_a)", converted to a concentration of c_j in $\mu g_{water}/m^3$ (or H—absolute humidity in g_w/m^3 which is a measure of water-vapor density). In some cases, it is necessary to convert the pollution-derived parameter to VOC concentration (conversion of the odor intensity *OI* to VOC concentration is described later in this section).

From the concentration values, total air pollution can be calculated, and, subsequently, also the energy needed to ventilate the indoor air pollution. When the concentration levels of pollutants are variable and are increasing due to the presence of emissions, then the formulas given in Reference [36] and the amended standard EN 16798-1:2019 are used to calculate the required ACH ventilation rate. When the level of contamination is set (or quasi-fixed) and the volume and other parameters of the ventilated room are known, it is possible to calculate ventilation-time curves, i.e., maximum ventilation curves for ACH ventilation rate to reach concentration levels ELV, LCI or the olfactory threshold level, according to Reference [40] or another adequate equation.

There are two outputs of each IAQ submodel as described below:

1. The weights of the weighting system for the model $\sum IAQ_{quality}$ and hypothetically $W_{1,\ldots 5}$ for the model $\sum IAQ_{comfort}$ or $W_{1,\ldots,7}$ for the model $\sum IAQ_{comfort/health}$ (in a hypothetical model with a set adjustment method). These should reflect the energy load of the IAQ expressed by the theoretically assumed increase in the current concentration of pollutant c_j relative to the reference concentration of $c_{j,ref}$, which determines the level of this concentration intended to be obtained by ventilation.

2. The $PD\%$ with the IAQ as a function of air pollution concentration. These values, determined in panel tests, reflect the impact of the interaction of air with a given pollutant at the actual level of concentration, estimated via panelists' sensations/perceptions ($PD = f(c_j)$ in %).

Examples of measurable physical parameters for the purpose of IAQ and IEQ calculations (see case study) are given in the following section.

For the construction of the combined model $\sum IAQindex$ with a weighting scheme useful for aggregating sub-indices, we proposed the model presented in Figure 5. In this scheme, the combined IAQ model is shown as the basic assumption for the aggregation of all sub-indices. First, the model was cut by a cross-connected vertical connection regarding the inputs of submodels—the calculation of the sum of the masses of all pollutants $\sum c_j$ in the ventilated space, using the values for the inputs of all submodels of IAQ concentration values of contaminants. The sum of the concentrations of all air pollutants expressed as mass units of pollution per m^3 of volume (which can be read after multiplication by V (m^3) as the mass to be displaced by ventilation), is the basis for calculating the "air change rate per hour" (ACH), the minimum air exchange rate needed to reduce the observed mass level of air pollution in a ventilated room (see Reference [40]). The second connection concerns the submodel outputs—the conversion of excess concentration to a dimensionless value, which allows to for the weighting scheme of the $\sum IAQ_{index}$ combined model and the weights of individual IAQ submodels to be determined.

Additional assumptions were as follows:

i. The sum $\sum \Delta c_{1,\ldots 7}$, which admittedly constitutes an excess mass increase of the sum of pollutants described by the submodels, was treated only as a "virtual energy load of the building" for ventilation conducted for the elimination of pollutants, and therefore, when adjusting the weights, the possibility of dividing the excess concentration by the sum of concentrations should be considered.

ii. The percentage of persons dissatisfied $PD(IAQ_{component})$ determined experimentally in sensory studies using panelists' sense of air quality during their exposure to an internal environment deteriorated by a given contamination component ($PD = f(c_j)$), was derived from the literature or direct experiments.

Values of weights $W_{1,\ldots j}$ in sets of three, five or more components of the three $\sum IAQ_{index}$ models (see Equations (12)–(14), were adjusted to a value of unity by dividing $\Delta c_{1,\ldots,j}$ of each component by the sum of excess concentrations $\sum \Delta c_{1,\ldots j}$ in μ/m^3. We proposed the use of c_{ref} values, apart from the values of c_{LCI} [18], c_{ELV}, and the threshold values c_{th} for odorous compounds, were as follows.

i. For IAQ(CO_2) and IAQ(HCHO), the c_{ELV} concentrations were derived from EN 16798-1:2019 [10];

ii. For IAQ(TVOC) and IAQ($VOC_{odorous}$), the threshold concentrations, c_{th}, for identified odorous compounds or mixtures are from Reference [44];

iii. For the IAQ(h), the water-vapor concentration H (g_w/m^3), recalculated from the moisture content x ($g_w/kg_{dry\ air}$) using the gas constant for the water vapor and the actual temperature, the value of h up to the critical value for "high enthalpy of humid air" was evaluated using the formula:

$$h = 1.006t_a + x\cdot(2501 + 1.805t_a) \tag{15}$$

where h is the specific enthalpy of humid air (kJ/kg), which must be >55 kJ/kg. The EN 16798-1:2019 standard [10] recommends a limit for the dehumidification of air of $12g_w/kg_{dry\ air}$ (this value must be converted to a c value in g_w/m^3).

i. For IAQ($VOC_{non-odorous}$), the c_{ELV} value in cases where no established LCI values were derived from the EN 16798-1:2019 standard.

ii. For IAQ($PM_{2.5}$, PM_{10}), the c_{ELV} values were derived from the WHO [7] or other organizations (Tables 1 and 2).

The proposed reference values of pollutants forming the sub-indices of the IAQ model are given in Table 2. With reference to the concentration values of c_{LCI}, it should be noted that according to References [18], this value is typically acquired by dividing occupational exposure limits by a safety factor (100 or 1000). Concentration c_{LCI} is taken from Lowest Concentration of Interest (EU-LCI) from European Commission lists.

However, the model values for exposure limit values (ELVs) of indoor air pollution, in accordance with the recommendations of the health-based ventilation guidelines [5], should be adopted in accordance with the current WHO guidelines given in the periodically issued *WHO Air Quality Guidelines* [7].

2.5. Selection of Submodels for Pollutant Components

Our overall selection of physical subcomponent equations and dependences for $\%PD = f(c_j)$ is presented in Table 3. The models presented were used to determine the IEQ for the sample building. The highlighted pollutants were taken into account in the case study building assessment.

Table 2. Exemplary pollution concentration reference values for low-pollution building.

Component of Pollution P_j	Reference Concentration	Reference Value	Reference/Recommendation List
CO_2 (outdoor 350 ppm)	c_{ELV}	380 ppm	EN 16798-1:2019
TVOC	c_{ELV} c_{th}	<300 µg/m³ 50 µg/m³	EN 16798-1:2019 MV. Jokl [45]
HCHO	c_{ELV} or c_{th}	30 µg/m³ (30 min) 100 µg/m³ (30 min) 9 µg/m³ (1 year) 300 µg/m³ 60 µg/m³	EN 16798-1:2019 WHO Guidelines for IAQ (2010) IEA-AIVC Report. Annex 68 (2017) AIHA Odor Thresholds for Chemicals with Established Health Standards (2013)
$VOC_{odorous}$ (as in EN 16516) Naphthalene Nitrogen dioxide Ammonia Ozone Carcinogenic VOCs	c_{th} or c_{ELV}	10 µg/m³ (1 year) 20 µg/m³ (1 year) 70 µg/m³ (1 year) 100 µg/m³ (8 h) 5 µg/m³	AIHA Odor Thresholds for Chemicals (2013) EN 16798-1:2019
Moisture content x in indoor air (enthalpy >55 kJ/kg, $t_a = 23\,°C$ and 60% RH) x [$g_{water}/kg_{dry\ air}$] $\Rightarrow$ H [g_{water}/m^3]	> unacceptable value of moisture content x at 23 °C.	x > 12 g/kg RH > 60%	EN 16798-1:2019
PM_{10}	c_{ELV}	20 µg/m³	WHO Air Quality Guidelines (2010)
$PM_{2.5}$	c_{ELV} $c_{DIRECTIVE}$	10 µg/m³ 2 25 µg/m³ (1 year)	EU Directive 2008/50/EC introduced additional $PM_{2.5}$ objectives targeting the exposure of the population to fine particles.
$VOC_{non-odorous}$ (EN 16516 includes pollutants with limit values on concentration that have been identified): Carbon monoxide Carcinogenic [1] VOCs	c_{ELV} or c_{LCI}	7 mg/m³ (24 h) 5 mg/m³	EN 16798-1:2019 EN 16516; WHO Guidelines for IAQ (2010)

[1] Limit values and carcinogenic effect: the level of PAHs, particles, benzene, and trichloroethylene should always be kept as low as possible.

Table 3. Selection of physical equations for IAQ_{index} components and dependences for $PD = f(c_j)$.

Sub Component	Input Parameters	Sensory Equations to Calculate the %PD Value	References
CO_2	$c(CO_2)$ (in ppm)	$PD_{IAQ(CO_2)} = 395 \cdot \exp(-15.15 \cdot C_{CO_2}^{-0.25})$	[36]
		$c_{IAQ(CO_2)} = 55\,833 \cdot (\ln(PD) - 5.98)^{-4}$	[45]
		$IAQ = 6 \times 10^{-0.07} c_{CO_2}^2 - 0.0025 \times c_{CO_2} + 1.9416$	[46]
		$PMV_{CO_2} = 6.364 \times \log \frac{c_{co_2}}{485}$	[47]
		$PD = 100 - 95 \cdot \exp(-0.03353 \cdot PMV^4 - 0.2179 \cdot PMV^2)$	[48]
$TVOC$ [1]	$c(TVOC)$ $\Rightarrow$ PD	$c_{TVOC} = 16{,}000(\ln(PD) - 5.988)^{-4}$; $\quad c_{TVOC}(\mu g/m^3)$	[49]
		$PD_{(IAQ)TVOC} = 405 \cdot \exp(-11.3 c_{TVOC}^{-0.25})$	[22,45]
	$IAQI \Rightarrow PD^*$	In Taiwan EPA Indoor Air Quality Index System (IAQI). Nine major indoor air pollutants are included: PM10, PM2.5, CO_2, CO, O_3, HCHO, TVOC, bacteria, and fungi. We proposed the use of a calibration curve $IAQI = f(c_{TVOC})$ for conversion to PD^* %	[30]
HCHO	$c(HCHO)$	$PMV_{HCHO} = 2\log \frac{c_{HCHO}}{0.01}$, $\quad c_{HCHO}(mg/m^3)$	[47]
	$(pi)^2 \Rightarrow$ $OI^3 \Rightarrow$ PD	$PD = 100 - 95 \cdot \exp(-0.03353 \cdot PMV^4 - 0.2179 \cdot PMV^2)$ $$PD_{IAQ(OI)} = \frac{1}{1 + \exp(2.14 \cdot OI - 3.81)}$$	[48] [50,51]
	$c(HCHO)$ $\Rightarrow$ $IAQI \Rightarrow PD^*$	In the Taiwan EPA Indoor Air Quality Index[6] System (IAQI). Nine major indoor air pollutants are included: PM10, PM2.5, CO_2, CO, O_3, HCHO, TVOC, bacteria and fungi. We proposed the use of the calibration curve $IAQI = f(c_{HCHO})$ for conversion[7] to PD^*%	[30]

Table 3. *Cont.*

Sub Component	Input Parameters	Sensory Equations to Calculate the %PD Value	References
	$c_{VOC} \Rightarrow OI^4$	Formulae for the conversion of odorant concentration c_j to odor intensity OI for recalculation (4) of the conversion of odor intensity to odorant concentration.	[52]
	$c_{VOC} \Rightarrow OI^3$	Conversion of the chemical concentrations c_j (mg m^{-3}) into odor concentrations c_{OD} (ou$_E$ m^{-3}) and odor intensities OI; $c_{OD,0}$ – unity of odor concentration ($c_{OD,0}$ = 1ou$_E$ m^{-3}) $$OI_j = k_j \log c_{OD,j} + 0.5$$	[53]
SVOC and VOC$_{odoros}$	$OI^3 \Rightarrow ACC_{VOC} \Rightarrow PD$	The mean of acceptability votes as a function of the mean of intensity votes. $ACC = -0.45\,OI + 0.93$, $R^2 = 0.979$ ACC in scale: from -1 to $+1$ OI odor intensity in scale: from 0 to 5 $$PD_{IAQ(OI)} = \left(\frac{\exp(-0.18-5.283ACC)}{1+\exp(-0.18-5.283ACC)}\right)3100$$	[54] [55]
	OI^2 in scale (pi) recalculation to $\Rightarrow OI^3$	The perceived intensity in pi units is determined by comparing the intensity of the sample with different specified intensities of the reference substance, (e.g., acetone). Concentrations for 1 to n (pi) follow a linear gradation of the acetone concentration. Confidence intervals should be within ±2 pi. Recalculation (pi) to OI	[51]
			[34]

Risk Assessment and Indoor Air Quality

Table 3. Cont.

Sub Component	Input Parameters	Sensory Equations to Calculate the %PD Value	References
Moisture x in high enthalpy h of moist air [5] (in room temperature t_a when air enthalpy $h > 55$ kJ/kg)	x (g_w/kg_a) $\Rightarrow$ H (g_w/m^3) (H-absolute humidity)	The recommended criteria for dimensioning of humidification and dehumidification. It is recommended to limit the absolute humidity value to $x = 12 g_w/kg_a$ or H(g_w/m^3)	[10]
		$h = 1.006t + 0.622(2501 + 1.84t) \cdot \dfrac{0.01 \cdot RH \cdot exp(23.58 - 4043/(t+273.15-37.58))}{Patm + 0.01 \cdot RH \cdot exp(23.58 - 4043/(t+273.15-37.58))}$	[56]
		IAQ acceptability equation $ACC = ah + b$ where a and b are different for different pollutants	
		$PD = \dfrac{100}{1+exp(-3.58+0.18(30-t_a)+0.14(42.5-0.01 p_v))}$ %	[23]
		$ACC = 5.63 + 0.46 \ln RH - 1.32 \ln h$ uncertainty of acceptability ACC is ± 0.12.	[57]
		$PD_{IAQ(h)} = \left(\dfrac{exp(-0.18-5.283ACC)}{1+exp(-0.18-5.283ACC)}\right)3100$	[55]
Particulate matter [6] PM$_{2.5}$ PM$_{10}$	$c_{PM2.5}$ c_{PM10}	IAQI (Indoor Air Quality index) [6] is the proposal for a system comparable to the World AQI EPA system $c_{PM2.5} \Rightarrow IAQI \Rightarrow PD*\%$ $c_{PM10} \Rightarrow IAQI \Rightarrow PD*\%$	[30]
Air pollutants [7] VOC$_{non\text{-}odorous}$ CO, NO$_2$	c_{VOC}	IAQI [6] value for c_{VOC} was calculated with System IAQI [27] by interpolation and after which the index IAQI must be converted [7] to $PD*\%$	[30]

[1] TVOC, according to Reference [45], represents a narrow chromatographic picture that excludes, for example, the lower aldehydes, e.g., formaldehyde. [2] The measurement of the intensity of the odors in the building in which emissions from construction materials occur can be performed by a panel of participants, where the room is treated as a "test room for background odor" according to Section 6.8.1 of ISO 16000-28:2013, "Indoor Air—Part 28: Determination of odor emissions from building products using test chambers" (2013) [51]. The assessment of the 90% confidence level is possible through the use of a 15-pi odor intensity scale with a reading uncertainty of ± 2 pi. [3] OI is perceived odor intensity on a six-level scale from 0 to 5 (no odor = 0, slight odor = 1, moderate odor = 2, strong odor = 3, very strong odor = 4, and overpowering odor = 5). [4] Based on the study Kim and Kim (2014) [52] selected 22 odorants with a similar chemical structure (structural formula) and determined an equation to convert the concentration x in ppm to intensify y, with odor OI on a scale of zero to five.

The study included odors from the distribution of food. These 22 odorants can be divided into five chemical groups: (1) reduced sulfur compounds, (2) carbonyls, (3) nitrogenous compounds, (4) VOCs, and (5) volatile fatty acids. For example, for the group of reduced sulfur compounds for compound number 1, the conversion equation for H_2S is $Y = 0.950logX + 4.14$, for the carbonyl compounds group for compound number 10, ammonium NH_3, it is $Y = 670logX + 2.38$, and for the VOC group for styrene it is $Y = 1.420logX + 3.10$. [5] If one considers a cooling system that removes heat from a space but does not remove moisture unless condensation occurs, such as radiant cooling without dehumidification in a ventilation system, the importance of humidity is very clear. The sensible cooling of air in a room (no change in absolute humidity) from 25 °C and 60% RH to 20 °C (process a–b for x = 0.012 kg_w/kg_a).This value, coupled with a high temperature, $t_a = 25$ °C, is accepted as the critical limit value for dehumidification by EN16798-1:2019 [10] and can be converted into an absolute humidity H (g_w/m^3). These processes are expected to significantly increase thermal comfort and air quality. Nevertheless, the same change in enthalpy of the air h can be achieved by simply reducing the humidity by 10% RH and keeping the temperature constant (processes a–c). Since IAQ is a function of enthalpy, these are expected to be the same. Here, a change in humidity of 10% RH at constant temperature is equivalent to a change in temperature of 5 or 6 °C at constant moisture content in air x ($kg_w/kg_{dry air}$). [6] The IAQI system [30] adopts the methodology of AQI to set up the range of IAQI values from zero to 500, including 50, 100, 150, 200, 300, and 500. The IAQI values of 100 and 150 correspond to the concentrations in the Taiwan IAQG standard. Other IAQI values between 50 and 200 correspond to the concentration rankings given by several reference resources including the US EPA AQI system [31]. All on-site concentrations of indoor air pollutants (HCHO, TVOC, PM10, PM2.5, CO, CO_2, O_3, bacteria, fungi, SO_2, and NO_2) are combined using the IAQI system. The IAQI values are calculated on the basis of the concentration value c using the interpolation method for each air pollutant. In the IAQI system, the index range 0–50 is "good" with a significance level of "little or no risk", 51–100 is "moderate" where "sensitive persons or those with respiratory symptoms are concerned", 101–150 is "unhealthy for sensitive groups", 151–200 is "unhealthy for all individuals", 201–300 is "very unhealthy—more serious health effects for everyone for short-term exposure", and 301–500 is "hazardous" with "health warning of emergency conditions for everyone". Therefore, the comfort scale for IAQI is adequate for index values from zero to 200 and our proposal is to use the "hypothetical" scale of PD^* of 0–100% in this range, converted from IAQI. [7] The method of conversion of the IAQI scale to the PD^* scale is based on experiences gained during research [27]. The IAQI values in a health-risk scale can be given in % for persons giving a verbal answer of "no risk", "moderate", and "unhealthy" as their health risk evaluation. Therefore, when determining, for instance, the concentration function values of the $\%PD_{IAQI}(TVOC)$ and $\%PD_{IAQI}(HCHO)$ components, it is necessary to recalculate the scale IAQI = f(c) in the range from zero to 200 to the scale PD^* from 0 to 100%. There are break points in the new $PD^*\%$ scale: for IAQI values 0–50, $PD^*\%$ is 0–25, for IAQI values 51–100, $PD^*\%$ is 26–50, for IAQI values 101–150, $PD^*\%$ is 51–75, and for IAQI values 151–200, $PD^*\%$ is 76–100.

The air quality indexes (i.e., AQI [31] and IAQI [30]) are piecewise linear functions of the pollutant concentrations. At the boundary between AQI categories, there is a discontinuous jump of one AQI unit. To convert from concentration c_j to I_j (in the converted scale index I_j will be PD_j, Equation (16) is used.

$$I_j = ((I_{high} - I_{low})/(c_{high} - c_{low})) \cdot (c_p - c_{low}) + I_{low} \tag{16}$$

where I_j is the air quality index I in the $PD^*\%$ scale, c_{low} is the pollutant concentration break point, which is $\leq c_j$, c_{high} is the pollutant concentration break point, which is $\geq c_j$, I_{low} is the index break point corresponding to c_{low}, I_{high} is the index break point corresponding to c_{high} and c_p is the truncated (to an integer) actual concentration for the pollutant. Little data exist on the AQI's metrological reliability for AQI and IAQI. Only in the EPA Air Program undertaken at Cornell University [26] is there a previous review of the quality assurance requirements for AQI.

2.6. The Representative VOCs for Indoor Environment

The time when IAQ studies focused on a class of contaminants referred to as volatile organic compounds (VOCs) is bygone. The analytical methodology available was the primary basis for this focus, but the recent broadening of analytical methods has led to growing realization that other compounds (i.e., SVOCs) beyond traditional VOCs are implicated in IAQ problems. The choice of VOCs remains a challenge in IAQ assessment. Moreover, VOCs is somewhat vague term, the definition of which is not universally agreed upon. It has been defined in terms of vapor pressures and boiling points, as well as molecular chain lengths detectable by chromatographic techniques. Due to the complexity of VOC emission profiles, it is tempting to simplify the analysis and reporting of emissions by grouping all detected compounds together. The first problem with this approach is that individual compounds have highly variable health and/or comfort effects, the result being that concentration alone is not predictive of IAQ effects. Levels of concern vary by orders of magnitude, so a collective concentration will not correlate with IAQ. Second, VOC detection and quantification are highly method dependent. A given sampling and analysis system cannot capture or respond to all the VOCs present in any indoor environment or in the test chamber for a given emitting material. Thus, the term "total" is misleading. The important aspect of IAQ submodel selection is the strategy defined by the US EPA as "VOCs—Total versus Target: Irritancy, Odor and Health Impact". The representative 90 target VOCs were presented by Canada's National Research Council's Institute for Research in Construction (NRG-IRC) in collaboration with several academic and governmental partners, including Health Canada. The compounds were selected based on health impact, occurrence in indoor air, known emission from building materials, as well as suitability for detection and quantification by gas chromatography-mass spectrometry (GC-MS) or high-performance liquid chromatography (HPLC). Our list of target VOCs was actually representative for indoor environment and recommended by the HealthVent project and is provided in Table 1.

2.7. Steps of ΣIAQ_{index} Calculation

After selection of the IAQ model type ($\Sigma IAQ_{quality}$ or $\Sigma IAQ_{comfort}$), the IAQ_{index} evaluation was carried out using the complex model ΣIAQ from Figure 5, which should contain the following stages.

(a) Calculation of the total concentration of pollutants in the ventilated space or the total mass of air pollutants per m^3, the level of which is to be reduced by the ventilation process (taking into account the ventilated volume of the room and the emissions present).
(b) Selection of the IAQ_{index} model shape from the models defined by Equations (12)–(14), with the provision that due to the multiplication of the submodels for $IAQ(VOC_{odorous})$, the number of subcomponents of the $IAQ_{comfort}$ model will be more than five.
(c) Processing the input data of the submodels to obtain the concentration value c_j, e.g., converting the measured OI value into a concentration value for a given pollutant c_j in $\mu g/m^3$.

(d) Calculation of the excess concentration values for each identified contaminant (Table 2) $\Delta c_j = c_j - c_{ref}$.

(e) Calculation of the sum of excess concentrations (see Table 2), $\sum \Delta c_j$.

(f) Calculation of adjusted weights Wj for the selected model equations. $IAQ_{quality}$ and $IAQ_{comfort}$, are determined on the basis of arithmetic means or by adjusting all the values of Δc_j in a given model using Equation (10), only for groups of pollutants with similar concentration values.

(g) Calculation of the value of the ventilating air flow for the environment described in the IAQ_{index} model in accordance with the requirements of the standard EN 16798-1:2019 (a method using the criteria for the ventilation required for the individual substance emitted) [10].

(h) Calculation of given IAQ environmental input parameters, including concentrations of pollutants c_j assigned to submodels. The PD values from their sensory equations (Table 3) are presented as the dependence of the percentage of persons dissatisfied $PD = f(c_j, \ldots)$, from one of the formulas from Table 3, in order to determine this function.

(i) Selection of the $\sum IAQ_{quality}$ model equation (with weights W_1, W_2, and W_3) or the $IAQ_{comfort}$ model equation (with weights $W_1, \ldots W_5$ or more) and calculation from Equation (13) of the value with adjusted weights, followed by multiplication of IAQ submodels (Equation (8)) and insertion as a term of the IEQ_{index} in Equation (18) [34].

When selecting the $\sum IAQ_{comfort/health}$ model type, an IAQ_{index} evaluation is carried out using the combined model $\sum IAQ$ from Figure 5, which should contain the following steps.

(a) Calculation of the total concentration of pollutants $\sum c_j$ in the ventilated space or the total mass of air pollutants per m^3, the level of which is to be reduced by the ventilation process (taking into account the ventilated volume of the room and the emissions present).

(b) Choosing the IAQ_{index} model (12) from among the models defined by Equations (12)–(14), with the provision that by multiplying the $IAQ(VOC_{non-odorous})$ submodels, the number of subcomponents of the $IAQ_{comfort/health}$ model will be more than seven.

(c) Processing of submodel input data to obtain concentration values c_j in $\mu g/m^{-3}$.

(d) Calculation of excess concentration values for each pollutant identified (Table 2) using $\Delta c_j = c_j - c_{ref}$.

(e) Calculation of the sum of excess concentration values for submodels 1–7 via $\sum \Delta c_{1 \ldots 7}$

(f) Determination of the adjusted weights $W_{1, \ldots 7}$ for the equation of the $IAQ_{comfort/health}$ model on the basis of arithmetic means or by adjusting all the values of Δc_j in a given model using Equation (10), only for groups of pollutants with similar concentration values.

(g) Calculation of the values of the ventilating air stream from the total concentration of indoor air pollutants $\sum c_j$ (for instance, Reference [42]), for the environment described by the $\sum IAQ_{index}$ model (Figure 5) in accordance with the requirements of the EN 16798-1:2019 standard for the individual substances emitted and using an alternative method when the concentration in the room has stabilized.

(h) Calculation of the given IAQ input parameters, including concentrations of all pollutants, c_j, assigned to the submodels. The PD values are taken from their sensory equations (Table 3) depending on the percentage of persons dissatisfied, $PD = f(c_j, \ldots)$, or selected from the formulas for determining this function given in Table 3.

(i) Development of $IAQ_{PM2.5}$, IAQ_{PM10}, and $IAQ_{non-odorous}$ submodels. When it is planned to use the indoor air quality index scale IAQI or a similar scale, it is necessary to convert these to PD^* values in %, in two steps: (1) by reading from the standard curves of the $IAQI = f(c_j)$ all IAQI values for the determined (measured) $VOC_{non-odorous}$ concentration values and using the converted scale calibration curve, $PD = f(IAQI)$, by reading from the recalibration curve (Equation (16)), the $PD = f(c_j)$ values on the dissatisfaction rating scale from zero to 100 in %, according to Footnote 7 in Table 3.

(j)   Calculation of IAQ values for *Pj* pollutant submodels from Equation (8) and insertion into the model equation ΣIAQ~comfort/health~ (14) (with weights $W_1, W_2, \ldots, W_7$ or more).

2.8. The IEQ Assessment Equation with ΣIAQ As a Subcomponent

The proposed IAQ model can be a substantial component of the IEQ model; for example, in the case study shown later in the article. The indoor environmental quality index refers to the quality of a building's environment with respect to the occupants' satisfaction in %. The morphology of the IEQ~index~ model used to assess buildings, to determine as an IEQ component the IAQ~index~ and to determine other subcomponents TC~index~—thermal comfort, ACc~index~—acoustic comfort and L~index~—light quality based on measurements of physical properties in each of the submodels—in accordance with the scheme of the Piasecki–Kostyrko model, is presented in Figure 6 [22].

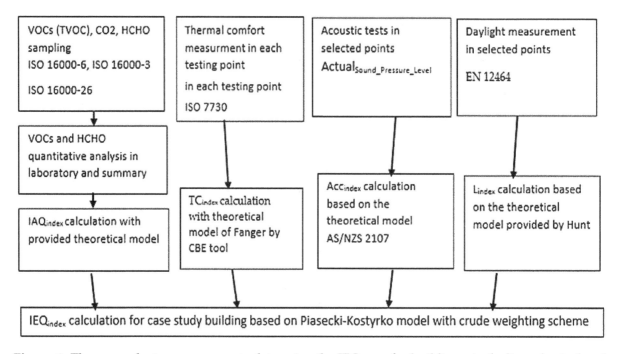

Figure 6. The research steps necessary to determine the IEQ~index~ for buildings, including physical and design parameters of buildings and subcomponent models.

The EN 16798-1:2019 is the reference for IEQ model creation [22,33]. The standard allows complex indoor information to be presented as one overall indicator of indoor environmental quality of the building—IEQ~index~. The model reliability, including the uncertainties of measurements and data for this model, was discussed clearly in Reference [34], where the authors also presented the internal incongruity in the IEQ model structure and the justification for using the crude weights method for each subcomponent. Originally, the IEQ model was expressed as a polynomial equation consisting of four terms by Wong [43]. The IEQ~index~ is composed of the following subcomponents (SI_i): thermal comfort (TC_{index}), indoor air quality (IAQ_{index}), acoustics (ACc_{index}), and lighting quality (L_{index}). Multiplying their weights, W_i, leads to Equation (17).

$$IEQ_{index} = \Sigma\ W_i \cdot SI_i \tag{17}$$

The authors adopted the crude weighting system, where all elements are weighted in the same way (0.25 for W_1–W_4), as shown in Equation (18).

$$IEQ_{index} = 0.25 \cdot TC_{index} + 0.25 \cdot \Sigma IAQ_{index} + 0.25 \cdot ACc_{index} + 0.25 \cdot L_{index} \tag{18}$$

As a consequence of the equation, the subcomponents SI_i (the predicted percentage of those satisfied) can be calculated using Equation (19).

$$SI_i = 100 - PD(SI_i) \qquad (19)$$

where PD is the predicted percentage dissatisfied (PPD) and $PD(SI_i)$ is the percentage of persons dissatisfied with the IEQ subcomponent (SI_i) level. The authors' simulations for IEQ_{index} sub-indices and preliminary metrological analysis of the overall IEQ model fitting were performed with Monte Carlo tests.

It is easy to show that the standard deviations of these values are equal:

$$SD(SI_i) = SD(PD(SI_i)) \qquad (20)$$

2.9. A Case Study of a Building

The experimental part of this study was performed simultaneously with the BREEAM certification process, including determination of the three primary IAQ pollutants: formaldehyde concentration, CO_2, and VOCs in the indoor air [22]. The building is a high tower, made of a convex concrete–steel structure with a glass facade. The basic information on the assessed building is presented in Table 4. At the time of the test, the building had a standard empty office without furniture (so-called pre-occupancy stage). The walls were plastered and painted, the suspended ceilings were in place, and the floors were finished with synthetic carpets. All building installations were active, including the mechanical ventilation controlled by the Building Management System BMS system with zonal CO_2 concentration sensors. The building was tested a few days after the formal end of finishing works. The tests were made on the 55th and 47th floor.

Table 4. Information on the building in the case study.

Office Building Certificate	Facade View	Indoor View	Life-Stage	Number of Floors	Net Area (m²)	IAQ Assessed	
						Area (m²)	Number of Floors
BREEA Mexcellent			pre-occupant	49	59,000	3000	2

Measurement points in the building were determined based on the analysis of frequencies of designed occupancies of the room and interior finish standards (open spaces). The sampling plan was prepared with the BREEAM assessors conducting the certification process of the facility. The main focus was on the IAQ index of open spaces in which the largest number of people may reside, and these represent the largest occupied usable floor space. According to the detailed design project documents, the building emphasizes the use of materials with known and low emission levels (BREEAM certified).

2.10. The Equipment, Measurements, and Experimental Approach

Standardized CEN and ISO analytical methods were used to determine the VOC concentrations and CO_2 and formaldehyde concentrations in the indoor air of the building. Selection of the sampling points was made with the BREEAM assessor in two representative office zones per tested floor and a minimum of two floors. The building was tested three days after formal final finishing works at the pre-occupancy stage with no users inside. For this office building, the tests were conducted on the 55th and 47th floors. Air samples were collected using an active sampling procedure with an electronic mass flow controller, which controlled the air flow (10 dm³/h for VOC tests and up to 30 dm³/h for formaldehyde tests). Indoor samples were set up in selected representative office locations, approximately 1.5 m above the floor, away from windows, doors, potential emission sources, and direct sunlight. Air samples were tested in accordance with the ISO 16000-6:2011 and ISO 16000-3:2011 standards. The VOCs were assessed using tubes filled with Tenax adsorbent. Then, they were thermally desorbed using a thermal desorption apparatus (TD-20, Shimadzu, Tokyo, Japan). The process of separation and analysis of volatile compounds was achieved using a gas chromatograph equipped with a mass spectrometer (GC/MS) (model: GCMS-QP2010, Shimadzu, Tokyo, Japan). The following GC oven temperature program was applied: initial temperature 40 °C for five min, 10 °C per min to 260 °C, and the final temperature of 260 °C for 1 min. The 1:10 split ratio injection mode was applied. The method used has a limit of quantification of 2 µg/m^3. The volatile compounds were identified by comparing the retention times of chromatographic peaks with the retention times of reference compounds and by searching the NIST data base (National Institute of Standards and Technology, Gaithersburg, MD, USA) mass spectral database. Identified compounds were quantified using a relative identification factor obtained from standard solution calibration curves. TVOC was calculated by summing identified and unidentified compounds eluting between n-hexane and n-hexadecane. In order to determine volatile aldehydes, air samples were taken via cassettes using a solid absorbent silica gel coated with 2,4-dinitrilophenyl hydrazine (2,4-DNPH), and then subjected to a laboratory test using high-performance liquid chromatography (HPLC) with UV-Vis detection (Dionex 170S, Dionex, Sunnyvale, CA, USA) and an isocratic pump (Dionex P580A, Dionex, Sunnyvale, CA, USA). The described method has a limit of quantification at 2 µg/m^3.

Other IEQ$_{index}$ components were tested as follows. The acoustic tests confirming the designed values were carried out by the measurement of the equivalent sound levels, LAeq, in the selected locations. The measurements were carried out during the daytime (starting at 11:00). The following equipment was used for the measurements: Brüel&Kjær 4231 acoustic calibrator (Brüel&Kjær, Nærum, Denmark), Nor-121 analyzer (Norsonic, Tranby, Norway), Brüel&Kjær 4165 measuring microphones (Brüel&Kjær, Nærum, Denmark), analyzer with microphone Norsonic-140 (Norsonic, Tranby, Norway). Before the tests were carried out, the calibration of the measuring path was conducted in accordance with the instructions to "check the acoustic measurement channel". The test results were evaluated in relation to the requirements considering permissible sound levels A in rooms intended for human dwellings. Thermal environmental measurements were provided using the microclimate multifunctional instrument HD32.1 and the tests were in accordance with ISO 7726 and ISO 7730. VOCs were tested simultaneously at all points. Visual comfort (Hea 01) was confirmed by using a MAVOLUX 5032C instrument (USB version) with a 3C15683 detector (Gossen, Nürnberg, Germany), in accordance with EN 12464 provisions.

2.11. Additional Explanations

The adaptation of the IAQ model to a practical casestudy was mainly for illustrative purposes in the context of the presented IAQ calculation/aggregation method. We did not focus deeply on discussing the technical or environmental issues of the presented building. Other IEQ subcomponents, such as thermal, acoustic, and visual satisfaction (in %), used to determine the IEQ index, were experimentally determined and partly presented in References [22,33]. Authors do not focus on these results in this article, as they have already been discussed in other papers [22].

3. Results

3.1. Results for theIAQ$_{index}$ and IEQ$_{index}$ Prediction

A previous publication of ours [22] reported on IEQ and IAQ building assessments for a larger number of BREEAM buildings, where IEQ was assessed without calculating the combined $\sum$IAQ index. The combined model of the $\sum$IAQ$_{index}$ presented in this paper had not yet been previously developed, and we were limited in determining the IEQ$_{index}$, thus we only took into account two of the most well-known pollutants (i.e., CO_2 and TVOC) separately. The assessment of the IEQ index was made by adaptation of the measured parameters (complying with the draft EN 16798-1:2019 standard for indoor environments) as the input values for the submodels of the IEQ_{index}. The input values for the case study are presented in Table 5, which provides the input data for determining the IEQ_{index} sub-indices of thermal comfort (TC_{index}), indoor air quality (IAQ_{index}), acoustics (ACc_{index}), and lighting quality (L_{index}) for an office building (47th floor) three days after completion of the finishing work before users were allowed in the building (i.e., pre-occupancy stage).

Table 5. Physical parameters [1] and IEQ_{index} results calculated using Equation (9) separately for an IAQ_{index} with internal air pollution of CO_2 and an IAQ_{index} with internal TVOC air pollution, assuming a realistic uncertainty of parameter measurement for the case study of a building (47th floor; open space) three days after the completion of finishing works.

Sub-Index	Sub-Index $PD(SI_i)$ Models	Input Values	Sub-Index (Satisfied) and ±SD
TC_{index}	PMV (Fanger-CBE-ISO 7730) $PMV = f(t_a, t_r, v_a, p_a, M, I_{cl,})$ $PD_{TC} = f(PMV)$	I_{cl} 0.55 clo t_a 24.0 °C t_r 24.5 °C v_a 0.15 m/s RH 45% M 1.1 met	90% ± 3.2%
IAQ_{index}	$PD_{IAQ(CO_2)} =$ $395 \cdot \exp(-15.15 \cdot C_{CO2}^{-0.25})$	450 ppm	85.2% ± 0.6%
	$PD_{IAQ(TVOC)} =$ $405 \cdot \exp(-11.3 \cdot C_{TVOC}^{-0.25})$	787 µg/m³	52.0% ± 18.0%
ACc_{index}	$PD_{ACc} =$ $2 \cdot (Actual_{Sound\_Pressure\_Level}(dB(A))$ $-Design_{Sound\_Pressure\_Level}(dB(A)))$ Actual (background) noise level Design sound level	55 dB(A) 45 dB(A)	80% ± 6.7%
L_{index}	$PD_L = -0.0175 + 1.0361/\{1 +$ $\exp(+4.0835 \cdot (\log_{10}(E_{min}) -$ $1.8223))\}$	450 lux	98.4% ± 9.0%
$IEQ_{(CO_2)}$ First variant with c_{CO2} as an IAQ_{index} parameter IEQ_{CO_2} = 92.2% ± 5.8% [1]			
IEQ_{TVOC}	Second variant with c_{TVOC} as an IAQ_{index} parameter	IEQ_{TVOC} = 80.1% ± 10.7%	

[1] The IEQ and its measurement's uncertainty (with subcomponent standard deviation values) were calculated for IEQ physical parameter values, where t_a is the air temperature (°C), t_r is the mean radiant temperature (°C), v_a is the relative air velocity (m/s), p_a is the water-vapor partial pressure (Pa), M is the metabolic rate (met), and $I_{cl,}$ is the clothing insulation (clo). In addition, c_{CO2} is the concentration in ppm, c_{TVOC} is the highest observed TVOC concentration in µg/m³, actual noise is in dB(A), and E_{min} is the minimum daylight illuminance (lux).

3.2. Results for the ΣIAQ_{index} and IEQ_{index} Assessment Including Identified Pollutants (CO_2, TVOC, and HCHO)

The example of a modified calculation of the collective submodel $\Sigma IAQ_{quality}$ for three basic pollutants, as a component of the IEQ model for determining one project value for this indicator, is provided in two variants. The first variant uses sub-indices of IAQ for two pollutants, CO_2 and TVOC, which are described in Table 5, as well the sub-index of the third pollutant, HCHO (according to Reference [47]), where these differences in the approaches mean one must combine them into one submodel ΣIAQ in order to be used in IEQ calculation. The second variant uses submodels of IAQ for TVOC and HCHO pollutants based on the IAQI system [30] and then converts them into percentages of persons dissatisfied PD* in %. According to the diagram of the model ΣIAQ from Figure 5 and using Equation (12) of the $\Sigma IAQ_{quality}$ model, the submodel weights are calculated as follows.

W_{CO_2} for the submodel IAQ(CO_2) = 0.5 is a component of the polynomial:

$$\sum IAQ_{quality} = 0.5 \cdot IAQ(CO_2) + 0.5 \cdot IAQ(VOC) \tag{21}$$

W_{VOC} for submodel IAQ(VOC) = 0.5 is a weight for combined submodel of the polynomial:

$$IAQ(VOC) = W_{TVOC} \cdot IAQ(TVOC) + W_{HCHO} \cdot IAQ(HCHO) \tag{22}$$

with the terms W_{TVOC} and W_{HCHO} calculated from Equation (14) using the measured values c_j (actual concentration of TVOC and HCHO) and the reference values c_{ref} (Table 6).

Table 6. Calculation of the weights of the W_{TVOC} and W_{HCHO} values for the two sub-indices of the combined IAQ(VOC) model.

Sub-Index	Input Value c_j	Input [1] Value c_{ref}	Excess Concentration Δc_j	W_j
IAQ(TVOC)	787 µg/m³	300 µg/m³	487 µg/m³	0.96
IAQ(HCHO)	18 µg/m³	0	18 µg/m³	0.04

[1] EN 16798-1:2019 for a very low-pollution building.

In our case study, the value of the submodel IAQ weight for the model (HCHO) ought to also be calculated from the measured value and the reference value c_{ref}. However, formaldehyde is an unusual pollutant because, although it belongs to $VOC_{odorous}$ compounds, the concentrations found in buildings are many times lower than the HCHO threshold c_{th} = 300 µg/m³ according to the WHO [7] and lower than the threshold concentrations of HCHO from 60 µg/m³ to 70 µg/m³ issued in 2013 by the American Industrial Hygiene Association [44]. The permissible value of c_{ref} = 100 µg/m³ is also higher than the formaldehyde concentration found in buildings, according to Reference [5] and the standard EN16798-1:2019 [10]. Therefore, the authors propose that in such a case (to avoid a negative value of Δc_j), the value modelling the reference should be taken as zero. Then, the form of the adjusted W_{HCHO} weight in the model described by Equation (11) for air with three pollutants, would be as follows.

$$W_{HCHO} = \frac{\Delta c_j}{\sum\limits_{j=2...3} \Delta c_{2...3}} = \frac{(c_{HCHO} - 0)}{(c_{TVOC} - c_{ELV}) + (c_{HCHO} - 0)} \tag{23}$$

The results of the weights assessment for the two variants of the $\sum IAQ_{quality}$ model are presented in Table 6.

According to the diagram of the model $\sum IAQ$ from Figure 5 and Equation (8), we proposed sensory equations for the percentage of persons dissatisfied %PD* in two variants.

The submodel $\sum IAQ$'s first variant includes the following:

1. The IAQ submodels used so far in References [22,33] for CO_2 and TVOC pollutants, as shown in Table 5;

2. The IAQ submodel for formaldehyde, using two types of equations depending on the range of HCHO concentrations measured in the building. Formaldehyde concentrations in the air with values above the threshold concentration, c_{th}, for its odor, i.e., above 60 or even 300 µg/m³, can be used to create IAQ submodels for rooms with volatile and aromatic VOC compounds as well as for the HCHO equation [50].

$$PD_{HCHO} = \frac{\exp(2.14 \cdot OI - 3.81)}{\exp(2.14 \cdot OI - 3.81) + 1} \tag{24}$$

However, in the case study building, the maximum concentration of HCHO was 18 µg/m³ and, therefore, its concentration in the air was several times lower than the concentration of the odor threshold, c_{th} [44]. The intensity of the formaldehyde odor was undetectable under these conditions, and the sensory equation $PD = f(OI)$, which is appropriate for sensory detection of IEQ, is not applicable for odors below the threshold. Therefore, for small concentrations, we proposed the use of the equation taken from the work of Zhu and Li [47] based on the analysis of "health effects on the human body", derived from "indoor air quality comfort evaluation experiments and the literature".

$$PMV_{HCHO} = 2log\frac{c_{HCHO}}{0.01} \qquad (25)$$

This equation links the value of the new unit "the effect of formaldehyde on human comfort", called PMV_{HCHO}, with its c_{HCHO} concentration ($\mu g/m^3$) in the air. It covers the range from 10 $\mu g/m^3$ to 320 $\mu g/m^3$ and, as declared by the authors, this value has the same nature as PMV thermal comfort, which can be converted into a $PD\%$ unit according to the formula in Reference [48], experimentally confirmed for nearly zero energy buildings (NZEBs) by Reference [58].

$$PD_{HCHO} = 100 - 95 \cdot \exp(-0.03353 \cdot PMV^4 - 0.2179 \cdot PMV^2) \qquad (26)$$

The submodel $\sum IAQ$'s second variant includes the following.

i.   The $IAQ(CO_2)$ submodel used so far for CO_2 pollution, as shown in Table 5.

ii.   The IAQ submodels for TVOC and HCHO types of pollution used as indoor air quality index ratio values borrowed from the IAQI system [30], which are then converted into percentages of persons dissatisfied (PD* in %) in the following way

(a)   The reference curves of $IAQI = f(c_j)$ [30] for two dependencies of the IAQI index on TVOC and HCHO contamination values must be reconstructed. On the y-axis are the IAQI index values from zero to 200 in the range from "no risk" to "unhealthy" and on the x-axis, the c_j values are presented.

(b)   In accordance with the measured values of c_{TVOC} and c_{HCHO}, the values $IAQI_{TVOC}$ and $IAQI_{HCHO}$ are determined from the functions $IAQI_{TVOC} = f(c)$ and $IAQI_{HCHO} = f(c)$.

(c)   Based on the IAQI system parameters [30] given in Footnote 6 of Table 3, which are presented as data for the functions for indexes, the $IAQI_{TVOC}$ and $IAQI_{HCHO}$ values appropriate for the break pointsi n perceived pollution concentration values are calculated in a range from zero (good) to 200 (unhealthy) using the ordinal scale $IAQI = f(c)$ [27]. This function was converted to $IAQI_{TVOC}$ and $IAQI_{HCHO}$ scales using the concentration function scales $PD*(TVOC)$ and $PD*(HCHO)$ in the $PD*$ range from 0 to 100% (Figure 8).

(d)   The data used for calculation and conversion of IAQI and $PD*$ scales are presented in Table 7.

(e)   Based on data determined for the new converted scales (Table 7) for the percentage of persons dissatisfied ($PD*(TVOC)$ and $PD*(HCHO)$ concentration functions), the $PD*(c)$ working graphs were drawn (Figure 7). (To harmonize the concentration scale of pollutants on the y-axis, c_{HCHO} values multiplied by 10 were applied.)

(f)   The interpolation of the percentage of persons dissatisfied ($PD*(TVOC)$ and $PD*(HCHO)$) in % for the measured values of pollution concentrations in the air must be made using the graph from Figure 7 or using Equation (16) [31].

Table 7. Recalculation and conversion of IAQI value scales.

PD* Value	IAQI Value	Evaluation	c_{TVOC}-TVOC (1 h)		c_{HCHO}-HCHO (1 h)	
%	-	-	ppm	$\mu g/m^3$	ppm	$\mu g/m^3$
0	0	No risk	0	0	0	0
25	50	Good	0.3 [1]	300	0.01 [2]	12.3
50	100	Moderate	0.9	900	0.04	49.1
75	150	Unhealthy for sensitive	3.0	3000	0.10	122.8
100	200	Unhealthy	4.6	4600	0.75	921.0

[1] Conversion factors TVOC: 0.3 ppm corresponds to 300 $\mu g/m^3$ [44]. [2] Conversion factors HCHO: 1 ppm corresponds to 1228 $\mu g/m^3$ [44].

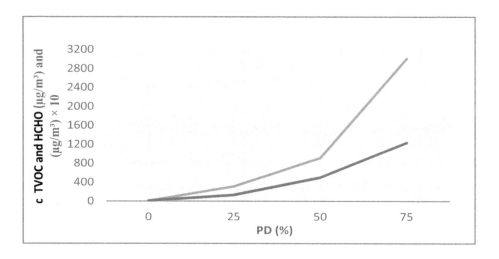

Figure 7. The percentage of persons dissatisfied function of TVOC (blue line) and HCHO (brown line) versus concentration.

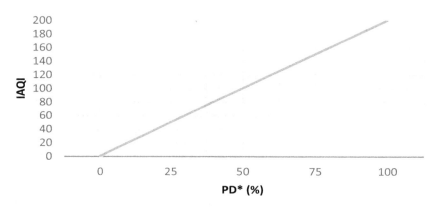

Figure 8. Percentage of persons dissatisfied, %PD*, in relation to IAQI values.

In the context of the results for the overall IEQ model index when treating both the main pollutants CO_2 and TVOC separately, we present in Table 8a the transformed IEQ calculations with the sub-indices ΣIAQ. The results for the individual case study building IAQ subcomponents are taken from Table 5 (for c_{TVOC} = 787 $\mu g/m^3$ and c_{HCHO} = 18 $\mu g/m^3$ [22]). The IEQ index with ΣIAQ$_{quality}$ values was calculated using two variants—the first conventional and the second borrowed from the IAQI scale [30] (Table 8a).

Standard deviations for each concentration are provided in Table 8b.

Table 8. (a) Physical parameters (Footnote [1] in Table 5) and IEQ results calculated from Equation (9) with $\sum IAQ$; assuming realistic uncertainty of parameter measurements for the case study building (47th floor; open space) a few days after completion of the finishing works. (b) Measured pollutant concentrations c and standard deviations SD(c).

(a)

Sub-Index	Sub-Index $PD(S_1)$ Models	Input Values	Sub-Index (Satisfied) and ±SD
TC_{index}	PMV (Fanger-CBE-ISO 7730) $PMV = f(t_a, t_r, v_a, p_a, M, I_{cl,dyn})$ $PD_{TC} = f(PMV)$	I_{cl} 0.55 clo t_a 24°C t_r 24.5°C v_a 0.15 m/s RH 45% M 1.1 met	90.0% ± 3.2%
$\sum IAQ_{index}(1)$ Sub-indices First variant	$PD_{IAQ(CO_2)} = 395 \cdot \exp(-15.15 \cdot C_{CO2}^{-0.25})$ $PD_{IAQ(TVOC)} = 405 \cdot \exp(-11.3 \cdot C_{TVOC}^{-0.25})$ $PMV_{HCHO} = 2log\frac{c_{HCHO}}{0.01}$ $PD_{HCHO} = 100 - 95 \cdot \exp(-0.03353 \cdot PMV^4 - 0.2179 \cdot PMV^2)$	$c = 450$ ppm $c = 787$ µg/m³ $c = 0.018$ mg/m³	85.2% ± 0.6% 52.0% ± 18.0% 65.8% ± 10.7%%
$\sum IAQ_{index}(1)$	$IAQ_{VOC} = 0.96 \cdot IAQ_{variant1}(TVOC) + 0.04 \cdot IAQ_{variant1}(HCHO)$ $\sum IAQ_{index}(1) = 0.5 \cdot IAQ(CO_2) + 0.5 \cdot IAQ_{variant1}(VOC)$		53.0% ± 17.3% 69.1% ± 9.0%
$\sum IAQ_{index}(2)$ Sub-indices [1] Second Variant	$PD_{IAQ(CO_2)} = 395 \cdot \exp(-15.15 \cdot C_{CO2}^{-0.25})$ PD^*_{TVOC} read from the graph with $PD^*_{TVOC} = f(c_{TVOC})$ or determined from Equation (16) PD^*_{HCHO} read from the graph with $PD^*_{HCHO} = f(c_{HCHO})$ or determined from Equation (16) SD for c_{TVOC} and c_{HCHO} at "break points" ±12%	$c = 450$ ppm $c = 787$ µg/m³ ±18% $c = 18$µg/m³ ±12%	85.2% ± 0.6% 54% ± 13.8% 71.1%± 11.0%
$\sum IAQ_{index}(2)$	$IAQ_{VOC} = 0.96 \cdot IAQ_{variant2}(TVOC)$ $+ 0.04 \cdot IAQ_{variant2}(HCHO)$ $\sum IAQ_{index}(2) = 0.5 \cdot IAQ(CO_2)$ $+ 0.5 \cdot IAQ_{variant\,2}(VOC)$		54.7% ± 13% 70.0% ± 6.5%
ACC_{index}	$PD_{ACc} = 2 \cdot (Actual_{Sound\_Pressure\_Level}(dB(A)) - Design_{Sound\_Pressure\_Level}(dB(A))$ Actual (background) noise Design sound level	55 dB(A) 45 dB(A)	80.0% ± 6.7%
L_{index}	$PD_L = -0.0175 + 1.0361/\{1 + \exp(+4.0835 \cdot (\log_{10}(E_{min}) - 1.8223))\}$	450 lux	98.4% ± 9.0%

Table 8. *Cont.*

Sub-Index	Sub-Index $PD(S_I)$ Models	Input Values	Sub-Index (Satisfied) and ±SD
$IEQ_{index}(1)$ with $\sum IAQ_{index}(1)$ meas. $IEQ_{index}(1) \pm SD$	$IEQ_{index} \pm SD = W_1 \cdot TC_{index} + W_2 \cdot \sum IAQ_{index}(1) + W_3 \cdot Ac_{cindex} + W_4 \cdot L_{index}$		$84.4\% \pm 3.7\%$ $u_{meas} = 2 \cdot 3.7 = \pm 7.4\%$
overall $IEQ_{index}(1) \pm u_{overall}$ [2]	$\pm u_{overall}(IEQ) = (\sum (SD_{real}(PD(SI_i))^2 + \sum (SD_{vote} PD(SI_i))^2)^{-2}$ $IEQ_{index}(1) \pm u_{overall}$		$u_{oveall} = \pm 16.24\%$ $84.4\% \pm 16.24\%$
$IEQ_{index}(2)$ with $\sum IAQ_{index}(2)$ meas$IEQ_{index}(2) \pm SD$	$IEQ_{index} \pm SD =$ $W_1 \cdot TC_{index} + W_2 \cdot \sum IAQ_{index}(2) + W_3 \cdot ACC_{index} + W_4 \cdot L_{index}$		$84.6\% \pm 3.3\%$ $u_{meas} = 2 \cdot 3.3 = \pm 6.6\%$
overall $IEQ_{index}(2) \pm u_{overall}$ [2]	$\pm u_{overall}(IEQ) = (\sum (SD_{real}(PD(SI_i))^2 + \sum (SD_{vote} PD(SI_i))^2)^{-2}$ $IEQ_{index}(2) \pm u_{overall}$		$u_{overall} = \pm 16.15\%$ $84.6\% \pm 16.15\%$

(b)

	c_{mes}	$SD(c_{meas})$	c_H	$SD(c_H)$	c_L	$SD(c_L)$	$PD*_H$	SD	$PD*_L$	SD
TVOC	787	$18\% \Rightarrow 141.7$	900	$12\% \Rightarrow 108$	300	$12\% \Rightarrow 36$	50	12%	25	12%
HCHO	18	$12\% \Rightarrow 2.16$	49.1	$12\% \Rightarrow 5.89$	12.3	$12 \Rightarrow 1.48$	50	12%	25	12%

$$PD* = \frac{(PD*_H - PD*_L)}{(c_H - c_L)} \times (c_{meas} - c_L) + PD*_H \qquad (27)$$

[1] The method of calculation of mean values $PD*$ and $\pm SD(PD*)$ for TVOC and HCHO is based on Equation (16) and takes the form (27).

Concentrations c and $SD(c)$ are in $\mu g/m^3$; $PD*_H$, $PD*_L$, and $SD(PD*)$ are in %, and standard deviations of the HCHO concentration of 12% was adopted on the basis of reports from IAQ research conducted as part of BREEAM in 2016 [22]. The assumptions were that c_H and $PD*_H$ are the coordinates of the upper break point (i.e., high break point) of the converted scale $PD* = f(c)$, and c_L and $PD*_L$ are the coordinates of the lower break point (i.e, low break point) of the converted scale $PD* = f(c)$. Standard deviations were assumed for c_{meas}, c_{TVOC}, and c_{HCHO} as well as for $PD* = \pm 12\%$, as this is half of the transformed segment of the scale, which according to Table 7 covers a range of 25% of $PD*$, with one perceived category of air quality, e.g., "no risk" or "moderate". Therefore, the maximum standard deviation was 12.5% and, according to the literature on AQI and IAQI values, it should be rounded to a total value. [2] $SD_{vote}(PD(SI_i))$ from the $\pm u_{overall}(IEQ)$ equation was the standard deviation of a probability distribution of an each. $(SI_i)_{vote}$ and was calculated primary using the $PD(SI_i)$ equation calibration curve [34].

4. Discussion

4.1. Discussion of the $\sum IAQ_{index}$ Theoretical Model

For years, the authors, as accredited laboratory personnel, have conducted IAQ pollution tests in indoor environments for various applications. Based on our experience, it was concluded that the general approach to assessing combined IAQ has not yet been systematized and that there is a global tendency to assess individual IAQ parameters separately or to group them without a justified aggregation method. This is not a good situation from the point of view of building users' needs. This, in our opinion, may lead to incorrect IAQ interpretations in specific building situations. In the context of analyzing the problem in this paper, authors presented a summary of the state-of-the-art methods and also provided a new approach for solving some of these problems. As presented, it is possible to create a $\sum IAQ$ index aggregating the results of indoor air analyses, taking into account various representative pollutants. Three levels of comprehensive air quality assessments (with three, five or seven subcomponents), depending on the application of the assessment, were proposed, together with step-by-step procedures. This may be practical, as shown in the evaluation of a case study on a building. We originally selected the main IAQ subcomponent equations and user satisfaction dependences, $\%PD = f(c_j)$, and provided them all in one place (Table 3). We then proposed and justified the weighting schemes for the IAQ total equation. In most of the studies in the literature, the weighting schemes used for IEQ or IAQ assessments are not physically justified or explained. There are known methods of weighting sub-indices, but the problem that was solved in this paper was an effective system for weight adjustments. For the construction of the combined model IAQ_{index}, with a weighting scheme useful for aggregating sub-indices, we proposed the model scheme presented in Figure 5. According to the results, the advantage of the complex model $\sum IAQ_{index}$, in which the input quantities always constitute concentrations of given pollutants, is the ability to use these concentrations to calculate excess pollution concentrations from Equation (10) and generate weighting schemes $W_{1, \ldots n}$ for all three models by adjusting the weights based on the concentration values of excess air pollutants to a value ≤ 1.0 for each IAQ_{index} model. The Δc_j values determine the masses of pollutants that must be removed by ventilation to eliminate the target pollutant effect. They can be determined as differences between the current concentrations of pollutants and the concentration of pollutants at the reference or standard level (e.g., c_{ELV} or c_{LCI}), and in the case of $VOC_{odorous}$, the odor threshold c_{th}. The presented approach may allow planning of air quality for the building.

As discussed, it is important to identify those VOCs with comfort, health, and impacts and focus on the IAQ sub-model choice aspect, briefly defined as the strategy "VOCs—Total vs. Target: Comfort, Irritancy, Odor, and Health Impact". The model from Figure 5 has uniform inputs, i.e., concentration levels c_j and two outputs: (1) weighed (adjusted) and (2) sensory equations, $PD^* = f(c_{TVOC})$, constituting the IAQ submodel equations. These second outputs of submodels (PD^* values) are coefficients of satisfaction from the comfort sensation or lack of "health risk". These are the terms of the equation describing "combined $\sum IAQ$", which meets the requirements of the abovementioned strategy of selecting IAQ sub-models related to IAQ components that have the most impact on the resulting IEQ perception.

Models for subcomponents of IAQ not perceived by humans but influencing health are recommended to be used from the index set in the AQI system [26–28], which was adapted by the authors to assess the quality of indoor air based on, and in accordance with, the concepts of the air quality assessment system used globally by the American EPA. In the context of the subcomponent of the TVOC concentration in Figure 7, the authors provided the relationships of $PD^* = f(c_{TVOC})$ based on Jokl research [49] and resulting from the IAQI scale [30] as converted by the authors. The relationship between PD^* and TVOC concentration in both approaches is strongly correlated, as shown in Figure 9.

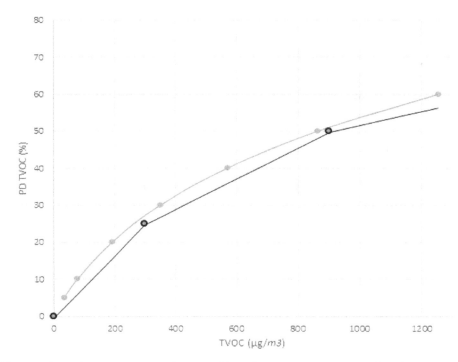

Figure 9. PD_{TVOC} based on conversion from IAQI to the PD* scale and study of Weber-Fechner theory.

The curves obtained from the conversion confirm Jokl's predictions provided in [49] and previously accepted idea. However, the final confirmation of these curves will be done experimentally as panel tests, as planned by the authors for the near future.

4.2. Discussion of Results for the Case Study on a Building

The experimental study was performed in a BREEAM certified building, and included the determination of formaldehyde concentration, CO_2, and VOCs in the indoor air. The example calculation of the combined $\sum IAQ$ model for three basic pollutants, as components of the IEQ_{index} model, is presented in two variants but the calculated PD^*_{TVOC} values obtained with both calculation methodswere very similar. The first variant of $\sum IAQ$ calculation used $\%PD = f(c_j)$ curves in % sub-indices of IAQ for three pollutants, and the differences in approach in Tables 5 and 8a meant combining them into one IEQ submodel: the $\sum IAQ$ model intended for the IEQ calculation. The second variant used submodels of IAQ for TVOC and HCHO pollutants based on the IAQI system [30] which were then converted into percentages of persons dissatisfied (PD^* in %).

The first conclusion is that CO_2 concentration cannot be used separately for the IAQ_{index} assessment, especially at the pre-occupancy stage (Table 5). The building was polluted with VOC emissions and HCHO from the construction products directly after finishing works were completed.

The authors confirmed that all three pollutions should be a simultaneously integrated part of the IAQ model, because the importance of TVOC is much greater, representing the main source of pollution—the construction and finishing materials.

According to the results, we recognized two variants of the combined $\sum IAQ_{index}$ calculation. For the first variant [22], the combined $\sum IAQ$ index of satisfied users was 69.1% and, for the second variant (new approach with converted AQI index), the $\sum IAQ$ index was 70.0% satisfied. The results ofthe $IEQ_{index}(1)$ (for Variant 1) were within the interval of combined overall uncertainty,±16.24%, and the results of the $IEQ_{index}(2)$ (for Variant 2) were that the overall uncertainty was ±16.15%. Therefore, the result was convergent, which confirms the credibility of the proposed approach.

The results obtained also showed that, in the period immediately after completion of finishing works in indoor spaces, there may be a temporarily increased concentration of TVOC, which systematically decreases over time, as we have shown in other papers. In the case of a building, the research showed that tests carried out immediately after finishing works gave results that significantly exceeded the BREEAM limits for TVOC at 300 $\mu g/m^3$ (twice as high). It should be expected that an acceptable level should be reached after a minimum of one month from the completion of the work.

For correctness of the obtained calculations, the authors are conducting a model credibility analysis that will be provided in the next article—Indoor Air Quality Model Part II: The Combined Model $\sum IAQ_{index}$ Reliability Analysis. The model uncertainty estimate may be compromised because the model reproduces the discomfort level associated with the dominant component.

Author Contributions: Conceptualization, M.P. and K.B.K.; methodology, M.P. and K.B.K.; formal analysis, M.P. and K.B.K.; investigation, M.P. and K.B.K.; resources, M.P. and K.B.K.; data curation, M.P. and K.B.K.; writing—original draft preparation, M.P. and K.B.K.; writing—review and editing, M.P. and K.B.K.; visualization, M.P. and K.B.K.; supervision, K.B.K.; project administration, M.P.; funding acquisition, M.P.

References

1. Fanger, P.O. What is IAQ? *Indoor Air* **2006**, *16*, 328–334. [CrossRef] [PubMed]
2. Jones, L.L.; Hashim, A.; McKeever, T.; Cook, D.G.; Britton, J.; Leonardi-Bee, J. Parental and household smoking and the increased risk of bronchitis, bronchiolitis and other lower respiratory infections in infancy: Systematic review and meta-analysis. *Respir. Res.* **2011**, *12*, 5. [CrossRef] [PubMed]
3. ASHRAE. *ASHRAE Guideline No. 10. Interactions Affecting the Achievement of Acceptable Indoor Environments*; ASHRAE: New York, NY, USA, 2011.
4. Kostyrko, K.; Kozicki, M. Trends in odor measurments and detecting MVOC content in the interiors of buildings. In *Conference on Metrology MKM18*; The Scientific Papers of Faculty of Electrical and Control Engineering; Gdańsk University of Technology: Gdansk, Poland, 2018; Volume 59, p. 97.
5. Carrer, P.; de Oliveira Fernandes, E.; Santos, H.; Hänninen, O.; Kephalopoulos, S.; Wargocki, P. On the development of health-based ventilation guidelines: Principles and framework. *Int. J. Environ. Res. Public Health* **2018**, *15*, 1360. [CrossRef] [PubMed]
6. WHO Regional Office for Europe. *Evolution of WHO Air Quality Guidelines: Past, Present and Future*; WHO Regional Office for Europe: Copenhagen, Denmark, 2017.
7. WHO Regional Office for Europe. WHO guidelines for indoor air quality. *Nutr. J.* **2010**, *9*, 454.
8. General Administration of Quality Supervision, Inspection and Quarantine of the People's Republic of China. *GB/T 18883-2002, Indoor Air Quality Standard*; General Administration of Quality Supervision, Inspection and Quarantine of the People's Republic of China: Beijing, China, 2002. (In Chinese)
9. Asikainen, A.; Carrer, P.; Kephalopoulos, S.; Fernandes, E.D.O.; Wargocki, P.; Hänninen, O. Reducing burden of disease from residential indoor air exposures in Europe (HEALTH-VENT project). *Environ. Health A Glob. Access Sci. Source* **2016**, *15* (Suppl. 1), 35.
10. CEN. Part 1: Indoor environmental input parameters for design and assessment of energy performance of buildings addressing indoor air quality, thermal environment, lighting and acoustics. In *EN 16798 Energy Performance of Buildings–Ventilation of Buildings*; CEN: Brussels, Belgium, 2019.
11. Wang, Z.; Bai, Z.; Yu, H.; Zhang, J.; Zhu, T. Regulatory standards related to building energy conservation and indoor-air-quality during rapid urbanization in China. *Energy Build.* **2004**, *36*, 1299–1308. [CrossRef]
12. *China National StandardsGB 3095-2012 Ambient Air Quality*; The Standardization Administration of China (SAC): Beijing, China, 2012.
13. The Government of the Hong Kong Special Administrative Region IAQ Management Group. *A Guide on Indoor Air Quality. Certification Scheme for Offices and Public Places*; The Government of the Hong Kong Special Administrative Region IAQ Management Group: Hong Kong, China, 2019.
14. Piasecki, M.; Kostyrko, K. Indoor Environmental Quality Assessment Model IEQ Developed in ITB. Part 1. Choice of the Indoor Environmental Quality Subcomponent Models. *Ciepłownictwo Ogrzew. Went.* **2018**, *49*, 223–232.
15. Sekhar, S.C.; Tham, K.W.; Cheong, K.W. Indoor air quality and energy performance of air-conditioned office buildings in Singapore. *Indoor Air* **2003**, *13*, 315–331. [CrossRef]
16. Moschandreas, D.J.; Sofuoglu, S.C. The indoor environmental index and its relationship with symptoms of office building occupants. *J. Air Waste Manag. Assoc.* **2004**, *54*, 1440–1451. [CrossRef]

17. Sofuoglu, S.C.; Moschandreas, D.J. The link between symptoms of office building occupants and in-office air pollution: The Indoor Air Pollution Index. *Indoor Air* **2003**, *13*, 332–343. [CrossRef]

18. IEA. *Indoor Air Quality Design and Control in Low-Energy Residential Buildings-Annex 68 Subtask 1: Defining the Metrics in the Search of Indices to Evaluate the Indoor Air Quality of Low-Energy Residential Buildings*; AIVC Contributed Report, 17; Abadie, M.O., Wargocki, P., Eds.; IEA: Paris, France, 2017.

19. Chen, C.; Chiang, C.; Shao, W.; Huang, L. Hazards of VOCs emissions from building material and ventilation effectiveness in Taiwan. In Proceedings of the Indoor Air Conference, Copenhagen, Denmark, 17–22 August 2008; Volume 1.

20. Zhu, C.; Li, N. Study on Grey Clustering Model of Indoor Air Quality Indicators. *Procedia Eng.* **2017**, *205*, 2246–2253. [CrossRef]

21. Salisa, L.C.R.; Abadiea, M.; Wargocki, P.; Rode, C. Towards the definition of indicators for assessment of indoor air quality and energy performance in low-energy residential buildings. *Energy Build.* **2017**, *152*, 492–502. [CrossRef]

22. Piasecki, M.; Kozicki, M.; Firlag, S.; Goljan, A.; Kostyrko, K. The approach of including TVOCs concentration in the indoor environmental quality model (IEQ)- case studies of BREEAM certified office buildings. *Sustainability* **2018**, *10*, 3902. [CrossRef]

23. Toftum, J.; Jørgensen, A.S.; Fanger, P.O. Upper limits of air humidity for preventing warm respiratory discomfort. *Energy Build.* **1998**, *28*, 15–23. [CrossRef]

24. Wang, Z.; Yu, Z. PM2.5 and Ventilation in a Passive Residential Building. *Procedia Eng.* **2017**, *205*, 3646–3653. [CrossRef]

25. Morawska, L.; Ayoko, G.A.; Bae, G.N.; Buonanno, G.; Chao, C.Y.H.; Clifford, S.; Fu, S.C.; Hänninen, O.; He, C.; Isaxon, C.; et al. Airborne particles in indoor environment of homes, schools, offices and aged care facilities: The main routes of exposure. *Environ. Int.* **2017**, *108*, 75–83. [CrossRef]

26. Cornell's Legal Information Institute. Available online: https://www.law.cornell.edu/sfr/text/40/appendix-A_to_part_58 (accessed on 20 May 2019).

27. Saad, S.M.; Shakaff, A.Y.M.; Saad, A.R.M.; Yusof, A.M.; Andrew, A.M.; Zakaria, A.; Adom, A.H. Development of indoor environmental index: Air quality index and thermal comfort index. In *AIP Conference Proceedings*; American Institute of Physics: Melville, NY, USA, 2017; Volume 1808, p. 020043.

28. Mintz, D. *Technical Assistance Document for the Reporting of Daily Air Quality–The Air Quality Index (AQI)*; U.S. EPA: Washington, DC, USA, 2013.

29. United States EPA-Environmental Protection Agency Office of Air Quality Planning and Standards Air Quality Assessment Division Research Triangle Park, NC. Available online: https://airnow.gov/index.cfm?action=airnow.calculator (accessed on 25 June 2019).

30. Wang, H.; Tseng, C.; Hsieh, T. Developing an indoor air quality index system based on the health risk assessment. In Proceedings of the Indoor Air Conference, Copenhagen, Denmark, 17–22 August 2008; Volume 1.

31. United States EPA-Environmental Protection Agency Office of Air Quality Planning and Standards Air Quality Assessment Division Research Triangle Park, NC. *EPA-454/B-18-007- Technical Assistance Document for the Reporting of Daily Air Quality*; United States EPA-Environmental Protection Agency Office of Air Quality Planning and Standards: Research Triangle Park, NC, USA, 2018.

32. Sakellaris, I.A.; Saraga, D.E.; Mandin, C.; Roda, C.; Fossati, S.; de Kluizenaar, Y.; Carrer, P.; Dimitroulopoulou, S.; Mihucz, V.G.; Szigeti, T.; et al. Perceived indoor environment and occupants' comfort in European "modern" office buildings. The OFFICAIR Study. *Int. J. Environ. Res. Public Health* **2016**, *13*, 444. [CrossRef]

33. Piasecki, M.; Kostyrko, K.; Pykacz, S. Indoor environmental quality assessment: Part 1: Choice of the indoor environmental quality subcomponent models. *J. Build. Phys.* **2017**, *41*, 264–289. [CrossRef]

34. Piasecki, M.; Kostyrko, K.B. Indoor environmental quality assessment, part 2: Model reliability analysis. *J. Build. Phys.* **2018**, *42*, 288–315. [CrossRef]

35. Ye, W.; Won, D.; Zhang, X. A Simple VOC Prioritization Method to Determine Ventilation Rate for Indoor Environment Based on Building Material Emissions. *Procedia Eng.* **2015**, *121*, 1697–1704. [CrossRef]

36. Bienfait, D. *ECA Report No. 11 Guidelines for Ventilation Requirements in Buildings*; Commission of the European Communities: Bruxelles, Belgium, 1992.

37. Ye, W.; Won, D.; Zhang, X. A practical method and its applications to prioritize volatile organic compounds emitted from building materials based on ventilation rate requirements and ozone-initiated reactions. *Indoor Built Environ.* **2017**, *92–93*, 324–333. [CrossRef]
38. Persily, A.K. Evaluating building IAQ and ventilation with indoor carbon dioxide. *ASHRAE Trans.* **1997**, *103*, 4072.
39. Gyot, G.; Sherman, M.H.; Walker, I.S.; Clark, J.D. *Residential Smart Ventilation: A Review*; LBNL 2001056; Ernest Orlando Lawrence Berkeley National Laboratory: Berkeley, CA, USA, 2017.
40. Engineering ToolBox. Available online: https://www.engineeringtoolbox.com (accessed on 5 May 2019).
41. Noguchi, M.; Mizukoshi, A.; Yanagisawa, Y.; Yamasaki, A. Measurements of volatile organic compounds in a newly built daycare center. *Int. J. Environ. Res. Public Health* **2016**, *13*, 736. [CrossRef] [PubMed]
42. Haag, W. Rapid continuous TVOC measurements using PPB-level PIDs. In Proceedings of the 10th International Conference on Indoor Air Quality and Climate, Beijing, China, 4–9 September 2005; pp. 938–941.
43. Wong, L.T.; Mui, K.W.; Hui, P.S. A multivariate-logistic model for acceptance of indoor environmental quality (IEQ) in offices. *Build. Environ.* **2008**, *43*, 1–6. [CrossRef]
44. American Industrial Hygiene Association (AIHA). *Odor Thresholds for Chemicals with Established Health Standards*; American Industrial Hygiene Association: Falls Church, VA, USA, 2013.
45. Jokl, M.V. Evaluation of indoor air quality using the decibel concept based on carbon dioxide and TVOC. *Build. Environ.* **2000**, *35*, 677–697. [CrossRef]
46. Tahsildoost, M.; Zomorodian, Z.S. Indoor environment quality assessment in classrooms: An integrated approach. *J. Build. Phys.* **2018**, *42*, 336–362. [CrossRef]
47. Zhu, C.; Li, N. Study on indoor air quality evaluation index based on comfort evaluation experiment. *Procedia Eng.* **2017**, *205*, 2246–2253. [CrossRef]
48. European Committe for Standardization. *CEN/CR 1752:1998 Ventilation for Buildings. Design Criteria for the Indoor Environment*; International Organization for Standardization: Geneva, Switzerland, 1998.
49. Jokl, M.V. Indoor Air Quality Assessment Based on Human Physiology–Part 1. New Criteria Proposal. *Acta Polytech.* **2003**, *43*, 31–37.
50. Wargocki, P.; Knudsen, H.N.; Krzyżanowska, J. Some methodological aspects of sensory testing of indoor air quality. In Proceedings of the CLIMA 2010, Antalya, Turkey, 9–12 May 2010.
51. ISO. *ISO 16000-28 Determination of Odor Emissions from Building Products Using Test Chambers*; ISO: Geneva, Switzerland, 2013.
52. Kim, K.H.; Kim, Y.H. Composition of key offensive odorants released from fresh food materials. *Atmos. Environ.* **2014**, *205*, 2246–2253. [CrossRef]
53. Wu, C.; Liu, J.; Zhao, P.; Piringer, M.; Schauberger, G. Conversion of the chemical concentration of odorous mixtures into odor concentration and odor intensity: A comparison of methods. *Atmos. Environ.* **2016**, *127*, 283–292. [CrossRef]
54. Yeganeh, B.; Haghighat, F.; Gunnarsen, L.; Afshari, A.; Knudsen, H. Evaluation of building materials individually and in combination using odor threshold. *Indoor Built Environ.* **2006**, *15*, 583–593. [CrossRef]
55. Gunnarsen, L.; Fanger, P.O. Adaptation to indoor air pollution. *J. Environ. Int.* **1992**, *18*, 43–54. [CrossRef]
56. Fang, L.; Clausen, G.; Fanger, P.O. Impact of temperature and humidity on the perception of indoor air quality. *Indoor Air* **1998**, *8*, 80–90. [CrossRef]
57. Polednik, B. *Zanieczyszczenia a Jakość Powietrza Wewnętrznego w Wybranych Pomieszczeniach*; Biblioteka Cyfrowa Politechniki Lubelskiej: Lublin, Poland, 2013. (In Polish)
58. Piasecki, M.; Fedorczak-Cisak, M.; Furtak, M.; Biskupski, J. Experimental Confirmation of the Reliability of Fanger's Thermal Comfort Model—Case Study of a Near-Zero Energy Building (NZEB) Office Building. *Sustainability* **2019**, *11*, 2461. [CrossRef]

12

Indoor Air Quality Improvement by Simple Ventilated Practice and *Sansevieria Trifasciata*

Kanittha Pamonpol [1],*, Thanita Areerob [2] and Kritana Prueksakorn [2,3,]*

[1] Environmental Science and Technology Program, Faculty of Science and Technology, Valaya Alongkorn Rajabhat University under the Royal Patronage, 1 Moo 20 Paholyothin Road, Klong Nueng, Khlong Luang, Pathum Thani 13180, Thailand

[2] Faculty of Technology and Environment, Prince of Songkla University, Phuket Campus 83120, Thailand; Thanita.a@phuket.psu.ac.th

[3] Andaman Environment and Natural Disaster Research Center, Faculty of Technology and Environment, Prince of Songkla University, Phuket Campus 83120, Thailand

* Correspondence: kanittha@vru.ac.th (K.Pa.); k.prueksakorn@gmail.com (K.Pr.)

Abstract: Optimum thermal comfort and good indoor air quality (IAQ) is important for occupants. In tropical region offices, an air conditioner is indispensable due to extreme high temperatures. However, the poor ventilation causes health issues. Therefore, the purpose of this study was to propose an improving IAQ method with low energy consumption. Temperature, relative humidity, and CO_2 and CO concentration were monitored in a poorly ventilated office for one year to observe seasonal variation. The results showed that the maximum CO_2 concentration was above the recommended level for comfort. Simple ventilated practices and placing a number of *Sansevieria trifasciata* (*S. trifasciata*) plants were applied to improve the IAQ with the focus on decreasing CO_2 concentration as well as achieving energy saving. Reductions of 19.9%, 22.5%, and 58.2% of the CO_2 concentration were achieved by ventilation through the door during lunchtime, morning, and full working period, respectively. Placing *S. trifasciata* in the office could reduce the CO_2 concentration by 10.47%–19.29%. A computer simulation was created to observe the efficiency of simple practices to find the optimum conditions. An electricity cost saving of 24.3% was projected for the most feasible option with a consequent reduction in global warming potential, which also resulted in improved IAQ.

Keywords: computational fluid dynamics; CO_2 concentration; indoor air quality; *Sansevieria trifasciata*; ventilation

1. Introduction

In tropical regions, where the mean annual temperature (T) exceed 30 °C, [1,2], air conditioning (AC) is often used in buildings to make the conditions more comfortable for occupants and it can be found in commercial buildings, government buildings, factories, universities, schools, and homes. AC is often used in closed rooms with low ventilation in order to prevent ambient air pollution [3], caused by high occupancy [4] as well as to maintain low T and save energy [5]. The desire to save energy and a lack of awareness regarding health and safety issues relating to indoor air quality (IAQ) have resulted in rooms being designed in tropical countries, including Thailand, which allow for all doors and windows to be closed, causing poor ventilation, particularly in hotels [6]. Gases that are generated in closed rooms, which cannot be effectively ventilated can increase to harmful levels, resulting in negative health effects [7], especially for office workers, who spend most of their working hours inside buildings [8].

Keeping windows open or using a ventilation fan can improve IAQ, but energy consumption is thereby increased, since AC systems have to work harder to maintain the indoor T within a comfortable range. Therefore, an important aspect of room design is how to sustain good IAQ and thermal comfort with low energy consumption. An additional benefit from low energy use is the reduction of greenhouse gas (GHG) emissions due to human activities, which are a cause of climate change. It is now generally accepted that GHG emissions have a considerable impact by raising ambient T [9], which results in higher energy use for air conditioning to make rooms comfortable for their occupants, thus creating a vicious circle.

However, poor indoor air quality is not only associated with closed rooms, but it can also result from errors in the design of ventilation systems, which can introduce pollutants from outside into indoor areas [10]. Major indoor air pollutants that have been studied include $PM_{2.5}$, PM_{10}, O_3, CO_2, CO, NO_2, NH_3, volatile organic compounds (VOCs), and aldehydes, which can be derived from a number of potential sources [11–15]. Moreover, in addition to physical pollutants, biological pollutants, such as bacteria, fungi, and mold, can be suspended in indoor air in the form of particles and they are considered to be indoor air pollutants [16,17], and people in buildings affected by indoor air pollutants are at risk of acquiring sick building syndrome (SBS). The symptoms of SBS are various and non-specific, but they include tiredness, feeling unwell, itching skin, high blood pressure, and heart rate, and even difficulty in concentrating. Sometimes, these effects are rapidly relieved after leaving the building [11,15,18], but this may not be an option for those affected.

In this study, the air quality and comfort parameters that were selected for study were relative humidity (RH), T and CO_2 concentration, and how they affect the proper design of rooms [19]. Also included is the CO concentration as a representative outdoor air pollutant generated by incomplete combustion of fossil fuels [9], being mostly derived from automotive exhaust fumes from roads and parking areas around buildings.

A high RH content in the ambient air can result in the low evaporation of perspiration from the surface of human skin with a consequent reduction in the excretion of substances by evaporation [20]. Further, exposure to high T has been found to not only affect work performance, but also to result in symptoms, such as mental fatigue and changes in blood pressure [21]. In addition, inhaling excessive amounts of CO_2 above 10,000 ppm can cause a condition that is known as acidosis (low pH of blood: <7.35), in which the body's defense mechanisms are stimulated, resulting in, e.g., an increase in breathing rate and volume and high blood pressure and heart rate [12]. Exposure to CO_2 levels of approximately 50,000 ppm can lead to the failure of the central nervous system (brain and spinal cord), possibly causing death [22]. On the other hand, breathing high levels of CO can lead to death due to tissue hypoxia, as CO can bind to hemoglobin more effectively than oxygen [23].

The results of research that was conducted by NASA's environmental scientists into the improvement of IAQ while using plants was published in 1989, with a number of houseplants being tested as a means of treating indoor air pollution by removing trace organic pollutants from the air in closed environments in energy-efficient buildings. The organic chemicals tested consisted of benzene, trichloroethylene, and formaldehyde, and the scientific names of the plants investigated were *Chamaedorea seifritzii, Aglaonema modestum, Hedera helix, Ficus benjamina, Gerbera jemesonii, Deacaena deremensis, Deacaena marginata, Dracaena massangeana, Sansevieria laurentii, Spathiphyllum, Chrysanthemum morifollum,* and *Dracaena deremensis* [24]. Among the plants that are generally found in tropical regions is *Sansevieria laurentii* (mother-in-law's tongue), which is a size that is suitable for a small office and it was selected in this study to test its effect on IAQ improvement.

IAQ has been an important issue in Europe and America since the 18th century [4]. However, there has been relatively limited research on the topic in Association of South East Asean Naitons (ASEAN) countries [15,25] and few long-term studies have been conducted [26]. The main aim of this research was to improve the IAQ of an air-conditioned office in Thailand with a poor ventilation system by practicing simple ventilation operations and locating the mother-in-law's tongue plants in the office, with the second aim of achieving lower energy consumption. Another beneficial outcome of this study was finding ways of reducing GHG emissions that are associated with the use of electricity. In this paper, alternative energy-saving scenarios are presented to demonstrate the effectiveness of simple operations in reducing GHG emissions and improving IAQ for office workers.

2. Experiments

2.1. Studied Site

The study site was an air-conditioned office in Valaya Alongkorn Rajabhat University under the Royal Patronage (VRU), which is located in Pathum Thani province, a suburban area 15 km north of Bangkok, Thailand at 14°8.004′ north latitude and 100°36.961′ east longitude, with the site, on average, 5 m above sea level. The office was located on the second floor of a four-story building that was surrounded by a parking area. The room dimensions were: Length × Width × Height of around 4 m × 10 m × 3 m for six occupants. The office was air conditioned with the only means of ventilation being a door, leading to open corridor, which was generally closed to maintain a low T and save energy. The air conditioner was a wall mounted type with cooling capacity 36,000 Btu/hour (Daisenko International Co., LTD., Thailand). Figure 1 presents a diagram of the office.

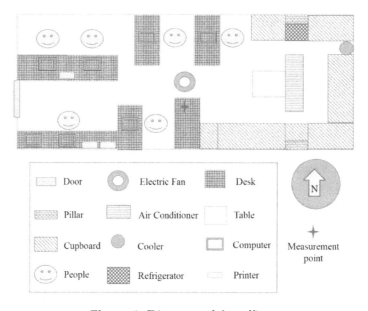

Figure 1. Diagram of the office.

2.2. Measurement of IAQ Parameters

Data were collected in the air-conditioned office for approximately one year from May 2017 to May 2018 to indicate the quality of the indoor air by the recommended levels of the American Society of Heating, Refrigerating, and Air-conditioning Engineers (ASHRAE) [27]. RH, T, CO, and CO_2 were monitored every minute using a FLUKE 975 AirMeter, a portable device for the measurement of IAQ. The specifications of the measurement device are, as follows; CO_2: accuracy ± 2.75%, range 0 to 5000 ppm; CO: accuracy ± 5% or ± 3 ppm at 20 °C and 50% RH, range 0 to 500 ppm; T: accuracy ± 0.5 °C from 5 °C to 40 °C, range −20 °C to 50 °C; and, RH: accuracy ± 2%, range 10% to 90% RH [28]. The device was installed on the desk in the middle of the room, at the same height as the breathing zone during working hours.

2.3. Simple Ventilation Practices for Improving IAQ

After obtaining the results of the one-year observation, four different systems were implemented in June 2018 during working hours (9:00–17:00) to discover the simplest and most efficient means of removing stale air from inside the room, and introducing fresh air from outside. The experimental condition was conducted in real practice where people in the office were working and doing activities as usual. The four systems tested were as follows: Case 1: the AC was turned on all day (normal case) from 1–7 June, Case 2: the AC was turned off during the lunch hour (12:00–13:00) from 8–14 June, Case 3: the AC was turned off for half a day (9:00–13:00) from 15–21 June, and Case 4: the AC was turned off all day (9:00–17:00) from 22–28 June. The statistical analysis was analyzed by one way ANOVA to consider among four cases at confidential level 95% ($p < 0.05$) by Statistic 8 Software (Version 8, USA).

For Case 3, it was decided to turn off the AC in the morning, because the high T in the afternoon [2] had a negative effect on work efficiency. Turning off the AC all day (Case 4) could not realistically be applied, since it would probably result in problems, such as heat strain. This system was included to establish the maximum rate of full-day ventilation with the AC turned off, the door constantly opened, and an electric fan mounted on the ceiling turned on.

An assessment of the envelope air permeability of the room was obtained by the infiltration rate, which was calculated by the following equation:

$$Q = -\frac{V}{t} \times \ln\left[\frac{C_t - C_{ext}}{C_0 - C_{ext}}\right]$$

where, Q is infiltration rate of air entering the room, V is volume of air in the office (m^3), t is time interval (s), C_t is indoor concentration of CO_2 at time t (ppm), C_{ext} is concentration of CO_2 in the ambient air (ppm), and C_0 is indoor CO_2 concentration at time 0 (ppm) [29,30].

The volume of air in the office (V) was calculated from the size of the room (4 m x 10 m x 3 m), Interval (t) was 3600 seconds from hourly average data, C_{ext} was average monitoring outdoor CO_2 concentration at 430 ppm, C_t was the monitored CO_2 concentration at time t (i.e., 18:00), and C_0 was the monitored CO_2 concentration at one hour before t (i.e., 17:00). The frequency of measurement was every one minute, so the raw data were calculated to hourly data for both indoor and outdoor CO_2 concentration. The indoor CO_2 concentration was obtained by monitoring inside the office at 1 m height or nose level while people were sitting. The condition in the room was no plant and no ventilation for a long period. The door was opened when people came in and went out in a short time, not over one minute. The outdoor CO_2 concentration was monitored at 1.5 m above ground level in the ambient air.

2.4. Sansevieria Trifasciata for IAQ Improvement

Another option tested for improving IAQ was locating the mother-in-law's tongue plants in the office to reduce the CO_2 concentration in the ambient air through their photosynthesis. In this experiment, the S. trifasciata was put in a pot that contained soil. The plant was watered twice a week. The experiments were conducted by monitoring the air quality for six conditions, as follows: 0, 2, 3, 4, 5, and 6 Mother-in-law's tongue plants with three replicates for each case. The plants were placed on the floor near the desks where people worked, as shown in Figure 1. The number of plants was limited by the space of the rooms in which they were located. The IAQ was monitored from March to April 2019 for 24 hours each day to observe the amount of CO_2 that the plants consumed for photosynthesis during the daytime and to establish the amount that they released through respiration at night. RH, T, CO, and CO_2 were monitored by indoor air meter (FLUKE 975 AirMeter, USA) every minute. We monitored CO_2 in a real situation to represent working activities or business as usual in tropical areas. Only the room temperature was controlled by air conditioner to comfort people at 25-degree Celsius, which was the general setting temperature in tropical countries.

2.5. Numerical Study

In addition to the measurement of the IAQ, a simulation was performed to estimate the efficiency of the office ventilation while using the computational fluid dynamics (CFD) software system, ANSYS Airpak 3.0.16 (Fluent Inc., Lebanon, NH, USA). Airpak simulation software has been broadly applied for numerical simulation of the indoor air alteration under conditions [31–34]. Based on the finite volume method, Airpak uses the FLUENT CFD solver engine for the thermal and fluid-flow calculations to solve equations for the conservation of mass, energy, and momentum of air. The two-equation K-epsilon turbulence model was chosen to solve turbulent flow equations. The number of cells in the domain was approximately 1.5 million, while using hexa-unstructured geometry to discretize. For this function, all of the element types were used to fit the mesh to the geometry. The simulation was iterated to a convergence level of 10^{-3} until the solutions were stable. Additionally, a mesh refinement study was conducted for quantifying and minimizing the error due to discretization. Four different mesh systems, i.e., coarser, course, medium, and fine were generated, to perform the test.

The investigated parameter was the mean age of the air, indicating the average time taken for the air to pass through the room, with a shorter time denoting higher air freshness [35]. A three-dimensional (3D) simulation of an experimental room with the same dimensions as that shown in Figure 1 was constructed using the ANSYS Airpak software and it is illustrated in Figure 2. Section 3.4 presents the results of numerical study.

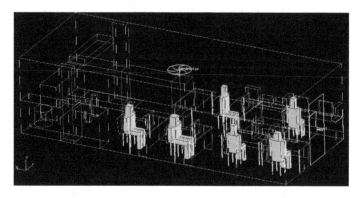

Figure 2. Model of the room in Airpak.

2.6. Estimation of Mitigation of Electricity Use and GHG Emissions

Electricity usage directly impacts the increases in GHG emissions. Reducing electricity consumption not only helps decrease GHG emissions, but also reduces the cost of electricity. Table 1 presents the possible options for electricity saving scenarios.

Table 1. Options for electricity saving scenario.

Appliances	Unit	Power (watt)	Working Time (hour)		
			Case A	Case B	Case C
Computer	6 [a]	450	8	8	7 [b]
Printer [c]	1	10	1	1	1
Printer (standby)	1	2.1	7	0 [d]	0 [d]
Refrigerator	1	90	24	8 [e]	8 [e]
AC	1	1000	8	7 [f]	4 [g]
Fan	1	39	0	1 [h]	4 [h]
Water dispenser	1	100	8	8	8
Light	3	28	8	7 [i]	7 [i]

Notes: Superscripts represent assumptions as follows: [a] the spare computer near the door was not in used; [b] all computers turned off during lunchtime; [c] not used continuously for printing; used for approximately one hour per day; [d] turned on only when in use; [e] no food kept in the office refrigerator overnight; [f] turned on from 09.00–12.00 and 13.00–17.00 (turned off during lunchtime); [g] turned on from 13.00-17.00; [h] turned on instead of AC; [i] turned off during lunchtime.

The different options that are shown in Table 1 were designed in collaboration with the room occupants, who were interested in the effect on the cost of electricity and global warming potential (GWP) if they all agreed to try them. The duration of working without AC for Cases A, B, and C were aligned with Cases 1, 2, and 3 in Section 2.3, respectively. Case 4 (no AC) was not re-assessed, since its results could be estimated from the other cases and its application was, in any event, not realistic. Case A was a typical case, whereas Cases B and C for other appliances represented situations that were not convenient, but feasible in practice.

3. Results and Discussion

3.1. Results of Monthly Monitoring Data

The data from 24-hour monitoring of IAQ at one-minute intervals were converted into monthly results, as presented in Figure 3. The average RH of the office was 67.50% ± 3.98%, which was a little over the range of the recommended standard of 65% or less [36]. Thailand is located in a tropical zone and Thais are accustomed to a hot-humid climate, so the ambient humidity can be higher than the recommended level for the USA [37]. However, a T of 26 °C and an RH of 50–60% were preferred, according to the results of a survey of thermal comfort for air conditioned buildings in Thailand [25]. During the period studied, the Bangkok's mean annual RH was 74% [38], which was consistent with the wide range of outdoor RH between 34 and 78% RH, detected at 17 locations referred by the study of Ongwandee et al. [25]. Therefore, opening the office door might only be helpful in reducing the RH at some times.

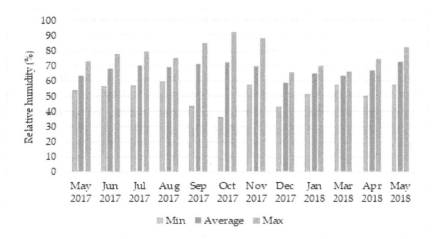

Figure 3. Monthly relative humidity (RH) data.

Figure 3 shows that the greatest variation in RH was apparent for the minimum values during September and October 2017, because of the influence of outdoor air that is caused by various groups visiting the office at that time. Reducing the RH by introducing ambient air is only practicable while taking that the RH value can vary diurnally or hourly into consideration.

The room T was maintained close to the comfortable standard (23 to 26 °C) [39], consistent with the preferred environmental conditions for Thailand established in the thermal comfort survey (26 °C at 50–60% RH) [25], through the use of the AC, as shown in Figure 4. The annual country average outside T was 27.61 ± 1.31 °C and high temperature was found to be higher than usual during August to November in 2017, because of rain, so the winter started late in December [40]. The outdoor T was usually higher, reaching more than 40 °C in the afternoon. The highest indoor T was detected during September and October 2017, which was consistent with the variation in the minimum RH data, and it was also possibly due to the number of visitors entering and leaving the office during that period.

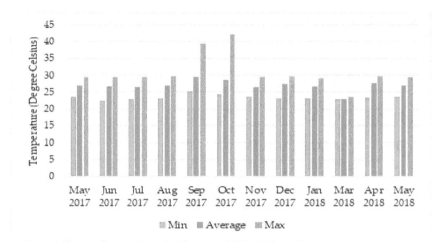

Figure 4. Monthly Temperature data.

The CO concentrations were measured and the data were analyzed by converting from ppm at the local T, to the standard ppm at 25 °C. Figure 5 shows that the average CO concentration was 1.32 ± 0.29 ppm, while the maximum concentrations were 4.57 ppm, 4.91 ppm in September, and October 2017, respectively. The CO detected must have originated from the ambient air outside the room with the probability that this was associated with the parking outside the room with the probability that this was associated with the parking area surrounding the building, since there was no source of CO generation in the room and the minimum values were close to zero. Moreover, the findings are also consistent with the findings related to T and RH in September and October. Further, while fluctuations can be observed between the minimum, maximum, and average levels of CO, these three parameters were closest in March 2018, because there were no events scheduled in that month. Nevertheless, although the ambient air outside the office probably influenced the concentration of CO, the level was not a significant factor in the IAQ, because it was lower than the indoor air quality standard, (9 ppm for eight hours and 35 ppm for one hour) [37].

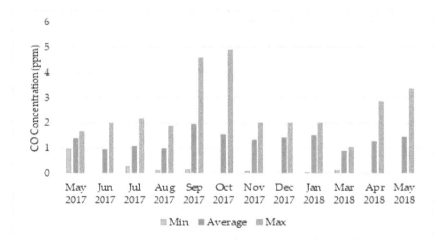

Figure 5. Monthly CO data.

Figure 6 shows that, from May 2017 to May 2018, the maximum, average (± standard deviation), and minimum indoor CO_2 concentrations were 1456.79 ppm, 600.67 ± 42.80 ppm, and 387.32 ppm, respectively. The maximum level of CO_2 was found to be above the comfortable level of 1000 ppm—as recommended by various standards [4,11,12] in every month, except December 2017 and April 2018 (maximum values, 987 ppm and 977 ppm, respectively). The maximum 24-hour CO_2 concentration was found in January 2018, on a day when all staff members (six people) were in the office together with an additional four people attending a long meeting. This emphasizes that human respiration was a significant source of indoor CO_2 (generally two pounds of CO_2 per day) [22]. There were no significant fluctuation in the level of CO_2 detected during September and October 2017, which implied that the exchange of air between inside and outside the office only occurred in the area near the door, and the stale air inside was not effectively removed, due to the lack of a ventilation system to support the exchange process.

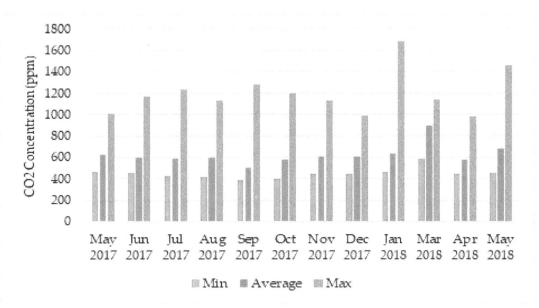

Figure 6. Monthly CO_2 data.

3.2. Results of Simple Ventilated Practices for The Improvement of IAQ

Based on the measurements for the entire year, CO_2 was the parameter that most obviously exceeded the comfortable standard and there was no obvious method of solving this problem, which did not involve reconstruction and the installation of a ventilation system. Therefore, this was the chosen parameter for the experiment to test the performance of simple ventilation practices. The results for the IAQ improvement in June 2017 between Cases A, B, and C that are shown in Table 1 corresponding to Cases 2–4 in Section 2.3, respectively, and Case 1 (normal case) are presented in Table 2.

Table 2. Measurement results in the operational practices for the improvement of indoor air quality (IAQ).

Parameters	Case 1 (Normal)	Case 2 (AC Turned off 12:00–13:00)	Case 3 (AC Turned off 9:00–13:00)	Case 4 (AC Turned off 9:00–17:00)
Maximum CO_2 concentration (ppm)	1167.09	934.56	904.21	488.07
Average CO_2 (± SD) concentration (ppm)	785.05 (67.94)	759.30 (111.66)	700.75 (164.60)	472.10 (10.02)
Minimum CO_2 concentration (ppm)	556.13	565.06	526.60	457.17
T (°C) Average Standard = 23–26 °C [38]	25.3	26.1	27.9	33.8
RH (%) Average Standard = 65% [30]	68.7	71.7	71.6	69.4
CO (ppm) Average Standard = 9 ppm [36]	1.81	1.92	4.61	5.83

Reductions of 19.9%, 22.5%, and 58.2% of the maximum CO_2 concentrations were found by turning off the AC during lunchtime (12:00–13:00, Case 2); in the morning (9:00–13:00, Case 3); and. during the full working time (9:00–17:00, Case 4), respectively, as can be deduced from Table 2. In all cases, the concentrations of CO_2 were reduced below the comfortable standard (1000 ppm) [4,11,12]. Cases 2 and 3 were able to reduce the maximum concentration of CO_2 by almost the same amount. The concentration of CO_2 was being reduced when the air conditioning was more time inactive because the air inside and outside the room could be exchanged due to the door leading to the opened corridor. More time opening the door resulted in more time for air to be exchanged. Higher CO_2 in the room was released into the ambient air then plants could lower CO_2 in the ambient air through photosynthesis process.

The measured values are in agreement with research that was conducted in 21 offices, in Taiwan in which the CO_2 levels were measured while using a Q-TRAK indoor air quality tester (Model 7575, TSI Corporation, Bangkok, Thailand)) with an average level of 708.2 ± 190.5 ppm, a maximum level of 1193.6 ppm, and a minimum level 464.0 ppm [11]. The normal case results were consistent with the indoor CO_2 level in Thai classrooms, which was measured by the similar method using Indoor Air Quality Meters (IAQ-CALC^{TM}) Model 8760/876, a real time monitoring device that used a dual wavelength non-dispersive infrared sensor (NDIR) for CO_2 and Electro chemical sensor for CO by frequency of five minutes. The average CO_2 concentration in classroom 1 (carpet) was 711 ± 272 ppm and classroom 2 (wooden) was 1332 ± 609 ppm [41].

The statistical analysis results by One-Way ANOVA found that the result was significant at $p = 0.002223$ ($p < 0.05$) and the f-ratio value was 5.18726. This result means concentrations of CO_2 were different among four cases.

The T and RH were also measured to check whether they aligned with the comfort standards. Figures 7 and 8 present the results of T and RH, respectively.

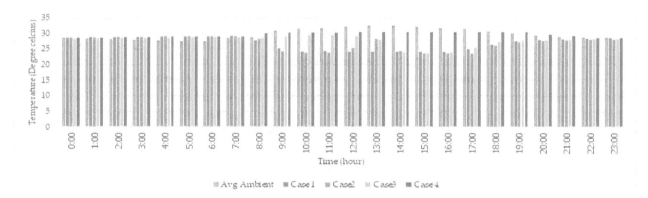

Figure 7. Diurnal temperature in the ambient air and in the room during experiments.

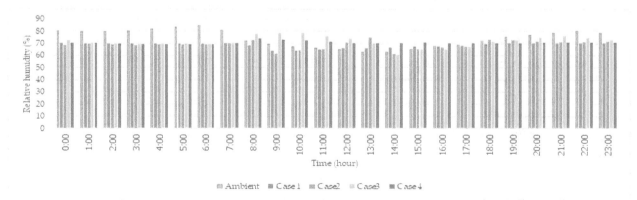

Figure 8. Diurnal relative humidity in the ambient air and in the room during experiments.

From Figure 7, case 2 would be the optimum choice if the occupants of the room preferred not to work under hot condition. The T in Case 3 exceeded the comfortable standard, but it did not reach 28 °C, the level with which 80% of Thai workers have been found to be comfortable [42]. Thus, the adoption of Case 3 would have to be based upon the agreement of all the occupants of the room. The values of T between Cases 3 and 4 are visibly different because the ambient T in the afternoon is higher than that in the morning. Certainly, the greatest saving of electricity can be achieved by not using AC at all during working hours, but it is not a realistic option with a working temperature of over 30 °C for 7–8 hours [35], which would be likely to affect work performance. From Figure 8, the RH was lower in the air conditioned room, so the skin would be dried when you stayed for a long time. If the door was opened in Case 2–4, the RH was increased in the room to be more comfortable. The monitored wind velocity in front of the room was 0.05 ± 0.04 m/s, which was calm wind that mostly blew from southwest and west direction. The door was in the west, so the mild wind entered the room to comfort people and exchanged the air between inside and outside the room to reduce CO_2.

Energy saving from turning off air conditioner was calculated. The air conditioner size 36,000 BTU/Hour (international) was converted into 10.55056 kWh. For case 2, the air conditioner was turned off one hour for 238 working days (not including special holiday 24 days and weekend 104 days in Thailand 2020). The conversion factor of Thailand grid mixed electricity year 2016–2018 was 0.5986 kg CO_2eq/kWh (LCIA method IPCC 2013 GWP 100a V1.03, Thai National LCI Database, TIIS-MTEC-NSTDA (with TGO, Electricity 2016–2018) updated in December 2019) [43]. Therefore, we can reduce CO_2eq emission for 1503.10 CO_2eq per year.

3.3. Results of Using Sansevieria trifasciata for IAQ Improvement

Table 3 presents the results of monitoring the parameters that are relevant to IAQ during the experimental placement of mother-in-law's tongue plants in the poorly ventilated room.

Table 3. Results of indoor air quality improvement using *Sansevieria trifasciata*.

Number of Plants		0 Plants	2 Plants	3 Plants	4 Plants	5 Plants	6 Plants
CO_2 (ppm)	Min	491.15	468.55	470.87	475.83	467.92	465.08
	Avg	681.32 ± 136.59	549.91 ± 52.46	531.52 ± 58.40	583.48 ± 103.08	588.22 ± 115.67	609.98 ± 60.81
	Max	1133.30	1029.27	770.12	902.67	1072.43	1037.10
CO (ppm)	Min	0.97	0.98	0.98	0.98	0.95	0.90
	Avg	1.05 ± 0.06	1.20 ± 0.16	1.17 ± 0.13	1.14 ± 0.14	1.01 ± 0.03	1.00 ± 0.02
	Max	1.30	1.90	1.92	1.57	1.25	1.17
RH (%)	Min	52.00	46.00	62.00	57.00	64.00	63.00
	Avg	64.23 ± 2.98	140.53 ± 2.21	66.00 ± 1.23	67.29 ± 1.67	69.11 ± 0.92	68.07 ± 1.75
	Max	72.00	72.00	71.00	70.00	76.00	73.00
T (°C)	Min	24.38	24.55	24.61	24.17	24.65	24.58
	Avg	26.61 ± 1.72	27.55 ± 1.72	28.02 ± 1.49	27.39 ± 1.98	27.20 ± 1.82	26.79 ± 1.73
	Max	28.67	29.72	29.62	29.64	29.50	29.64

Table 3 shows that the average concentration of CO_2 was decreased by placing 2, 3, 4, 5, and 6 mother-in-law's tongue plants in the office as compared to there being no plants in the room, as can be visually observed in Figure 9. The concentration of CO_2 in the room was not varied by the number of plants, but influenced by temperature of the room. This means if the door is opened, the temperature will be high and the CO_2 concentration will be reduced. Closed chamber is required to control infiltration and ventilation to only consider the influence of plant on indoor CO_2.

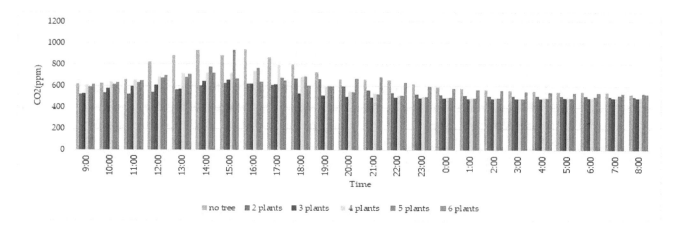

Figure 9. The trend of 24-hour CO_2 concentration and number of *Sansevieria trifasciata*.

From Figure 9, it can be seen that there was an increasing trend in CO_2 concentrations from 9:00 to 15:00, after which they declined to an ambient air concentration at 473.23 ± 8.66 ppm. The reductions in the percentage CO_2 concentrations with 2, 3, 4, 5, and 6 of *S. trifasciata* when compared to no plants being placed in the office were 19.29%, 21.99%, 14.36%, 13.66%, and 10.47%, respectively. The overall average CO_2 decreased by 15.95 ± 4.13%, which was slightly (±4%) lower than the reduction that was achieved in Case 2 by turning off the AC during lunchtime. The statistical analysis by One-Way ANOVA results was significant at $p = 0.009132$ ($p<0.05$) and f-ratio value is 3.53877. The results are different among treatments. However, the result of statistical analysis to compare between Case 2 (turned off AC at noon) and 3–6 indoor plants found $p = 0.061$, so it is not significant at the 95% confidence level. Therefore, reducing indoor CO_2 could be done by turning off air conditioner and opening the door during lunch hour, or by planting *S. trifasciata* in offices because there was no difference from statistical analysis results.

These results were consistent with those of other studies, which have found that plants could reduce indoor CO_2 concentrations [44,45]. Therefore, during the daytime, human respiration is clearly the key factor in increasing the CO_2 concentration, while ventilation is the main factor in decreasing the level of CO_2 in a room. Hence, the number of people in the room and their activities are the main drivers of the CO_2 concentration. Thus, the concentration of CO_2 is not directly related to the number of plants that are placed in the room. It was found that the respiration of plants during the night had no effect on the CO_2 concentration when compared with no plants in the room with a declining trend in CO_2 to the same level among different options from 0–6 mother-in-law's tongue plants being apparent.

The envelope air permeability of the room was the air passing through 2–3 millimeters around the door, which was caused by damage of the sealed material. The infiltration rate of the room was considered from the CO_2 concentration when there was no plant in the room to avoid the effect from photosynthesis or respiration of plant and microorganism in soil. The maximum rate was found at around 17:00–18:00 when people went back after finish working. The average maximum value of ventilation rate was 0.0152±0.0006 m³/s. The maximum infiltration was found at 0.0162 m³/s.

3.4. Results of The Simulation

As noted above, the stale air inside the room might not be properly ventilated and this section presents the results of the computer simulation generated to explain that hypothesis, as shown in Figure 10.

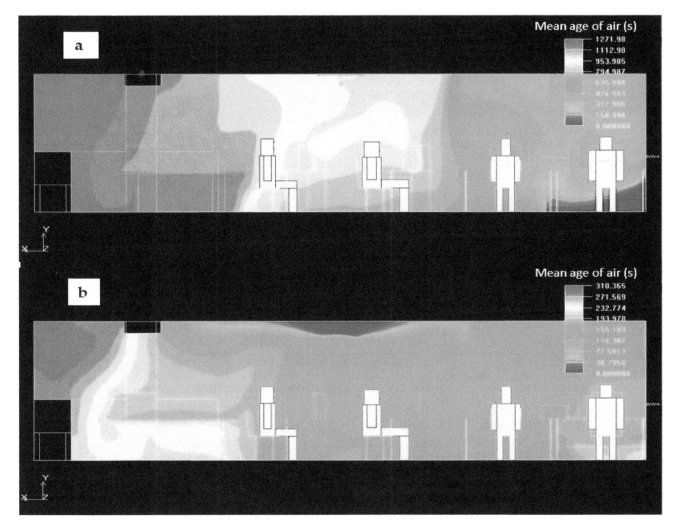

Figure 10. Assessment of air ventilation: mean age of air, (**a**) fan mounted on the ceiling was turned off (**b**) fan mounted on the ceiling was turned on.

Figure 10 shows the comparative results for the mean age of air when the fan mounted on the ceiling is turned off (Figure 10a) and turned on (Figure 10b). The electric fan was only used when the AC was switched off so there was no effect from the AC on the movement of air in this simulation. The average mean age of the air in Figure 10a,b were 747 s and 164 s, respectively. Thus, the efficiency of mixing the air by a 39-watt ceiling fan, from which the volume of air blown is 30 m³/min., is improved by approximately four times. Moreover, the exchange of air between the inside and outside of the room only occurred in the area near the door, and the stale air (i.e., that with the highest concentration of CO_2 from human exhalation) was retained inside of the room, and this is clearly illustrated in Figure 10a.

The fan needs to be used for mixing air and the door must be opened, as shown in both Figures 10a and 10b, since otherwise there will be no movement of air near the door, as presented in Figure 10a. Thus, the ventilation is undoubtedly much better when compared to when the AC is turned on when the door is closed. This finding should encourage room occupants to sometimes apply general ventilation to increase the IAQ.

A mesh refinement study was performed based on four grids, i.e., coarser, coarse, medium, and fine grid, with approximately 0.4–0.5 million, 0.8–0.9 million, 1.4–1.5 million, and 1.6–1.8 million cells, respectively, to minimize and ensure that the error was below the tolerance level. The average mass flow rate in the room was used for comparison of the meshes. The differences between fine grid and the other grids (coarser, coarse, and medium grids) are about 7.4%, 3.6%, and 1.7%, respectively.

The performance of the medium grid, selected for the simulation of this numerical study, was not significantly different to the fine one, and it was concluded as the suitable option.

3.5. Mitigation of GHG Emissions

Table 4 summarizes the reduction in electricity costs and hence GWP for the three experimental cases.

Table 4. Estimation of mitigations of greenhouse gas (GHG) emissions based on the electricity saving scenarios.

Appliances	Case A			Case B	Case C
	Electricity	cost[a]	GHGs[b]	%	%
PC	5356.8	604.4	3118.2	0.0	−8.1
Printer	5.0	0.6	2.9	0.0	0.0
Printer (standby)	3.1	0.4	1.8	0.0	0.0
Refrigerator	535.7	60.4	311.8	−4.3	−4.3
AC	1984.0	223.8	1154.9	−3.0	−12.0
Fan	0.0	0.0	0.0	0.1	0.5
Water dispenser	198.4	22.4	115.5	0.0	0.0
Light	166.7	18.8	97.0	−0.3	−0.3
Total	8249.6	930.7	4802.1	−7.5	−24.3

Notes: Cases A, B, and C are detailed in Table 1. Superscripts: [a] electricity cost per unit varies from month to month. The cost was averaged from the annual bill based on a figure of 3.96 Baht per kWh; [b] The GWP is calculated based on IPCC guidelines [46]. The emission factor for electricity production in Thailand is 0.5821 kgCO$_2$-eq [47].

When the feasible measures for the electrical appliances in the office were applied representing Case 2/B with the AC turned off from 12.00–13.00, or as in Case 3/C with the AC turned off from 9.00–13.00, the electricity cost and GWP can be reduced by 7.5% and 24.3%, respectively. AC is generally accepted as the largest consumer of electrical energy per unit and, thus, contributes most to the emission of GHGs. However, the total amount of electricity that is used by computers is higher due to the number of computers in the office (6 units). Overall therefore, since computers consume the greatest amount of energy, turning them off for only one hour per day during lunchtime can help to reduce the overall electricity consumption by almost 10%. However, this option was least convenient as compared to other choices, based on the opinion of the room occupants. The different practices relating to computers and the AC are the reason why the reductions that result from Cases B and C are so great.

Therefore, the methods that were investigated in this study to reduce electricity consumption and GWP would be feasible means of mitigating costs by reducing the usage of appliances in offices and could be adopted and implemented in energy saving plans in universities and other workplaces. However, it should be noted that the figures that are presented in Table 4 are only the estimated values, and based on the power ratings (wattage) of the various devices. The actual electricity consumption also depends on the settings actually used (such as standby mode for computers and T settings of the AC and refrigerator). In general, by observation, there is unlikely to be a significant effect from the adjustment of the T setting of the AC based on seasonal variation (i.e., the T setting is not varied during the year based on the outside temperature). More accurate data could be obtained if an electricity meter was installed in each room to monitor the actual electricity consumption in different scenarios.

4. Conclusions

This study conducted measurements of four IAQ parameters i.e., T, RH, CO, and CO$_2$, in a small room with six occupants and found that the levels of some of the parameters exceeded the recommended levels, particularly the level of CO$_2$, the source of which was human respiration. Therefore, the number of occupants in the room and its ventilation efficiency are the key factors for the CO$_2$ concentration. Poor conditions (i.e., a CO$_2$ concentration of over 1000 ppm) were not detected after the simple mitigation practices were implemented. However, T is a significant factor that must be taken into consideration in the adoption of measures to reduce electricity consumption. Without reconstruction of the office space and the installation of a ventilation system, Case 3/C was the best option with good IAQ, and it would achieve a reduction of electricity consumption of 24.3% based on the situation without taking any mitigating action. Moreover, this system would be feasible with the agreement of all members of staff working in the office. Another feasible option for CO_2 reduction was placing mother-in-law's tongue plants in the office. The average reduction in the CO_2 level based on using between two and six plants was almost 16%, which rendered the CO_2 concentration within the standard comfortable level of 1000 ppm. However, human activities are the key factor in the CO_2 concentration in a room, not the number of plants. The data that were derived from measuring the actual IAQ parameters in various scenarios and the results of the computer simulation are helpful in identifying and promoting simple practices that aimed at achieving good IAQ while reducing electricity costs and, hence, GWP in office situations. Further research should be directed towards measuring other IAQ parameters, and the causes of SBS, which are possibly associated with the CO, particulate matter, and VOCs in car exhaust fumes.

Author Contributions: Conceptualization, K.P. (Kanittha Pamonpol); methodology, K.P. (Kanittha Pamonpol) and K.P. (Kritana Prueksakorn); software model K.P. (Kritana Prueksakorn); validation K.P. (Kanittha Pamonpol) and K.P. (Kritana Prueksakorn); formal analysis, K.P. (Kanittha Pamonpol) and K.P. (Kritana Prueksakorn); investigation, K.P. (Kanittha Pamonpol); resources, K.P. (Kanittha Pamonpol); data curation, K.P. (Kanittha Pamonpol) and T.A.; writing-original draft preparation, K.P. (Kanittha Pamonpol) and K.P. (Kritana Prueksakorn); writing-review and editing, T.A.; visualization, K.P. (Kanittha Pamonpol) and K.P. (Kritana Prueksakorn); supervision, K.P. (Kanittha Pamonpol); funding acquisition, K.P. (Kanittha Pamonpol), K.P. (Kritana Prueksakorn), and T.A.; project administration, T.A. All authors have read and agreed to the published version of the manuscript.

Acknowledgments: The authors would like to express our appreciation to the Research and Development Institute, Valaya Alongkorn Rajabhat University under the Royal Patronage, Prince of Songkla University, Phuket Campus for the financial support. The support by assistants, i.e., Tip Sophea and Hong Anh Thi Nguyen are also acknowledged. The authors would like to thank Robert William Larsen for kind advice on English writing.

References

1.   NOAA. Global Climate Station Summary: Annual Mean Temperature 2010–2015. 2020. Available online: https://gis.ncdc.noaa.gov/maps/ncei/summaries/global (accessed on 13 February 2020).
2.   Thai Meteorological Department: Meteorological Development Bureau, The Climate of Thailand. 2015. Available online: https://www.tmd.go.th/en/archive/thailand_climate.pdf (accessed on 13 February 2020).
3.   Canha, N.; Lage, J.; Candeias, S.; Alves, C.; Almeida, S.M. Indoor air quality during sleep under different ventilation patterns. *Atmos. Pollut. Res.* **2017**, *8*, 1132–1142. [CrossRef]
4.   Persily, A.K. Indoor carbon dioxide concentrations in ventilation and indoor air quality standards. In Proceedings of the 36th Air Infiltration and Ventilation Centre Conference, Madrid, Spain, 23–24 September 2015.
5.   EERE. Common Air Conditioner Problems. Energy Saver, Office of Energy Efficiency & Renewable Energy, 2017. Available online: https://energy.gov/energysaver/common-air-conditioner-problems (accessed on 17 October 2018).
6.   Boonyayothin, V.; Hirnlabh, J.; Khummongkol, P.; Teekasap, S.; Shin, U.; Khedari, J. Ventilation Control Approach for Acceptable Indoor Air Quality and Enhancing Energy Saving in Thailand. *Int. J. Vent.* **2016**, *9*, 315–326. [CrossRef]
7.   Spiru, P.; Simona, P.L. A review on interactions between energy performance of the buildings, outdoor air pollution and the indoor air quality. *Energy Procedia.* **2017**, *128*, 179–186. [CrossRef]
8.   EPA. An Office Building Occupants Guide to Indoor Air Quality. United States Environmental Protection Agency, 2017. Available online: https://www.epa.gov/indoor-air-quality-iaq/office-building-occupants-guide-indoor-air-quality (accessed on 19 October 2019).

9. NASA. The Consequences of Climate Change. National Aeronautics and Space Administration, 2017. Available online: https://climate.nasa.gov/effects/ (accessed on 10 November 2019).

10. Junaid, M.; Syed, J.H.; Abbasi, N.A.; Hashmi, M.Z.; Malik, R.N.; Pei, D.S. Status of indoor air pollution (IAP) through particulate matter (PM) emissions and associated health concerns in South Asia. *Chemosphere* **2017**, *191*, 651–663. [CrossRef] [PubMed]

11. Jung, C.C.; Su, H.J.; Liang, H.H. Association between indoor air pollutant exposure and blood pressure and heart rate in subjects according to body mass index. *Sci. Total Environ.* **2016**, *539*, 271–276. [CrossRef] [PubMed]

12. Ramalho, O.; Wyart, G.; Mandin, C.; Blondeau, P.; Cabanes, P.-A.; Leclerc, N.; Mullot, J.-U.; Boulanger, G.; Redaelli, M. Association of carbon dioxide with indoor air pollutants and exceedance of health guideline values. *Build. Environ.* **2015**, *93*, 115–124. [CrossRef]

13. World Health Organization. Regional Office for Europe. In *WHO Guidelines for Indoor Air Quality: Selected Pollutant*; World Health Organization, Regional Office for Europe: Copenhagen, Denmark, 2010; ISBN 9789289002134.

14. Risuleo, R.S.; Molinari, M.; Bottegal, G.; Hjalmarsson, H.; Johansson, K.H. A benchmark for data-based office modeling: Challenges related to CO_2 dynamics. *IFAC-PapersOnLine.* **2015**, *48*, 1256–1261. [CrossRef]

15. Mandin, C.; Trantallidi, M.; Cattaneo, A.; Canha, N.; Mihucz, V.G.; Szigeti, T.; Mabilia, R.; Perreca, E.; Spinazzè, A.; Fossati, S.; et al. Assessment of indoor air quality in office buildings across Europe—The OFFICAIR study. *Sci. Total Environ.* **2017**, *579*, 169–178. [CrossRef]

16. Ye, W.; Zhang, X.; Gao, J.; Cao, G.; Zhou, X.; Su, X. Indoor air pollutants, ventilation rate determinants and potential control strategies in Chinese dwellings: A literature review. *Sci. Total Environ.* **2017**, *586*, 696–729. [CrossRef]

17. Arar, S.; Al-Hunaiti, A.; Masad, M.H.; Maragkidou, A.; Wraith, D.; Hussein, T. Elemental Contamination in Indoor Floor Dust and Its Correlation with PAHs, Fungi, and Gram+/− Bacteria. *Int. J. Environ. Res. Public Health* **2019**, *16*, 3552. [CrossRef]

18. Norhidayah, A.; Kuang, L.C.; Azhar, M.K.; Nurulwahida, S. Indoor air quality and sick building syndrome in three selected buildings. *Procedia Eng.* **2013**, *53*, 93–98. [CrossRef]

19. ASHRAE. Indoor air quality guide. In *Best Practices for Design, Construction, and Commissioning*; American Society of Heating, Refrigerating and Air-conditioning Engineers (ASHRAE): Atlanta, GA, USA, 2009; ISBN 9781933742595.

20. Jin, L.; Zhang, Y.; Zhang, Z. Human responses to high humidity in elevated temperatures for people in hot-humid climates. *Build. Environ.* **2017**, *114*, 257–266. [CrossRef]

21. Liu, W.; Zhong, W.; Wargocki, P. Performance, acute health symptoms and physiological responses during exposure to high air temperature and carbon dioxide concentration. *Build. Environ.* **2017**, *114*, 96–105. [CrossRef]

22. RFA. *Carbon Dioxide (CO_2) Safety Program*; Renewable Fuels Association: Vancouver, WA, USA, 2015.

23. Sohn, C.H.; Huh, J.W.; Seo, D.W.; Oh, B.J.; Lim, K.S.; Kim, W.Y. Aspiration pneumonia in carbon monoxide poisoning patients with loss of consciousness: Prevalence, outcomes, and risk factors. *Am. J. Med.* **2017**, *130*, 1465.e21–1465.e26. [CrossRef] [PubMed]

24. Wolverton, B.C.; Johnson, A.; Bounds, K. *Interior Landscape Plants for Indoor Air Pollution Abatement*; NASA: Washington, DC, USA, 1989; p. 22.

25. Ongwandee, M.; Moonrinta, R.; Panyametheekul, S.; Tangbanluekal, C.; Morrison, G. Investigation of volatile organic compounds in office buildings in Bangkok, Thailand: Concentrations, sources, and occupant symptoms. *Build. Environ.* **2011**, *46*, 1512–1522. [CrossRef]

26. Sun, S.; Zheng, X.; Villalba-Díez, J.; Ordieres-Meré, J. Indoor Air-Quality Data-Monitoring System: Long-Term Monitoring Benefits. *Sensor* **2019**, *19*, 4157. [CrossRef]

27. ASHRAE. Standards & Guidelines. American Society of Heating, Refrigerating and Air-Conditioning Engineers, 2017. Available online: https://www.ashrae.org/standards-research--technology/standards--guidelines (accessed on 30 September 2019).

28. Fluke Corporation. Fluke 975 Air Meter™, 2019. Available online: http://www.fluke.com/fluke/then/HVAC-IAQ-Tools/Air-Testers/Fluke-975.htm?PID=56156 (accessed on 3 August 2019).

29. Almeida, R.M.S.F.; de Freitas, V.P.; Delgado, J.M.P.Q. *School Buildings Rehablilitation: Indoor Environmental*

Quality and Enclosure Optimization; Springer International Publishing: Basel, Switzerland, 2015; pp. 35–83, ISSN 2191-530X; ISBN 978-3-319-15359-9. [CrossRef]

30. Gubb, C.; Blanusa, T.; Griffiths, A.; Pfrang, C. Can houseplants improve indoor air quality by removing CO_2 and increasing relative humidity? *Air Qual. Atmos. Health* **2018**, *11*, 1191–1201. [CrossRef]

31. Alonso, M.J.; Andreson, T.; Frydenlund, F.; Widell, K.N. Improvement of air flow distribution in a freezing tunnel using Airpak. *Procedia Food Sci.* **2011**, *1*, 1231–1238. [CrossRef]

32. Li, G.; She, C.; Wu, N.; Li, Z.; Wang, L. The solution and simulation of the condensation problem of the capillary network system in the children's hospital in Shenyang in Summer. *Procedia Eng.* **2015**, *121*, 1215–1221. [CrossRef]

33. Prueksakorn, K.; Piao, C.X.; Ha, H.; Kim, T. Computational and experimental investigation for an optimal design of industrial windows to allow natural ventilation during wind-driven rain. *Sustainability* **2015**, *7*, 10499–10520. [CrossRef]

34. Habchi, C.; Ghali, K.; Ghaddar, N. Coupling CFD and analytical modeling for investigation of monolayer particle suspension by transient flows. *Build. Environ.* **2016**, *105*, 1–12. [CrossRef]

35. Buratti, C.; Mariani, R.; Moretti, E. Mean age of air in a naturally ventilated office: Experimental data and simulations. *Energy Build.* **2011**, *43*, 2021–2027. [CrossRef]

36. ASHRAE. *ANSI/ASHRAE Standard 62.1-2016, Ventilation for Acceptable Indoor Air Quality*; American Society of Heating, Refrigerating and Air-conditioning Engineers, the American National Standards Institute: Atlanta, GA, USA, 2016; ISSN 10412336.

37. ASHRAE. *ANSI/ASHRAE Standard 55-2017, Thermal Environmental Conditions for Human Occupancy*; American Society of Heating, Refrigerating and Air-conditioning Engineers, the American National Standards Institute: Atlanta, GA, USA, 2017; ISSN 10412336.

38. Boonyou, S.; Jitkhajornwanich, K. Enhancement of natural ventilation in office buildings in Bangkok. In *Architecture, City, Environment: Proceedings of the PLEA 2000, Cambridge, UK, 2-5 July 2000*; James & James (Science Publishers) Ltd.: London, UK, 2000.

39. Busch, J.F. A tale of two populations: Thermal comfort in air-conditioned and naturally ventilated offices in Thailand. *Energy Build.* **1992**, *18*, 3–4. [CrossRef]

40. Thai Meteorological Department: Meteorological Development Bureau, Summary of the Climate in Thailand 2017. 2018. Available online: https://www.tmd.go.th/programs/uploads/yearlySummary/สรุปสภาวะอากาศปี%202561.pdf (accessed on 13 February 2020). (In Thai).

41. Klinmalee, A.; Srimongkol, K.; Kim Oanh, N.T. Indoor air pollution levels in public buildings in Thailand and exposure assessment. *Environ. Monit. Assess.* **2009**, *156*, 581–594. [CrossRef]

42. Baklanov, A.; Molina, L.T.; Gauss, M. Megacities, Air quality and Climate. *Atmos. Environ.* **2016**, *126*, 235–249. [CrossRef]

43. Thailand Greenhouse Gas Management Organization. Carbon Label and Carbon Footprint for Organization: Emission Factor. 2020. Available online: http://thaicarbonlabel.tgo.or.th/products_emission/products_emission.pnc (accessed on 28 February 2020). (In Thai).

44. Cetin, M.; Sevik, H. Measuring the impact of selected plants on indoor CO_2 concentration. *Pol. J. Environ. Stud.* **2016**, *25*, 973–979. [CrossRef]

45. Parhizkar, H.; Khoraskani, R.A.; Tahbaz, M. Double skin façade with Azolla; ventilation, Indoor Air Quality and Thermal Performance Assessment. *J. Clean. Prod.* **2020**, *249*, 119313. [CrossRef]

46. IPCC. Guidelines for National Greenhouse Gas Inventories. Intergovernmental Panel on Climate Change, 2006. Available online: http://www.ipcc-nggip.iges.or.jp/public/2006gl/ (accessed on 19 September 2019).

47. TGO. Emission Factor Collected from Secondary Data for Carbon Footprint for Organization. Thailand Greenhouse Gas Management Organisation, 2017. Available online: http://thaicarbonlabel.tgo.or.th/admin/uploadfiles/emission/ts_11335ee08a.pdf (accessed on 1 October 2019).

13

Keeping Doors Closed as One Reason for Fatigue in Teenagers

Anna Mainka * and Elwira Zajusz-Zubek

Department of Air Protection, Silesian University of Technology, 22B Konarskiego St., 44-100 Gliwice, Poland
* Correspondence: Anna.Mainka@polsl.pl.

Featured Application: Presented results are only a case study and the authors are not able to relate them to the population of children and teenagers in Poland; however, it is worth underlining the positive effects of opening doors during sleep, since the CO_2 concentration decreases 55–64% without a reduction in thermal comfort. This simple action applied by parents can decrease the contribution of low indoor environment quality (IEQ_{IV}) in children's and teenager's bedrooms to approximately 1% during the night.

Abstract: (1) Background: Healthy teenagers are often sleepy. This can be explained by their physiology and behavioral changes; however, the influence of CO_2 concentration above 1000 ppm should not be neglected with respect to sleep dissatisfaction. (2) Methods: CO_2 concentrations were measured in two similar bedrooms occupied by girls aged 9 and 13 years old. The scheme of measurements included random opening and closing of the bedroom doors for the night. Additionally, the girls evaluated their sleep satisfaction in a post-sleep questionnaire. (3) Results: During the night, the CO_2 concentration varied from 402 to 3320 ppm in the teenager's bedroom and from 458 to 2176 ppm in the child's bedroom. When the bedroom doors were open, inadequate indoor air quality (IEQ_{III} and IEQ_{IV} categories) was observed in both the teenager's and child's bedroom during 11% and 25% of the night, respectively; however, closing the doors increased the contribution of moderate (IEQ_{III}) and low (IEQ_{IV}) categories of air to 79% and 86%, respectively. The girls were dissatisfied only when the bedroom door was closed. The satisfied category of sleep was selected only by the younger girl. (4) Conclusions: Opening the bedroom door during the night can decrease the CO_2 concentration 55–64% without reducing thermal comfort.

Keywords: teenagers; children; bedroom; IEQ; CO_2

1. Introduction

There is a generally accepted opinion in society that teenagers are sleepy. This notion indicates that sleepiness is a negative issue; however, this particular group of young people has several reasons for being tired. From a physiological perspective, teenagers undergo the processes of puberty and rapid increases in height [1], and, from a psychological aspect, they have problems with low self-esteem, seeking their role in society, and getting easily involved in personal conflicts at school or within the family [1,2]. The duration of sleep varies among individuals, with an average teenager attending school requiring at least 8 h of sleep per night; however, the maturational changes combined with a cell phone or computer use within the hour prior to trying to fall asleep lead to shorter sleeping hours or sleep deprivation [3]. Sleep issues have drawn worldwide attention in recent years. Sleep problems threaten health and quality of life for up to 45% of the world's population, and 35% of individuals feel that they do not get enough sleep, which has negative effects on both physical and mental health [4]. Rafihi-Ferreira et al. [5] referenced several studies, in which the association between sleep quality and behavioral problems was investigated, demonstrated a relationship between externalizing and internalizing problems and sleep problems in children. This association becomes even more serious when considering that childhood sleep problems may persist and that they constitute a greater risk for other behavioral problems. In comparison with psychological state, body condition, and circadian rhythm, indoor environment quality (IEQ) including thermal conditions and indoor air quality (IAQ) can improve sleep through environmental control. Matricciani et al. [6] in the meta-review underlined that sleep duration is associated with adiposity in children, while sleep quality, timing and variability appear important for children's health, but further research is needed. Children and teenagers are more adversely affected by indoor air pollution than adults since they breathe a greater volume of air relative to their body weight, which may lead to a greater burden of pollutants on their bodies [7,8]. The quality of the indoor environment not only affects health and comfort, but it may also impair learning ability. Poor IEQ, as a result of inadequate thermal conditions, and IAQ can cause symptoms such as being too hot or too cold as well as feeling restless or sleepy [7].

A powerful remedial to improve IAQ is to increase the outdoor air supply rate; however, in recent years, effort has been focussed on decreasing energy consumption. Within houses, the building airtightness is maximized and the ventilation rates are minimized, leading to a general deterioration in indoor air quality. Consequently, by focussing on optimal indoor thermal comfort, air infiltration and dilution of indoor air pollutants are lowered.

Generally, in passive stack ventilation systems, to create optimal air quality conditions for the maintenance of health and comfort of the occupants, windows should be open slightly. In moderate climate zones such as Poland, this is the best solution from spring to fall seasons; however, during winter, unsealed windows decrease thermal comfort. The Upper Silesia region, in comparison with other Polish regions and European Union countries, is characterized by relatively high particulate matter concentrations. Although the last three decades of economic changes have forced the greatest drop in Polish industrial air pollution by closure of the old steelworks, cookeries, coal mines, and coking plants, emissions from small-scale combustion utilities, such as domestic boilers, together with re-suspension processes from urban surfaces and road traffic, have become particularly dangerous. Investigations performed during the winter season confirm that the hazard of domestic sources originates from the low quality of fuels (coal, biomass, culm, or even refuse) used for heating [9,10]. According to the dominant role of coal combustion in the region, residents are even advised not to open windows during winter, particularly at low wind speed, since it decreases IAQ.

IAQ is a multi-disciplinary phenomenon and is determined by many pathways in which chemical, biological, and physical contaminants eventually become a portion of the total indoor environmental composition. There is spatial and temporal heterogeneity of these contaminants, and the determination of exposure is difficult due to the diversity of time that occupants spend within the space [11,12]. Among indoor air pollutants, carbon dioxide (CO_2) is considered a useful and easily measurable indicator of the ventilation and air quality in indoor environments. Although this pollutant itself does not cause serious health issues at lower concentrations and, for short durations, higher concentrations can indicate a lower ventilation level and possible air contamination with other pollutants [13].

In residential non-smoking areas, with the exception of kitchens, occupants are a major source of CO_2 through exhalation. Mean concentrations of CO_2 during cooking times are significantly higher ($p < 0.01$) than those during non-cooking times, with no significant differences among fuel types [14]. While the CO_2 generation rate per person varies as a function of age, activity, and diet, on average, children aged 6–11 years old produce 448 mg/min/p and adults generate 763 mg/min/p [15]. Persily and de Jonge in [16] presented CO_2 generation rates according to age and gender. During sleep, the level of physical activity (MET) is 1.0; thus, the calculated metabolic rate in children aged 6–11 years old is 295 mg/min/p and 271 mg/min/p for boys and girls, respectively, while, for 11–16-year-old males and females, it is 401 mg/min/p and 342 mg/min/p, respectively. Teenagers generate more CO_2 according to their higher body weight, and they demand more privacy, therefore keeping their door closed all day and night. The metabolic activity itself influences air quality in bedrooms by reducing the concentration of oxygen and increasing the level of CO_2.

For CO_2, the exposure limits have been derived exclusively on the basis of health considerations. The Exposure Guidelines for Residential Indoor Air Quality [17] suggest that CO_2 concentrations above 1800 mg/m^3 (1000 ppm) are indicative of an inadequate supply of fresh air, although complaints have been documented at concentrations as low as 1100 mg/m^3 (600 ppm) [17]. Assumptions regarding these levels are in accordance with ASHRAE standards, which previously (62-1989) considered 1000 ppm as the highest acceptable concentration for a minimum sanitary requirement and 8 l/p/s as the minimum ventilation rate. More recent standards (ASHRAE 62-1999, 62-2001, and 62-2004) recommend that the indoor-outdoor differential concentration should not exceed 700 ppm [15]. Generally, a sufficient margin to protect against undesirable changes in the acid-base balance and subsequent adaptive changes such as the release of calcium from bones is a level of 6300 mg/m^3 (3500 ppm) [17]. However, a higher maximum exposure concentration is recommended for direct physiological effects of exposure to CO_2 as opposed to subjective symptoms. Subjective symptoms such as fatigue, headaches, and an increased perception of warmth and unpleasant odors have been associated with CO_2 levels of 900–5800 mg/m^3 (500–3200 ppm) [17].

Polish legal acts do not specify permissible concentrations of CO_2 in ambient air or in rooms intended for the permanent residence of individuals, i.e., apartments and houses. The regulations of the Ministry of Family, Work, and Social Policy define the highest permissible concentrations and intensities of agents harmful to health in the work environment [18]. The highest permissible CO_2 concentration is 9000 mg/m^3. Thus far, the Polish Committee for Standardisation followed general guidelines concerning the quality of air inside non-residential buildings (PN-EN 13779 standard [19]), but, in May 2019, the Polish Committee for Standardisation accepted EN-16798-1 (in English) [20] as a national standard and in CEN/TR 16798-2 Technical Report [21] including recommended criteria for the CO_2 calculation for demand-controlled ventilation in occupied living rooms and bedrooms. Design ΔCO_2 concentration for bedrooms (ppm above outdoors) are within the limits of ≤380 ppm (IEQ$_I$), 380–550 ppm (IEQ$_{II}$), 550–950 ppm (IEQ$_{III}$), and ≥950 ppm (IEQ$_{IV}$) corresponding to high (I), medium (II), moderate (III), and low (IV) indoor environment quality, respectively.

IEQ evaluation based on CO_2 concentration and possible effects on occupants comfort has been examined [11,22–27]. Polish research on CO_2 levels and possible occupant fatigue includes a few publications [9,28–31]; however, no research regarding IEQ in the bedrooms of Polish children and teenagers have been performed. Such studies can help to better define how behavioral patterns influence the possible exposure levels of air pollutants and occupant fatigue. The present paper presents the results of the measurement of CO_2 levels and the corresponding IAQ linked to the possible tiredness and lethargy of a teenager as compared with a child.

2. Materials and Methods

The concentration of CO_2 was continuously measured in two bedrooms: one occupied by a teenage girl aged 13 and a second by a younger girl aged 9. The choice of girls instead of boys was intentional since, among children and teenagers, more males than females exhibit excessive nightly use of computer games and consequent sleep disturbances [1]. Moreover, Karjalainen [32] suggested that female occupants should primarily be used as subjects when examining indoor thermal comfort requirements, since, if women are satisfied, then it is highly probable that men are also satisfied. Participation in the present study was voluntary. Prior to starting the measurements, the participants' parents gave informed consent.

The home of the girls was located in the suburbs of the western part of the Upper Silesia region. The distance from the urban area was approximately 10 km (Figure 1). The house was built in 2008 and was located in the third row of buildings, 100 m from the local street. It was a detached house (Figure 1) with an attic and no basement. The house area was 98 m^2 with a glass terrace of 12 m^2. The measurements were performed in the South and North bedrooms, which each had an area of approx.

12 m^2 and were occupied by the older and younger girl, respectively; detailed information has been added in Supplementary Materials Table S1.

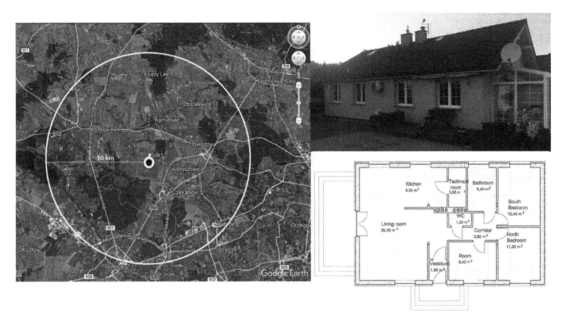

Figure 1. Location, south view and ground floor plan of the house.

The sampling strategy for CO_2 included indoor CO_2 concentration, measurement, and questionnaires. Precautions necessary to avoid measurements in air directly exhaled by building occupants were performed as described in [33]. One automatic portable monitor (model 77535, Az Instruments International Ltd., Hong Kong, China) connected to a PC with RS232 software installed was used in the present study. According to Mahyuddin et al. [34], one CO_2 sensor in a room with $<100 \text{ m}^2$ floor area had significant p-value relationships, and the breathing zone within the occupied space was considered to be between 1.0 and 1.2 m in the middle of a zone as a representative location. The monitor was equipped with a non-dispersive infrared sensor. The precision of measurements ranged between +0 and +10,000 ppm CO_2: ±100 ppm CO_2 or ±3% at a concentration below 100 ppm. The monitor also displayed and recorded in real-time the measurements of air temperature and relative humidity in the bedroom, allowing for logged data to be downloaded for analysis. The selected sampling interval was 60 s.

To estimate the parameters influencing CO_2 concentrations in the bedrooms during the measurements, parents were asked to note whether each daughter left the room for a longer period during the night. They provided a diary in which it was specified whether the windows and doors were closed, ajar, or fully open during the monitored days and nights. Additionally, the girls were asked to evaluate their sleep satisfaction as categorized into five levels [4]: very dissatisfied (1), dissatisfied (2), moderate (3), satisfied (4), and very satisfied (5). However, for the analysis, three levels were used: very dissatisfied and dissatisfied, defined as dissatisfied; moderately satisfied, defined as moderate; and satisfied and very satisfied, defined as satisfied. The evaluations of sleep quality were included in the post-sleep questionnaire. In total, 137 questionnaires were acquired, with 121 being valid.

During the study, the family was asked to maintain their regular routine regarding the opening of doors and windows. It should be underlined that, during daylight hours, the bedroom windows were fully open only twice for cleaning and ajar for a maximum of 1 h less than 10 times. The measurements were performed between 16 September 2018 and 2 March 2019. For the final analysis, 102 nights from 9:00 p.m. to 7:00 a.m. (51 nights per room) were included.

3. Results

As mentioned above, proper ventilation controls IAQ. Among the techniques adopted to evaluate ventilation rates, the one based on the measurement and analysis of the indoor CO_2 concentration and trends is the most common approach; however, it could be improved by the integration of an electronic nose for odor detection [35]. Acceptable ventilation conditions can be easily achieved in mechanically ventilated rooms, but it is not equally simple to maintain CO_2 and odor levels under control in indoor environments that are naturally ventilated [35]. Natural ventilation is typical for detached buildings in Poland, where the ventilation and the IAQ are controlled only by means of air infiltration through cracks and openings. The IAQ gets worse during the winter season when the desire for thermal comfort and acceptable IAQ are in conflict. For the characterization of IAQ by measuring CO_2 concentration, several indicators and criteria can be used [25]. We selected the average CO_2 concentration and a time fraction over a limit of ΔCO_2 values spent during the night according to the PN-EN 16798 standards [20,21].

Figure 2 shows the field measurements in the teenager's and child's bedrooms, including the CO_2 concentration range, when the doors were open and closed.

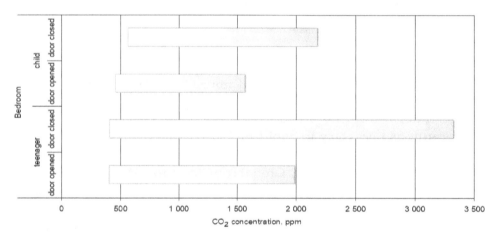

Figure 2. CO_2 concentration inside the teenager's and child's bedrooms during the night.

During the night, the short-term (60 s average) concentrations varied from 402 to 3320 ppm in the teenager's bedroom and from 458 to 2176 ppm in the child's bedroom. Thus, the highest concentrations exceeded the limits characteristic of subjective symptoms such as fatigue, headaches, and an increased perception of warmth and unpleasant odors [14]. If we compare the data separately by day of the week (Table 1), it is clear that, despite similar average concentrations, the maximum concentrations were significantly higher in the teenager's bedroom.

Table 1. Average, minimum, and maximum CO_2 concentrations according to the day of the week.

Parameters	CO$_2$ Concentrations, ppm					
	Teenager's Bedroom			Child's Bedroom		
	Average	Minimum	Maximum	Average	Minimum	Maximum
Monday	1127	402	3102	1828	1828	1828
Tuesday	998	597	2336	1114	554	1855
Wednesday	1088	452	2897	1144	458	1855
Thursday	1199	411	2695	1118	688	1778
Friday	1195	479	3320	1076	563	1757
Saturday	3061	3061	3061	1262	599	2176
Sunday	1191	405	3188	1993	1993	1993

Distinguishing between open and closed doors in the teenager's and child's bedrooms, the average nightly concentrations were not significantly different ($p = 0.76$; 1077 and 1103 ppm, respectively); however, there was a statistically significant difference in CO_2 concentration between bedrooms with open and closed doors (Figure 3). In the teenager's bedroom, opening the doors during the night decreased the average CO_2 concentration by 55%, and in the child's bedroom, the observed decrease was 64%.

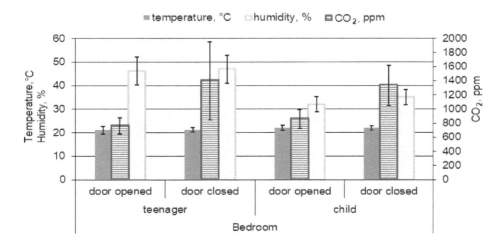

Figure 3. Mean temperature, relative air humidity, and CO_2 concentration inside the teenager's and child's bedrooms during the night.

Moreover, room temperature and humidity are important factors that may influence perceived environment quality. They both affect the thermal balance of the human body through effects on the skin and respiratory organs. Changes in air temperature trigger a sympathetic reflex via the skin that strengthens with lower air temperature. High temperatures and humidity require the human body to respond by increasing heat loss through the skin surface via blood circulation. The relation between indoor temperature and humidity depends on the season and indoor conditions [36]. In our study in the younger girl bedroom, the temperature and humidity correlation coefficient were -0.10 and -0.08 ($p < 0.05$) with opened or closed doors, respectively. In the older girl bedroom, the temperature and humidity were weakly positively correlated in the case of opened doors 0.26 and negatively correlated -0.33 in the case of closed doors. The mean room temperature differed between the teenager's and child's bedrooms significantly ($p < 0.01$). The mean temperature in the teenager's bedroom (21.1 ± 1.11 °C) was lower than that in the child's bedroom (22.05 ± 0.59 °C); however, the influence of opening and closing the doors to each bedroom during the night was not statistically significant ($p > 0.05$). The mean nightly temperature in the teenager's bedroom was 21.07 ± 1.39 °C and 21.01 ± 0.76 °C with open and closed doors, respectively, while, in the child's bedroom, the mean nightly temperature was 22.06 ± 0.62 °C and 22.03 ± 0.57 °C, respectively. In the case of humidity, a significant difference ($p < 0.01$) between the teenager's and child's bedrooms was also observed (46.46 ± 5.79% and 33.39 ± 3.41%, respectively). In the teenager's bedroom, the humidity did not differ between nights with respect to open and closed doors (46.04 ± 5.97% and 46.89 ± 5.68%, respectively), while, in child's bedroom, this difference was statistically significant ($p < 0.01$; 31.83 ± 3.20% and 34.88 ± 2.95%, respectively).

Based on WHO regulations [37] and the PN-EN 16798-1 and CEN/TR 16798-2 standards [20,21], the increase in CO_2 concentration in relation to that of outdoor air (ΔCO_2) was measured in both bedrooms. During the night, the general IEQ in the bedrooms included 32% in the high category (I), 28% in the medium category (II), 19% in the moderate category (III), and 21% in the low category (IV) in the older girl's bedroom and 14%, 29%, 32%, and 24%, respectively, in the younger girl's bedroom. This indicates that, for 10 h of rest at night, the teenager spent approximately 3 h in proper air quality, while the child spent only approximately 1.5 h. The indoor concentrations of CO_2 showed a higher contribution of inadequate air quality in the III and IV categories, corresponding to moderate and low IEQ, in the younger girl's bedroom. When the bedroom doors were open, inadequate IEQ was observed in the teenager's and child's bedrooms for 11% and 25% of the night, respectively (Figure 4); however, closing the doors increased the contribution of the moderate (III) and low (IV) categories to 79% and 86%, respectively.

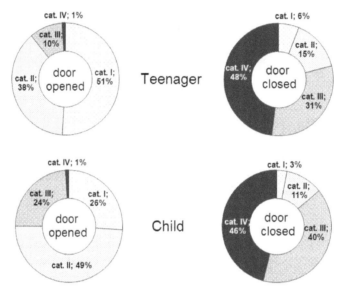

Figure 4. Categories of the indoor environment inside the teenager's and child's bedrooms during the night.

Despite the fact that the teenage girl spent an extra 1.5 h in the high and medium quality environment (IEQ$_I$ and IEQ$_{II}$), this did not correspond to sleep satisfaction. In the older girl's opinion, her sleep during study was never satisfying. During 51 nights of measurement, the teenage girl evaluated her sleep as moderate (3) for 36 nights, 13 nights as dissatisfied (2), and only two nights as very dissatisfied (1). The younger girl made a more positive assessment, evaluating seven nights as satisfied (4) and one night as very satisfied (5), eight nights categorized as dissatisfied (2), while one night as very dissatisfied (1), and 34 nights as moderate (3). Following the small contribution of very dissatisfied (1) and very satisfied (5) categories, we gathered categories (1) and (2) into one group defined as low sleep satisfaction, in addition to categories (4) and (5) in one group defined as high sleep satisfaction, as presented in Figure 5.

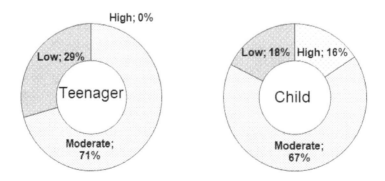

Figure 5. Low, moderate, and high sleep satisfaction of the teenager and child corresponding to categories: Very dissatisfied and dissatisfied (1) + (2); moderate (3), as well as satisfied and very satisfied (4) + (5).

Deeper analysis reveals that the girls were dissatisfied only when the bedroom door was closed. It is interesting that the average CO_2 concentration during the nights evaluated as dissatisfied had a similar average value of 1691 ppm and 1623 ppm in the older and younger girl's bedroom, respectively. Moreover, the average concentration ranges were not statistically different (1070 ppm and 1079 ppm to 2308 ppm and 1878 ppm) in the teenager's and child's bedroom, respectively. The moderate category was selected during the nights with open as well as closed doors. The average concentration of CO_2 during nights defined as moderate with open doors was 766 ppm and 903 ppm in the teenager's and the child's bedroom, respectively, while, during nights with closed doors, the CO_2 concentration was 968 ppm and 1185 ppm, respectively. The satisfied category of sleep was selected only by the younger girl during nights with open doors, corresponding to an average CO_2 concentration of 769 ppm.

4. Discussion

To ensure that the CO_2 concentration is at a lower level, it is necessary to supply an adequate amount of fresh air instead of used air. Bekö et al. [38] summarised that the most influential variables on the air change rate (ACR) are room volume, number of individuals sleeping in the bedroom, average window and door opening habits, location of the measured room (ground or higher floor), and year of building construction. The authors underlined that more door opening results in a higher air change rate. Our study was performed in two similar bedrooms occupied by girls aged 13 and 9 years old. Due to a low IAQ outside the building, the window opening was very limited; thus, we focussed on door opening and compared it with subjective evaluation of sleep satisfaction. Research by Bekö et al. [38] revealed that, when the bedroom doors are ajar or fully open instead of closed, the ACR increases from 0.48 to 0.55 (14%) and 0.71 (48%). Although we did not calculate the ACH, the observed decrease in CO_2 concentration while keeping the bedroom doors open during the night was 55% and 64% in the teenager's and the child's bedroom, respectively.

Generally, adults appear to be additional immune to the consequences of sleep deprivation, whereas children, teenagers and particularly young children tend to be a vulnerable subpopulation as they spend the majority of their time indoors at home while their respiratory and other systems are under development [8]. When an individual does not get enough sleep to feel awake and alert, they begin to experience symptoms of sleep deprivation such as yawning, irritability, fatigue, inability to concentrate, moodiness, forgetfulness, lack of motivation, depression, and poor perspective on life [3]. In our study, the average concentration during the night was similar in both bedrooms, at 1077 ppm (teenager) and 1103 ppm (child), which is comparable with the average CO_2 concentration in unoccupied bedrooms (999 ppm and 1236 ppm in South and North bedrooms, respectively) as reported by Bouvier et al. [25]. In occupied bedrooms, the authors in [25] reported higher (1585 ppm and 1760 ppm in South and North bedrooms, respectively) than our average concentration of CO_2, which was 766 ppm and 860 ppm with open doors and 1402 ppm and 1337 ppm with closed doors, in the teenager's and the child's bedroom, respectively (Figure 3). However, it is similar to the average CO_2 concentration of 716 ppm measured in 13 bedrooms located in Athens' residences [39].

During the night with the doors closed, in the teenager's and the child's bedroom, 48% and 43% of CO_2 concentration measurements were higher than 950 ppm above outdoor CO_2 concentration (398 ppm), which corresponds to category IV–low indoor environment quality (IEQ$_{IV}$), and, when open, it was < 1% in both bedrooms. The PN-EN 16798-1 standard [20] underlines that a lower level will not provide any health risk but may decrease comfort. At the same time, the categories are related to the levels of expectations. A normal level would be medium, but, for occupants with special needs (children, elderly, persons with disabilities, etc.), a higher level might be selected. During the research with opened doors, a high IEQ prevailed during 51% of the night in the teenager's bedroom, and during 26% of the night in the younger girl's bedroom. In comparison, the authors in [25] reported that time spent by an occupant in a CO_2 concentration over 1500 ppm in a mechanically ventilated detached house was approximately 30% of an entire day; thus, the IEQ in researched bedrooms could be satisfying with the door open; however, only the younger girl evaluated sleep satisfaction as high

(16%). The subjective assessment of sleep satisfaction points to a lower IEQ in the case of the older girl's bedroom, which can be supported by a greater CO_2 concentration range (Figure 2) in the teenager's bedroom. Nevertheless, the sharing of categories of environment calculated based on the PN-EN 16798-1 standard [20] does not explain the lower sleep satisfaction in the older girl's assessment. The reason can be seen in Table 1, which presents average CO_2 concentrations and ranges depending on the day of the week. In the teenager's bedroom, the maximum CO_2 concentrations were >3000 ppm on four days of the week (Mondays, Fridays, Saturdays, and Sundays), while, in the younger girl's bedroom, the maximum CO_2 concentration exceeded 2000 ppm only on Saturday nights. These results indicate the important role of short-term high concentrations in sleep satisfaction assessment.

Increasing the effectiveness of the building ventilation system is to decrease the number of exceedances of CO_2 concentration in naturally ventilated buildings. A cheap and simple solution is the application of ventilation grills to the window frames [40]. We would recommend simple passive grilles, where the airflow is due to the pressure drop between the indoor and the outdoor environment in addition to the typical pressure drop of the grille itself. Humidity-sensitive grills might not be very effective in this case because of lower humidity values (29.7–50.4%) in the bedrooms during the heating season (Supplementary Materials Table S1).

5. Conclusions

We investigated the variability of CO_2 concentration in naturally ventilated bedrooms occupied by a teenage girl and a female child according to their sleep satisfaction. Despite the fact that the average CO_2 concentration was 1077 ppm in the teenager's bedroom and 1103 ppm in the child's bedroom, the concentration ranges were much wider, from 402 to 3320 ppm and 458 to 2176 ppm, respectively. The average concentration during the night with the doors closed was 1402 ppm and 1336 ppm, respectively; however, the simple action of door opening decreased the CO_2 concentration by 55% and 64% in the older and younger girl's bedroom, respectively.

In the studied period of heating season (from 16 September 2018 to 2 March 2019) during the night (9:00 p.m. to 7:00 a.m.), the time spent by the teenage and younger girls at a concentration over 1348 ppm (IV—low IEQ category) was generally 21% and 24%, respectively. With the doors open, the low IEQ contributed to <1% of the time in both bedrooms; however, keeping the doors closed increased the contribution of low IEQ to 48% and 46% in the teenager's and the girl's bedroom, respectively. This highlights the strong influence of simple door opening during children's sleep.

The evaluation of sleep satisfaction highlights that the teenage girl was more dissatisfied in the mornings than was the younger girl, indicating the important role of short-term CO_2 concentrations >3000 ppm in sleep satisfaction assessment.

There are several limitations to this work. This study was limited by small sample size and was confined to one house. The suburban site where the house was located may be different from the general population of children and teenagers in Poland with respect to socioeconomic status, education, and other risk factors for adverse health outcomes, all of which may affect the level of IEQ of the home. The relationships presented here cannot be directly linked to other groups of children, and may not be applicable to other regions due to community differences and the air quality in the area. In Poland, during winter, suburban and rural areas according to individual coal heating systems tend to be more polluted than urban areas connected to collective heating systems, which influences the air quality in the area. Finally, these results apply to residential indoor exposure and may not apply to indoor exposure experienced in settings such as multi-family buildings. Of two bedroom IEQ measures during nights, we found that the parent's approval of children's and teenager's privacy by keeping the doors closed increased airtightness in the bedrooms, favoring low ventilation rates and poor indoor environment quality (IEQ). Caution should be taken regarding the fact that the measurement results may not be able to be interpreted as a human exposure level under regularly ventilated conditions since the data were obtained after the windows were closed. However, it could be useful for predicting exposure of CO_2 in children and teenagers during the night, since it is not only in Poland where occupants tend to close the windows and doors while sleeping. Examination of these relationships in other buildings, over a longer period of time, as well as involving wilder groups of participants, could help to explain what parameters affect indoor environment and sleep quality, which has effects on both physical and mental health.

Author Contributions: Conceptualization, A.M.; Methodology, A.M.; Investigation, A.M.; Data curation, A.M.; Writing—original draft preparation, A.M.; Writing—review and editing, E.Z.-Z.; Visualization, A.M.; Project administration, E.Z.-Z.

Acknowledgments: We would like to thank our colleague Walter Mucha for drawing the scheme of the building.

References

1. Landtblom, A.M.; Engström, M. The sleepy teenager-diagnostic challenges. *Front. Neurol.* **2014**, *5*, 1–5. [CrossRef]
2. Herrmann, J.; Koeppen, K.; Kessels, U. Do girls take school too seriously? Investigating gender differences in school burnout from a self-worth perspective. *Learn. Individ. Differ.* **2019**, *69*, 150–161. [CrossRef]
3. Preety, R.; Devi, R.G.; Priya, A.J. Sleep deprivation and cell phone usage among teenagers. *Drug Invent. Today* **2018**, *10*, 2073–2075.
4. Zhang, N.; Cao, B.; Zhu, Y. Indoor environment and sleep quality: A research based on online survey and field study. *Build. Environ.* **2018**, *137*, 198–207. [CrossRef]
5. Rafihi-Ferreira, R.E.; Laura, M.; Pires, N.; Ferreira, E.; Silvares, M. Behavioral intervention for sleep problems in childhood: A Brazilian randomized controlled trial. *Psicol. Reflexão Crítica* **2019**, *32*, 1–13. [CrossRef]
6. Matricciani, L.; Paquet, C.; Galland, B.; Short, M.; Olds, T. Children's sleep and health: A meta-review. *Sleep Med. Rev.* **2019**, *46*, 136–150. [CrossRef]
7. Hall, R.; Hardin, T.; Ellis, R. School indoor air quality best management practices manual. 2003. Available online: https://www.doh.wa.gov/portals/1/Documents/Pubs/333-044.pdf (accessed on 26 August 2019).
8. Stamatelopoulou, A.; Saraga, D.; Asimakopoulos, D.; Vasilakos, C.; Maggos, T. The link between residential air quality and children's health. *Fresenius Environ. Bull.* **2017**, *26*, 162–176.
9. Mainka, A.; Zajusz-Zubek, E. Indoor air quality in urban and rural preschools in Upper Silesia, Poland: Particulate patter and carbon dioxide. *Int. J. Environ. Res. Public Health* **2015**, *12*, 7697–7711. [CrossRef]
10. Mainka, A.; Zubek, E.Z.; Kaczmarek, K. PM10 composition in urban and rural nursery schools in Upper Silesia, Poland: A trace elements analysis. *Int. J. Environ. Pollut.* **2017**, *61*, 98. [CrossRef]
11. Tham, K.W. Indoor air quality and its effects on humans—A review of challenges and developments in the last 30 years. *Energy Build.* **2016**, *130*, 637–650. [CrossRef]
12. Monn, C. Exposure assessment of air pollutants: A review on spatial heterogeneity and indoor/outdoor/personal exposure to suspended particulate matter, nitrogen dioxide and ozone. *Atmos. Environ.* **2001**, *35*, 1–32. [CrossRef]
13. Pathirana, S.M.; Sheranie, M.D.M.; Halwatura, R.U. Indoor thermal comfort and carbon dioxide concentration: A comparative study of air conditioned and naturally ventilated houses in Sri Lanka. In Proceedings of the Moratuwa Engineering Research Conference (MERCon), Moratuwa, Sri Lanka, 29–31 May 2017; pp. 401–406.
14. Khalequzzaman, M.; Kamijima, M.; Sakai, K.; Chowdhury, N.A.; Hamajima, N.; Nakajima, T. Indoor air pollution and its impact on children under five years old in Bangladesh. *Indoor Air* **2007**, *17*, 297–304. [CrossRef]
15. Santamouris, M.; Synnefa, A.; Asssimakopoulos, M.; Livada, I.; Pavlou, K.; Papaglastra, M.; Gaitani, N.; Kolokotsa, D.; Assimakopoulos, V. Experimental investigation of the air flow and indoor carbon dioxide concentration in classrooms with intermittent natural ventilation. *Energy Build.* **2008**, *40*, 1833–1843. [CrossRef]
16. Persily, A.; de Jonge, L. Carbon dioxide generation rates for building occupants. *Indoor Air* **2017**, *27*, 868–879. [CrossRef]
17. Government of Canada: Health Canada Exposure Guidelines for Residential Indoor Air Quality. Available online: http://publications.gc.ca/collections/Collection/H46-2-90-156E.pdf (accessed on 26 August 2019).

18. Regulation of the Minister of Family, Labor and Social Policy Regarding the Highest Permissible Concentrations and Intensities of Harmful Factors in the Work Environment. Available online: http://www.ilo.org/dyn/natlex/natlex4.detail?p_lang=en&p_isn=99664 (accessed on 26 August 2019).

19. PN-EN13779. Ventilation for Non-Residential Buildings. Performance Requirements for Ventilation and Room-Conditioning Systems. Available online: http://www.cres.gr/greenbuilding/PDF/prend/set4/WI_25_Pre-FV_version_prEN_13779_Ventilation_for_non-resitential_buildings.pdf (accessed on 26 August 2019).

20. PN-EN 16798-1: 2019. Indoor environmental input parameters for design and assessment of energy performance of buildings addressing indoor air quality, thermal environment, lighting and acoustics. Available online: http://sklep.pkn.pl/pn-en-16798-1-2019-06e.html (accessed on 27 August 2019).

21. CEN/TR 16798-2. Technical Report: Energy Performance of Buildings-Ventilation for Buildings-Part 2: Interpretation of the Requirements in EN 16798-1-Indoor Environmental Input Parameters for Design and Assessment of Energy Performance of Buildings Addressing Indoor. Available online: https://epb.center/documents/centr-16798-2/ (accessed on 26 August 2019).

22. Bekö, G.; Lund, T.; Nors, F.; Toftum, J.; Clausen, G. Ventilation rates in the bedrooms of 500 Danish children. *Build. Environ.* **2010**, *45*, 2289–2295. [CrossRef]

23. Satish, U.; Mendell, M.J.; Shekhar, K.; Hotchi, T.; Sullivan, D. Is CO_2 an indoor pollutant? Direct effects of low-to-moderate CO_2 concentrations on human decision-making performance. *Environ. Health Perspect.* **2012**, *120*, 1671–1678. [CrossRef]

24. Norbäck, D.; Nordström, K.; Zhao, Z.; Nordstro, K.; Norba, D. Carbon dioxide (CO_2) demand-controlled ventilation in university computer classrooms and possible effects on headache, fatigue and perceived indoor environment: An intervention study. *Int. Arch. Occup. Environ. Health* **2013**, *86*, 199–209. [CrossRef]

25. Bouvier, J.L.; Bontemps, S.; Mora, L. Uncertainty and sensitivity analyses applied to a dynamic simulation of the carbon dioxide concentration in a detached house. *Int. J. Energy Environ. Eng.* **2019**, *10*, 47–65. [CrossRef]

26. Belmonte, J.F.; Barbosa, R.; Almeida, M.G. CO_2 concentrations in a multifamily building in Porto, Portugal: Occupants' exposure and differential performance of mechanical ventilation control strategies. *J. Build. Eng.* **2019**, *23*, 114–126. [CrossRef]

27. Bakó-Biró, Z.; Clements-Croome, D.J.; Kochhar, N.; Awbi, H.B.; Williams, M.J. Ventilation rates in schools and pupils' performance. *Build. Environ.* **2012**, *48*, 215–223. [CrossRef]

28. Telejko, M.; Zander-Świercz, E. Attempt to improve indoor air quality in kindergartens. *Procedia Eng.* **2016**, *161*, 1704–1709. [CrossRef]

29. Zender-Swiercz, E.; Telejko, M. Indoor air quality in kindergartens in Poland. In Proceedings of the IOP Conference Series: Materials Science and Engineering, 3rd World Multidisciplinary Civil Engineering, Architecture, Urban Planning Symposium (WMCAUS 2018), Prague, Czech Republic, 18–22 June 2018; Volume 471.

30. Krawczyk, D.A.; Rodero, A.; Gładyszewska-Fiedoruk, K.; Gajewski, A. CO_2 concentration in naturally ventilated classrooms located in different climates—Measurements and simulations. *Energy Build.* **2016**, *129*, 491–498. [CrossRef]

31. Mainka, A.; Zajusz-Zubek, E.; Kozielska, B.; Brągoszewska, E. Investigation of air pollutants in rural nursery school–a case study. In Proceedings of the E3S Web of Conferences, X-th Scientific Conference Air Protection in Theory and Practice, Zakopane, Poland, 18–21 October 2017; Volume 28, pp. 1–8.

32. Karjalainen, S. Thermal comfort and gender: A literature review. *Indoor Air* **2012**, *22*, 96–109. [CrossRef] [PubMed]

33. Seppanen, O.A.; Fisk, W.J.; Mendell, M.J. Association of ventilation rates and CO_2 concentrations with health and other responses. *Indoor Air* **1999**, *9*, 226–252. [CrossRef] [PubMed]

34. Mahyuddin, N.; Awbi, H.B. A Review of CO_2 measurement procedures in ventilation research. *Int. J. Vent.* **2012**, *10*, 353–370. [CrossRef]

35. Eusebio, L.; Derudi, M.; Capelli, L.; Nano, G.; Sironi, S. Assessment of the indoor odour impact in a naturally ventilated room. *Sensors* **2017**, *17*, 778. [CrossRef] [PubMed]

36. Nguyen, J.L.; Schwartz, J.; Dockery, D.W. The relationship between indoor and outdoor temperature, apparent temperature, relative humidity, and absolute humidity. *Indoor Air* **2014**, *24*, 103–112. [CrossRef]

37. WHO. *WHO Air Quality Guidelines for Europe*; WHO: Copenhagen, Denmark, 2000.

38. Bekö, G.; Toftum, J.; Clausen, G. Modeling ventilation rates in bedrooms based on building characteristics and occupant behavior. *Build. Environ.* **2011**, *46*, 2230–2237. [CrossRef]
39. Stamatelopoulou, A.; Asimakopoulos, D.N.; Maggos, T. Effects of PM, TVOCs and comfort parameters on indoor air quality of residences with young children. *Build. Environ.* **2019**, *150*, 233–244. [CrossRef]
40. D'Ambrosio Alfano, F.R.; Ficco, G.; Palella, B.I.; Riccio, G.; Ranesi, A. An experimental investigation on the air permeability of passive ventilation grilles. *Energy Procedia* **2015**, *78*, 2869–2874. [CrossRef]

14

Probability of Abnormal Indoor Air Exposure Categories Compared with Occupants' Symptoms, Health Information, and Psychosocial Work Environment

Katja Tähtinen [1,2,*], **Sanna Lappalainen** [2], **Kirsi Karvala** [2], **Marjaana Lahtinen** [2] and **Heidi Salonen** [1]

[1] Department of Civil Engineering, Aalto University, 02150 Espoo, Finland; heidi.salonen@aalto.fi
[2] Finnish Institute of Occupational Health, Healthy Workspaces, P.O. Box 18, 00032 Työterveyslaitos, Finland; sanna.lappalainen@ttl.fi (S.L.); kirsi.karvala@ttl.fi (K.K.); marjaana.lahtinen@ttl.fi (M.L.)
* Correspondence: katja.tahtinen@ttl.fi.

Abstract: Indoor air problems are complicated and need to be approached from many perspectives. In this research, we studied the association of four-level categorisation of the probability of abnormal indoor air (IA) exposure with the work environment-related symptoms, group-level health information and psychosocial work environment of employees. We also evaluated the multiprofessional IA group assessment of the current indoor air quality (IAQ) of the hospital premises. We found no statistical association between the four-level categorisation of the probability of abnormal IA exposure and the employees' perceived symptoms, health information, and perceived psychosocial work environment. However, the results showed a statistical association between perceived symptoms and man-made vitreous fibre sources in ventilation. Furthermore, extensive impurity sources in the premises increased the employees' contact with health services and their perceived symptoms. The employees perceived stress and symptoms in all categories of abnormal IA exposure, which may be related to IAQ or other factors affecting human experience. Prolonged process management may influence users' experiences of IAQ. The results suggest that an extensive impurity source in premises does not always associate with the prevalence of perceived symptoms. We conclude that indoor air questionnaires alone cannot determine the urgency of the measures required.

Keywords: perceived indoor air quality; building research; indoor air questionnaires; psychosocial work environment; categorisation; ventilation; mould; moisture; man-made mineral fibres

1. Introduction

To assess the health significance, urgency, and extent of required indoor air quality (IAQ) measures, property owners and occupational health and safety professionals need reliable information on the buildings' conditions and impurity sources. Information is also needed regarding the experiences and health of the users of the premises, and on the cooperation in indoor air (IA) solution processes. When all the factors affecting the IAQ problem have been properly assessed, the degree, timing, and possible prioritisation of measures can be decided on. Properly timed and targeted measures have important implications for the economy, health, and well-being. IAQ problems can be controlled and good IAQ achieved if (i) the factors affecting the indoor environment are under control, (ii) the indoor environment is perceived as good and healthy [1–3], and (iii) good practices are in place for maintaining the indoor environment and solving indoor air (IA) problems [3,4].

IAQ problems are often the result of many different factors and their interaction or complex combination. In addition to moisture and mould damage, several other factors and their interactions, such as material emissions [5], ventilation deficiencies [6], system impurities [7], outdoor and soil impurities [8], human activities in the premises [9], IA temperature, and dry air [10] can cause IAQ problems. Office-type buildings have no established measurement methods for all IA pollutants, and no health-based limit values for most of them [2,11].

Several IA pollutant sources can cause symptoms and harm to the users of premises. A recent review concluded that the greater the presence of moisture and mould damage in the premises, the greater the risk of respiratory health effects [11]. Man-made vitreous fibres (MMVF) in the ventilation system may cause upper respiratory irritation and skin symptoms among users of the premises [5,12–14]. Volatile organic compounds (VOC) from building materials may cause sensory irritation [5,7]. IAQ problems may also affect sick leaves and work efficiency [15].

It has been estimated that predictive property management reduces the IAQ-related symptoms of premises users [16], and some evidence shows that repairing moisture and mould damage and removing contaminants from buildings can reduce respiratory symptoms [17,18] and improve work efficiency [19].

In addition to IAQ factors, several other factors at workplaces, such as stress, poor cooperation, heavy workload, and individual factors may also affect perceived IAQ and play a role in IAQ problems [20,21]. These problems should be examined from a wider perspective, and experience of the users of the premises and the psychosocial environment should be considered [20–23]. It has been suggested that good practices for solving complex or prolonged IAQ problems are well organised and involve long-term multiprofessional cooperation between experts [23,24].

The aim of this study was to test the use of the holistic approach in determining the urgency of the measures required from the perspective of building health. It can be divided into the following sub-aims: (i) to evaluate the relation between the four-level categorised probability of abnormal IA exposure and employees' work environment-related symptoms, group-level health information, and psychosocial work environment, (ii) to assess the relation between ventilation system deficiencies and employees' work environment-related symptoms and (iii) to evaluate the impact of prolonged IAQ problem solution processes on perceived IAQ.

This paper uses the term probability of abnormal IA exposure, which means a comprehensive method of categorising the results of building and ventilation system research. The method used to assess the probability of abnormal IA exposure is presented in our earlier study [25].

2. Materials and Methods

2.1. Materials

This study is based on two research and development projects conducted at the Finnish Institute of Occupational Health (FIOH). These projects were carried out between February 2013 and April 2014, in 27 hospital buildings (studied area altogether about 130,000 m^2), which form a unified building complex located in two Finnish hospital districts. Background information on the buildings' earlier history and documents revealed that parts of the buildings had IAQ problems. We investigated altogether 111 building floors or sections and selected forty building floors or sections on which to focus in more detail, from the premises in which both the IA questionnaire and the assessment of probability of abnormal IA exposure were carried out. We also conducted the building investigations and abnormal IA exposure assessments were still carried out in building premises that were not workplaces or were not in use. The oldest building was built around 1902 and the newest in 2010. One half of the buildings (48%) were built between the 1940s and 1950s. All of them were of stone or different combinations of stone materials and were mostly multistorey and divided between many hospital department areas and hospital functions. Some of the buildings had been repaired in several different stages and these renovations varied greatly. The ventilation system of the buildings was mostly mechanical extract and supply ventilation. However, several different ventilation systems and machines served different parts of the building. The maintenance, repair, reliability, and age of the ventilation systems varied considerably across the floors or sections of even one building.

The same IA researcher group conducted all the building research and ventilation system assessments. All the data were analysed by the same multi-professional group of experts, which comprised IA researchers, a civil engineer, an occupational health physician, a microbiologist, and a ventilation and building health specialist.

2.2. Assessment of Probability of Abnormal IA Exposure

We carried out systematic building examinations that covered (i) structural and architectural plan surveys, (ii) maintenance staff interviews, (iii) examinations and openings of high-risk building structures, (iv) moisture- and mould-damaged range and severity authentications, (v) assessments of ventilation systems, (vi) assessments of air leaks from or through damaged structures, (vii) assessments of air pressure differences, and (viii) assessments of other IA pollutants or pollutant sources in the buildings [25] (Figure 1). We collected building investigation and IAQ measurement results and used a four-level categorisation method to assess the probability of abnormal IA exposure [25] (Figure 1).

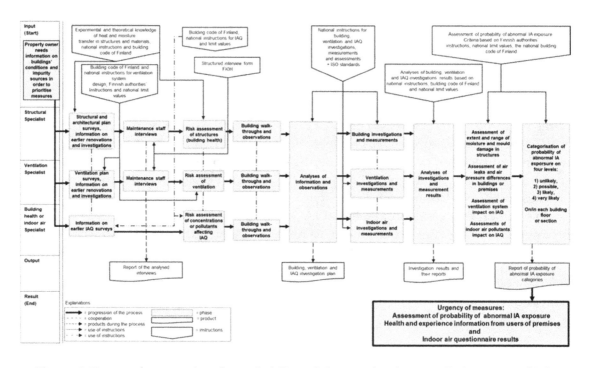

Figure 1. Process for assessing the probability of abnormal indoor air (IA) exposure [25].

We collected categorised parameters for the final assessment of the probability of abnormal IA exposure. This probability was categorised as: (1) probability of abnormal IA exposure unlikely, (2) probability of abnormal IA exposure possible, (3) probability of abnormal IA exposure likely, and (4) probability of abnormal IA exposure very likely [25] (Table 1). In cases of moisture and mould damage, air leaks from or through damaged structures to IA must be examined simultaneously with indoor negative pressure. In the main criteria for assessing the probability of abnormal IA exposure, the predominant IA impurity source is a determining criterion. The probability of abnormal IA exposure arises when the national limit values (IAQ, material samples, ventilation) are exceeded, structures or systems have found damaged, or IAQ pollutant sources that are known to affect indoor air quality and building health are found [25]. The national maximum limit values for IA concentrations, microbial growth on building material, MMVF and asbestos in dust, and specific detailed methods for evaluating building and ventilation conditions and IAQ are presented in our earlier research [25].

Table 1. Main criteria and categories for assessing probability of abnormal IA exposure in buildings.

Categories	Main Criteria for Assessing Probability of Abnormal IA Exposure in Buildings.
Unlikely	No moisture or mould damage in structures. No air leaks from or through damaged structures. Ventilation system can be controlled by indoor pressure difference from the building envelope. Room acoustic materials and ventilation system have no man-made vitreous fibres (MMVF) sources. Indoor air quality corresponds to national reference values and guidelines set for the premises.
Possible	Mould-damaged structure type is not widespread in building and repairs are easily definable (less than 1 m^2). A few or single air leaks from or through damaged structures or from surrounding premises. Room acoustic materials or ventilation system have MMVF sources and fibres may end up in the indoor air or on surfaces.[1] Concrete floor has extensive moisture, which can cause water vapour damage to permeable floor coating (emissions).[1] Indoor air quality does not correspond to national reference values or the guidelines set for the premises, and an indoor air impurity source has been identified.[1]
Likely	Building or premises have widespread mould-damaged structure. Repairs are significant and affect a large part of the (one) structure of the building or premises, e.g., whole base floor structure. There is recurrent damage in the type of structure. Air leaks from or through damaged structures or from surrounding premises and moisture or mould-damaged materials are regular and recurrent in the structure, occasionally there is negative pressure in the premises and/or air-tightness is risky. Indoor air quality does not correspond to national reference values or the guidelines set for the premises, and an indoor air impurity source has been identified.[1] Creosote has been used in the structure and air leaks into the indoor air from the structure. There is also a notable smell of creosote (e.g., naphthalene) in the indoor air.[1]
Very likely	The building or premises has a great deal of extensive mould damage in several structures. The extent of repairs is significant and affects several structures in the building or premises e.g., whole façade and whole base floor. There is recurrent damage in the type of the structures. Air leaks from or through damaged structures are regular and recurrent, negative pressure is significant in the premises and/or air-tightness is very risky. Indoor air quality does not correspond to national reference values or the guidelines set for the premises, and an indoor air impurity source has been identified.[1] Creosote has been used in the structures and air leaks into the indoor air from the structures. In addition, concentrations of polycyclic aromatic hydrocarbons (PAH) or separate components exceed the set national values and guidelines.[1] Dust sample tests have found asbestos fibres in the premises, and the pollution source has been defined.[1] Indoor radon concentrations exceed the set national values and guidelines (400 Bq/m^3 [26]).[1]

[1] The assessment must take into account the extent and impact of the problem and impurity source.

2.3. Employees' Experiences of Indoor Air Quality and Psychosocial Work Environment

To study the users', i.e., employees' experiences of the work premises, work environment-related symptoms and psychosocial work environment, we used FIOH's validated and frequently used IA questionnaire, which is based on Örebro's [27] indoor climate questionnaire [28,29]. To study perceived stress, we used a validated single-item measure of stress symptoms [30]: "Stress means a situation in which a person feels tense, restless, nervous or anxious or is unable to sleep at night because his/her mind is troubled all the time. Do you feel this kind of stress these days?". The response options were: (1) not at all, (2) just a little, (3) some, (4) quite a lot, and (5) very much. In the analyses, we combined the levels (1) not at all and (2) just a little into one level, and levels (4) quite a lot and (5) very much into one level.

We sent the questionnaire to 3608 hospital employees, of whom 2669 responded. The total response rate was 74%, with a range of 51% to 93%. The surveys were conducted in the spring from February to April and in the autumn in November, in 2013. We selected 40 IA questionnaire groups for the study, totaling 1558 respondents. The selected IA questionnaire groups were in premises in which the probability of abnormal IA exposure assessment had already been performed. The employees did not know the results of the assessment prior to responding to the IA questionnaires. This is a questionnaire-based study, in which participation was voluntary and performed no intervention on individuals, according to Finnish legislation it did not require ethics committee handling.

2.4. Group-Level Information from Occupational Health Services and Multiprofessional Indoor Air Group

We obtained information on the assessment of the group-level health of employees from occupational health services (OHS). The information covered employees' health from 43 building sections or floors. The group-level information from the OHS did not contain information on how many employees had work environment-related health symptoms in the building sections or on the floors. In the survey, we used short forms to ask about the following issues in relation to employees' health: (i) case of new onset asthma or aggravation of existing asthma, (ii) having to change workroom because of IAQ and work environment-related symptoms, (iii) increased amount of employee visits to OHS due to IAQ-related issues, and (iv) increased sickness absences due to respiratory symptoms. The hospital's multiprofessional IA group also provided information on the estimated duration of the IAQ problem solution on every building floor or section.

2.5. Statistical Analyses

Statistical analyses were carried out using IBM SPSS Statistics program 25.0 with a statistically significant level of $p < 0.05$. The statistical analysis used weighted averages of group response rates. The Mann–Whitney U test studied the differences between the probability of abnormal IA exposure categories (*unlikely, possible, likely,* and *very likely*) and the employees' complaints about their work environment-related symptoms and psychosocial work environment. This test also compared the difference between the two groups' (yes/no) ventilation adequacy, ventilation MMVF sources, ventilation moisture problems, and expired ventilation lifespan and the employees' complaints about their work environment-related symptoms. Fisher's exact test studied the relation between the weekly work environment-related symptoms experienced by the employees and the categorised group-level information on employee health. The group-level health information was categorised as 'yes' and 'no' as follows: case of new onset asthma or aggravation of existing asthma, having to change workroom or workplace because of work environment-related symptoms, increased amount of employee visits to OHS, and increased sickness absences due to respiratory symptoms.

3. Results

Probability of Abnormal IA Exposure and Employees' Experience

All building floors or sections (total 111) were investigated and we were able to assess the probability of abnormal IA exposure on or in 95 building floors or sections. In the case of forty building floors or sections, the assessment of the probability of abnormal IA exposure and IA questionnaire could both be conducted in the same areas (the IA questionnaire group was located in the area that was assessed as belonging to an abnormal IA exposure category). The probability of abnormal IA exposure was assessed as *unlikely* for 5% (n = 2), *possible* for 40% (n = 16), *likely* for 45% (n = 18), and *very likely* for 10% (n = 4) of the selected forty floors or sections of the buildings. In the *likely* and *very likely* categories, these floors or sections had wide moisture and mould damage in their structures together with air leaks from damaged materials to the IA and often had a detected MMVF source in the ventilation system as well as ventilation deficiencies (Table 2 and criteria in Table 1). These categories also had other impurity sources (Table 1), but moisture and mould damage in the building structures were dominant. The higher (more abnormal) the assessed category of probability of abnormal IA exposure, the more insufficient the ventilation was, the more often the lifespan of the ventilation system was exceeded and the more often MMVF sources were detected in the ventilation system from the categories likely and very likely (and not from unlikely category) (Table 2).

Table 2. Ventilation survey findings are included in all the sections or floors in which probability of abnormal IA exposure was assessed (95 floors or sections). (All building floors or sections (111) were investigated and we were able to assess the probability of abnormal IA exposure on or in 95 building floors or sections.).

Assessed Probability of Abnormal IA Exposure on/in Building Floors or Sections (n = 95)	Lifespan of Ventilation System Had Been Exceeded n (%)	Insufficient Ventilation n (%)	MMVF (Man-Made Vitreous Fibres) Source in Ventilation System n (%)	Moisture Problem in Ventilation System n (%)
Unlikely (n = 7)	7 (100)	3 (43)	3 (43)	0 (0)
Possible (n = 39)	23 (59)	22 (56)	12 (31)	6 (15)
Likely (n = 37)	26 (70)	24 (65)	26 (70)	15 (41)
Very likely (n = 12)	10 (83)	9 (75)	9 (75)	1 (8)

The perceived weekly work environment-related symptoms among the employees exceeded the corresponding number of weekly work environment-related symptoms in FIOH's reference data [28,29] (Figure 2). Even in premises in which the building research showed no source of contamination, some employees perceived weekly symptoms more than those in FIOH's reference data [28,29] (Figure 3). An analysis of the differences between the probability of abnormal IA exposure categories (*unlikely, possible, likely, very likely*) and the employees' weekly perceived work environment-related symptoms (headaches, concentration difficulties, irritation of the eyes and nose, irritation of the skin on the face and hands, hoarse throat, coughing, coughing at night, shortness of breath, wheezing, fever or chills, joint pain, and muscular pain revealed no statistically significant differences (Table 3).

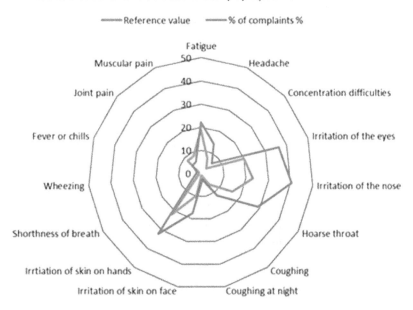

Figure 2. In the buildings assessed for probability of abnormal IA exposure, weekly work environment-related symptoms (n = 1558) were perceived more often than in the Finnish Institute of Occupational Health's (FIOH) comparable reference data [28].

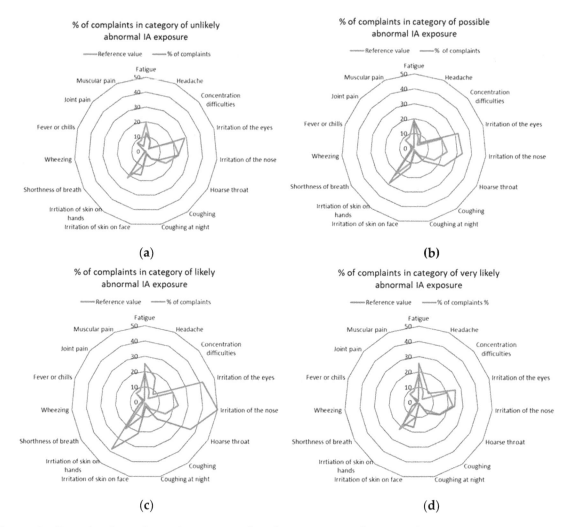

Figure 3. Perceived work environment-related symptoms of users of premises in assessment of probability of abnormal IA exposure categories, compared to FIOH reference data [28] (**a**) unlikely (n = 61), (**b**) possible (n = 618), (**c**) likely (n = 760), (**d**) very likely (n = 118).

Table 3. An analysis of the differences between the probability of abnormal IA exposure categories (unlikely, possible, likely, very likely) and the employees' weekly perceived work environment-related symptoms.

Variable (Weekly Symptoms)	Unlikely 1 n = 2 (p-Value)	Possible 2 n = 16 (p-Value)	Likely 3 n = 18 (p-Value)	Very Likely 4 n = 4
Fatigue	1 vs. 2 (0.673) 1 vs. 3 (0.257) 1 vs. 4 (0.355)	2 vs. 3 (0.220) 2 vs. 4 (0.570)	3 vs. 4 (0.609)	
Headache	1 vs. 2 (0.778) 1 vs. 3 (0.208) 1 vs. 4 (0.643)	2 vs. 3 (0.157) 2 vs. 4 (0.777)	3 vs. 4 (0.481)	
Concentration difficulties	1 vs. 2 (0.672) 1 vs. 3 (0.165) 1 vs. 4 (0.064)	2 vs. 3 (0.233) 2 vs. 4 (0.507)	3 vs. 4 (1.000)	
Irritation of the eyes	1 vs. 2 (0.888) 1 vs. 3 (0.378) 1 vs. 4 (0.814)	2 vs. 3 (0.262) 2 vs. 4 (0.777)	3 vs. 4 (0.125)	
Irritation of the nose	1 vs. 2 (0.779) 1 vs. 3 (0.130) 1 vs. 4 (0.643)	2 vs. 3 (0.073) 2 vs. 4 (0.925)	3 vs. 4 (0.061)	

Table 3. *Cont.*

Variable (Weekly Symptoms)	Unlikely 1 n = 2 (*p*-Value)	Possible 2 n = 16 (*p*-Value)	Likely 3 n = 18 (*p*-Value)	Very Likely 4 n = 4
Hoarse, dry throat	1 vs. 2 (0.399) 1 vs. 3 (0.130) 1 vs. 4 (1.000)	2 vs. 3 (0.101) 2 vs. 4 (0.508)	3 vs. 4 (0.061)	
Coughing	1 vs. 2 (0.481) 1 vs. 3 (0.378) 1 vs. 4 (0.355)	2 vs. 3 (0.147) 2 vs. 4 (0.636)	3 vs. 4 (0.287)	
Coughing at night	1 vs. 2 (0.941) 1 vs. 3 (0.792) 1 vs. 4 (0.411)	2 vs. 3 (0.652) 2 vs. 4 (0.289)	3 vs. 4 (0.237)	
Irritation of skin on face	1 vs. 2 (0.260) 1 vs. 3 (1.000) 1 vs. 4 (1.000)	2 vs. 3 (0.133) 2 vs. 4 (0.508)	3 vs. 4 (0.551)	
Irritation of skin on hands	1 vs. 2 (0.888) 1 vs. 3 (0.378) 1 vs. 4 (1.000)	2 vs. 3 (0.152) 2 vs. 4 (0.705)	3 vs. 4 (0.349)	
Shortness of breath	1 vs. 2 (0.562) 1 vs. 3 (0.509) 1 vs. 4 (0.355)	2 vs. 3 (0.508) 2 vs. 4 (0.502)	3 vs. 4 (0.663)	
Wheezing	1 vs. 2 (0.374) 1 vs. 3 (0.178) 1 vs. 4 (0.480)	2 vs. 3 (0.342) 2 vs. 4 (0.600)	3 vs. 4 (0.275)	
Fever or chills	1 vs. 2 (0.352) 1 vs. 3 (0.736) 1 vs. 4 (0.623)	2 vs. 3 (0.308) 2 vs. 4 (0.562)	3 vs. 4 (1.000)	
Muscular pain	1 vs. 2 (1.000) 1 vs. 3 (0.44) 1 vs. 4 (1.000)	2 vs. 3 (0.161) 2 vs. 4 (0.298)	3 vs. 4 (0.932)	
Joint pain	1 vs. 2 (0.324) 1 vs. 3 (0.900) 1 vs. 4 (0.643)	2 vs. 3 (0.283) 2 vs. 4 (0.570)	3 vs. 4 (0.898)	
Other work environment-related symptoms	1 vs. 2 (0.176) 1 vs. 3 (0.074) 1 vs. 4 (0.060)	2 vs. 3 (0.099) 2 vs. 4 (0.288)	3 vs. 4 (0.831)	

Statistically significant level of $p < 0.05$.

As regards the IA questionnaire results, most of the employees (88%) felt that their work was often stimulating and interesting, 74% believed they would receive help from their colleagues if needed, 21% often had the opportunity to influence their own work and working conditions, and 53% had no feelings of stress. A heavy workload was reported by 14% of the employees, which is below FIOH's reference value [28]. Stress was reported by 16% of the employees and 4% believed that they would not get help from colleagues if needed. Stress and lack of help from colleagues were perceived more often than in FIOH's comparable reference data [28]. Stress was perceived more often than in FIOH's reference data in every category of abnormal IA exposure (Table 4). Most often, stress was perceived in premises in which the probability of abnormal IA exposure was estimated as being *unlikely* (Table 4). An analysis of the differences between the probability of abnormal IA exposure categories (*unlikely, possible, likely, very likely*) and the employees' perceived psychosocial work environment and stress revealed no statistically significant differences.

Table 4. Perceived psychosocial work environment and stress according to probability of abnormal IA exposure categories (*unlikely, possible, likely, very likely*).

Question on Psychosocial Work Environment	Unlikely n = 61 (%)	Possible n = 619 (%)	Likely n = 763 (%)	Very Likely n = 118 (%)	FIOH's Reference Value %
Do you regard your work as interesting and stimulating?					
Yes, often	52 (88.1)	540 (87.5)	684 (89.9)	102 (86.4)	82
Yes, sometimes	7 (11.9)	66 (10.7)	61 (8.0)	11 (9.3)	16
No, seldom or rarely	0 (0)	11 (1.8)	16 (2.0)	5 (4.2)	0
Do you have too much work?					
Yes, often	12 (20.7)	95 (15.5)	98 (12.9)	15 (12.7)	20
Yes, sometimes	36 (62.1)	360 (58.7)	489 (64.4)	71 (60.2)	64
No, seldom or rarely	10 (17.2)	158 (25.8)	172 (22.7)	32 (27.1)	16
Do you have opportunities to influence your working conditions?					
Yes, often	38 (32.8)	120 (19.5)	161 (21.2)	38 (32.8)	21
Yes, sometimes	61 (52.6)	335 (54.3)	379 (49.9)	61 (52.6)	51
No, seldom or rarely	17 (14.7)	162 (26.3)	220 (29.0)	17 (14.7)	28
Do your fellow workers help you with work-related problems?					
Yes, often	33 (55.9)	463 (75.2)	569 (74.8)	83 (70.3)	79
Yes, sometimes	24 (40.7)	129 (20.9)	171 (22.5)	28 (23.7)	19
No, seldom or rarely	2 (3)	24 (3.9)	21 (2.8)	7 (5.9)	2
Do you feel stress?					
Quite a lot or very much	15 (26.3)	92 (15.1)	105 (15.0)	16 (14.0)	10
Some	12 (21.1)	181 (29.7)	243 (32.4)	40 (35.1)	28
Not at all or just a little	30 (52.6)	337 (55.2)	403 (53.7)	58 (50.9)	63

In addition, an analysis of the differences between the probability of abnormal IA exposure categories' (*unlikely, possible, likely, very likely*) and the employees' group-level health information (obtained from OHS) revealed no statistically significant differences. However, the more abnormal the probability of the IA exposure category, the more employees contacted OHS due to IAQ-related issues from the categories possible, likely and very likely (Table 5).

Table 5. IAQ-related group-level health information on employees (obtained from OHS), according to building floors or sections in which probability of abnormal IA exposure categories (*unlikely, possible, likely, very likely*) were assessed.

IAQ-Related Health Information	Unlikely n = 2 (%)	Possible n = 16 (%)	Likely n = 20 (%)	Very likely n = 5 (%)
Some employees have new asthma or aggravation of previous asthma [1]				
Yes	2 (100)	5 (31)	5 (25)	3 (60)
No	0 (0)	10 (63)	14 (70)	2 (40)
No information	0 (0)	1 (6)	1 (5)	0 (0)
Some employees have changed work premises or work places due to IAQ-related symptoms [1]				
Yes	1 (50)	4 (25)	4 (20)	2 (40)
No	1 (50)	11 (69)	16 (80)	3 (60)
No information	0 (0)	1 (6)	0 (0)	0 (0)
The amount of employee contacts with OHS due to IAQ-related issues has increased [1]				
Yes	2 (100)	5 (31)	10 (50)	3 (60)
No	0 (0)	10 (63)	10 (50)	2 (40)
No information	0 (0)	1 (6)	0 (0)	0 (0)
The amount of sickness absence due to respiratory symptoms has increased [1]				
Yes	1 (50)	4 (25)	8 (40)	1 (20)
No	0 (0)	7 (44)	11 (55)	3 (60)
No information	1 (50)	5 (31)	1 (5)	1 (20)

[1] Group-level health information does not contain information on how many employees have IAQ-related health symptoms on building floors or in sections or the assessed categories. IAQ: indoor air quality; OHS: occupational health services.

The results show a statistical association between detected MMVF sources in ventilation systems and perceived work environment-related symptoms and a statistical association between ventilation system age and perceived work environment-related symptoms (Table 6). The hospital's multiprofessional IA group estimated the duration of the IAQ problem solution process on every building floor or in each section, and the attempts to solve the indoor air problems in the *unlikely*, *possible*, *likely*, and *very likely* categories had lasted a year or more (Table 7). The hospital's multiprofessional IA group also estimated 'No IAQ problems' in premises in which the research group had assessed the probability of abnormal IA exposure as being *possible* and *likely* (Table 7).

Table 6. Statistical significance of differences (*p*-value) between weekly reported work environment-related symptoms and ventilation factors (yes/no) studied.

Perceived Work Environment-Related Symptoms Weekly	Respondents N = 1558 n	Technical Lifespan of Ventilation System Had Been Exceeded	Moisture Problem in Ventilation System	MMVF Source in Ventilation System	Insufficient Ventilation or Ventilation System Did Not Match Purposes of Facilities
Fatigue	532	NS	NS	0.005	NS
Headache	266	0.035	NS	0.002	NS
Concentration difficulties	164	0.049	NS	0.005	NS
Irritation of the eyes	559	0.014	NS	0.006	NS
Irritation of the nose	656	0.003	NS	0.001	NS
Hoarse, dry throat	475	0.001	NS	0.001	0.025
Coughing	225	0.016	NS	0.022	NS
Coughing at night	50	0.008	NS	NS	NS
Irritation of skin on face	373	NS	NS	NS	NS
Irritation of skin on hands	508	0.035	NS	0.003	NS
Shortness of breath	64	0.045	NS	NS	NS
Wheezing	32	0.001	NS	NS	0.044
Fever or chills	42	NS	NS	NS	NS
Muscular pain	160	0.027	NS	0.003	NS
Joint pain	193	0.039	NS	NS	NS
Other work environment-related symptoms	75	0.005	NS	0.001	0.027

NS: not significant. Statistically significant level of $p < 0.05$.

Table 7. Estimated duration of IAQ problems on each building floor or in each section studied.

Estimated Duration of IAQ Problems: Number of Cases	Unlikely (n = 2)	Possible (n = 16)	Likely (n = 18)	Very Likely (n = 4)
No IAQ problems	1	7	6	0
Duration of IAQ problems less than a year	0	2	4	0
Duration of IAQ problems one year or more	1	7	8	4

For each building floor or section, we looked at all the collected data (results of assessed probability of abnormal IA exposure, IA questionnaires results, group-level health information and IA group information) at the same time. The most urgent measures were required for floors or sections on/in which the probability of abnormal IA exposure was *likely* or *very likely*, the health information (Table 5) and IA questionnaire results (Figure 3a–d) indicated health problems, and solutions were delayed (Table 7).

4. Discussion

The detailed examination of the buildings and the four-level categorisation of the probability of abnormal IA exposure made it easier to organise the outcomes and obtain a clear picture of the many factors affecting the IAQ of the buildings and premises. The strength of our research was its multiprofessional approach, which took into account the employees' perceived work environment-related symptoms and health information, multiprofessional IA group information and the results of the technical building investigations.

According to previous studies, observed indoor mould and moisture damage indicates an increased health risk, and the greater the mould and moisture damage, the more prevalent respiratory symptoms in adults [11]. This study found no statistically significant differences between the four-level categories and employees' perceived work environment-related symptoms, perceived psychosocial environment, or OHS group-level information on employees' health. However, extensive impurity sources in the premises increased some employees' perceived work-related symptoms. Our earlier study [31] shows that all the symptoms perceived by the employees were very similar to the work-related symptoms examined in this study. The results showed a statistical association between MMVF sources in ventilation systems and perceived IAQ and work environment-related symptoms. Other studies have achieved similar results [5,14]. In addition, the age of the ventilation system was associated with perceived symptoms. The more the probability of abnormal IA exposure was estimated to differ, the more prevalent were ventilation system deficiencies and the MMVF sources in the ventilation system. Therefore, IAQ problems were usually affected by impurities from both the building and ventilation, which shows that many IAQ factors can affect perceived symptoms.

Employees perceived more work environment-related symptoms and stress and lack of social support from colleagues than those reported in FIOH's comparative data from (damaged and nondamaged) hospital buildings, a result which may be related to the poor condition of the buildings or other factors affecting the human experience. On the other hand, the amount of employee contact with OHS increased due to IAQ and work environment-related issues in all the categories of the probability of abnormal IA exposure. An earlier study has shown an association between symptoms and work strain [21], the psychosocial work environment [4,32], and individual factors (e.g., gender, age) [21,33]. The risk of experiencing the workplace as harmful has shown to be higher among employees who report mould problems than those who report ventilation problems in workplaces [34].

In our research, symptoms were common, stress was high, and the amount of contact with OHS in IAQ and work environment-related issues was great on floors or in sections in which the category of abnormal IA exposure was assessed as being *unlikely* and building floors or sections that were undamaged. Workers in non-damaged buildings have also shown to have IAQ and work environment-related symptoms [29,33]. Overall, the employees perceived stress in every category of abnormal IA exposure more than the amount reported in FIOH's reference data [28]. On the other hand, the employees often perceived their work as stimulating and interesting in all categories of the probability of abnormal IA exposure. An earlier study has also shown that hospital employees find their work more interesting and stimulating than office workers [28,32]. As the buildings and the factors affecting IAQ have been carefully studied, the building technology and IAQ alone may not explain the prevalence of the perceived work environment-related symptoms and the stress and the amount of IAQ- and work environment-related contact with OHS. These issues may also be partly affected by other factors influencing human experience, such as work-related and organizational factors not investigated in this study, or individual human factors.

Based on the IA group's information, attempts to solve the indoor air problems in several buildings had lasted more than a year. The IA group's evaluation of no IAQ problems in the premises often contradicted our building investigation results. One reason for these prolonged problem-solving processes may be that building-related problems or impurity sources have been unclear or building investigations incomplete. An earlier study also had similar findings [4]. Prolonged or unclear problem-solving processes may have increased health concerns and distrust of the problem-solving process and influenced the IAQ-related experience of employees. Thus, careful decision-making procedures are important, especially when people feel threatened by IAQ-related risks [34].

The employees did not know the results of the assessment of the probability of abnormal IA exposure prior to responding to the IA questionnaires, which contributes to the reliability of the study. The differences between previous studies and our research results may result from differences in the building research methods. Many studies are based on observations of visible damage or indications of damage, and hidden damage has remained unclear [11]. Our study was very detailed, and we also investigated hidden damage, in addition other pollutant sources. Possible limitations of our research may be that our data concerned only a small number of buildings with no IAQ problems or impurities affecting IAQ, which may have affected data distribution. Due to this, we had no reliable comparison survey of damaged and nondamaged buildings or premises. The method for assessing the probability of abnormal IA exposure is very pragmatic and is always based on strong technical expertise in building technology and IAQ. In addition, the criteria for the probability of abnormal IA exposure recommend taking into account many impurity sources in a building, based on national instructions, regulations and limit values in the field of the built environment and IAQ. However, the assessment also involves a researcher's subjective view. OHS' information was collected at group-level, and as the results were analysed categorically, they may not have provided sufficiently accurate results. In addition, the questions concerning the psychosocial work environment were quite limited in the IA questionnaire. They covered qualitative and quantitative workload, opportunities to influence one's working conditions, and social support at work. Although these are essential factors in the light of the stress theory, they fail to provide a comprehensive picture of the psychosocial work environment. The questionnaire did not include, for instance, factors such as organizational changes and questions concerning leadership.

The probability of abnormal IA exposure provides us with a holistic picture of the many factors affecting the IAQ of buildings and premises. In this case, the most urgent measures could have been identified more easily and holistically. Moreover, in the premises that needed the most urgent repairs, employee contact with OHS was increased, and the employees' perceived work environment-related symptoms indicated poor IAQ. On the other hand, in the premises in which no technical problems were found, the employees still perceived more work environment-related symptoms and stress than those reported in FIOH's reference data. Thus, IAQ problems should always be analysed from many perspectives; (i) the building's technical condition, (ii) perceived IAQ and psychosocial work environment, (iii) OHS information, and (iv) measures for solving IAQ problems, which may all affect the experience of IAQ problems. Indoor air questionnaires can serve as a parallel method with technical investigations in the building.

5. Conclusions

The four-level categorisation of the probability of abnormal IA exposure provides a comprehensive and systematic way of ranking building sectors from the perspective of building health. The method is based on national instructions for building and ventilation investigations, building codes, and limit values, and is therefore systematic and partially established. Thus, it can be applied in different environments. The method may also be used in other countries (with similar environments to that in Finland), if the national instructions, limit values, and building codes are taken into account and applied. The method enabled the holistic identification of the most urgent measures. This may help property owners allocate resources for proper repairs and also help OHS identify employees' IAQ and work environment-related symptoms. The results suggest that the extensive impurity source in premises does not always associate with the prevalence of perceived IAQ and work environment-related symptoms. Therefore, the solution to the IAQ problem is more specific when technical survey results, the health and experience information of the users of the premises, as well as the problem solution process are taken into account. The results also suggest that IA questionnaires alone cannot determine the urgency of the measures required. Possible limitations of our research are that the study was only conducted in hospital buildings and premises. Limitations may be that our data concerned only a small number of buildings with no IAQ problems or impurities affecting IAQ: this may have affected data distribution. The method for assessing the probability of abnormal IA exposure is, however, very pragmatic and always based on strong technical expertise in building technology and IAQ. Further studies should assess the probability of abnormal IA exposure in different work environments (e.g., offices), and the associations between the probability of abnormal IA exposure categories with perceived IAQ and the health of employees. They should also assess the impact of the IAQ solution process perceived IAQ. The results of further studies may possibly validate the method.

Author Contributions: K.T. wrote the paper and analysed the data; S.L. and K.K. contributed to the study design and data analysis and participated in writing the paper; M.L. contributed to the data analyses and participated in writing the paper; H.S. was the principal supervisor of the work and participated in writing the paper.

Acknowledgments: The authors acknowledge the contribution of Leena Aalto, Veli-Matti Pietarinen; Ulla-Maija Hellgren, Jouko Remes and Pauliina Toivio and all co-workers in building research.

Abbreviations

The following abbreviations are used in this manuscript:

FIOH	Finnish Institute of Occupational Health
IA	Indoor air
IAQ	Indoor air quality
OHS	Occupational health service
PAH	Polycyclic aromatic hydrocarbon
VOC	Volatile organic compound

References

1. Frontczak, M.; Wargocki, P. Literature survey on how different factors influence human comfort in indoor environments. *Build. Environ.* **2011**, *46*, 922–937. [CrossRef]
2. WHO. *Dampness and Mould—Who Guidelines for Indoor Air Quality*; World Health Organisation: Geneva, Switzerland, 2009.
3. Lappalainen, S.; Lahtinen, M.; Palomaki, E.; Korhonen, P.; Niemelä, R.; Reijula, K. Comprehensive procedure for solving indoor environment problems. In Proceedings of the NAM Nordic Work Environment Meeting, Espoo, Finland, 31 August–2 September 2009; p. 93.
4. Lahtinen, M.; Huuhtanen, P.; Kahkonen, E.; Reijula, K. Psychosocial dimensions of solving an indoor air problem. *Indoor Air* **2002**, *12*, 33–46. [CrossRef] [PubMed]
5. Salonen, H.; Lappalainen, S.; Riuttala, H.; Tossavainen, A.; Pasanen, P.; Reijula, K. Man-made vitreous fibers in office buildings in the helsinki area. *J. Occup. Environ. Hyg.* **2009**, *6*, 624–631. [CrossRef] [PubMed]
6. Sundell, J.; Levin, H.; Nazaroff, W.W.; Cain, W.S.; Fisk, W.J.; Grimsrud, D.T.; Gyntelberg, F.; Li, Y.; Persily, A.K.; Pickering, A.C.; et al. Ventilation rates and health: Multidisciplinary review of the scientific literature. *Indoor Air* **2011**, *21*, 191–204. [CrossRef] [PubMed]
7. Wolkoff, P.; Wilkins, C.K.; Clausen, P.A.; Nielsen, G.D. Organic compounds in office environments—Sensory irritation, odor, measurements and the role of reactive chemistry. *Indoor Air* **2006**, *16*, 7–19. [CrossRef] [PubMed]
8. Airaksinen, M.; Pasanen, P.; Kurnitski, J.; Seppänen, O. Microbial contamination of indoor air due to leakages from crawl space: A field study. *Indoor Air* **2004**, *14*, 55–64. [CrossRef] [PubMed]
9. Nevalainen, A.; Täubel, M.; Hyvärinen, A. Indoor fungi: Companions and contaminants. *Indoor Air* **2015**, *25*, 125–156. [CrossRef]
10. Nordström, K.; Norbäck, D.; Akselsson, R. Effect of air humidification on the sick building syndrome and perceived indoor air quality in hospitals: A four month longitudinal study. *Occup. Environ. Med.* **1994**, *51*, 683–688. [CrossRef]

11. Mendell, M.J.; Kumagai, K. Observation-based metrics for residential dampness and mold with dose–response relationships to health: A review. *Indoor Air* **2017**, *27*, 506–517. [CrossRef]

12. Hellgren, U.-M.; Hyvärinen, M.; Holopainen, R.; Reijula, K. Perceived indoor air quality, air-related symptoms and ventilation in finnish hospitals. *Int. J. Occup. Med. Environ. Health* **2011**, *24*, 48–56. [CrossRef]

13. Redlich, C.; Sparer, J.; Cullen, M.R. Sick-building syndrome. *Lancet* **1997**, *349*, 1013–1016. [CrossRef]

14. Schneider, T. Dust and fibers as a cause of indoor environment problems. *Scand. J. Work. Environ. Health* **2008**, *33*, 10–17.

15. Milton, D.K.; Glencross, P.M.; Walters, M.D. Risk of sick leave associated with outdoor air supply rate, humidification, and occupant complaints. *Indoor Air* **2000**, *10*, 212. [CrossRef] [PubMed]

16. Cox-Ganser, J.; Rao, C.; Park, J.; Schumpert, J.; Kreiss, K. Asthma and respiratory symptoms in hospital workers related to dampness and biological contaminants. *Indoor Air* **2009**, *19*, 280–290. [CrossRef] [PubMed]

17. Mendell, M.; Brennan, T.; Lee, H.; Odom, J.D.; Offerman, F.J.; Turk, B.H.; Wallingford, K.M.; Diamond, R.C.; Fisk, W.J. Causes and prevention of symptom complaints in office buildings. *Facilities* **2006**, *24*, 436–444. [CrossRef]

18. Sauni, R.; Verbeek, J.H.; Uitti, J.; Jauhiainen, M.; Kreiss, K.; Sigsgaard, T. Remediating buildings damaged by dampness and mould for preventing or reducing respiratory tract symptoms, infections and asthma. *Cochrane Database Syst. Rev.* **2015**, *76*, 1. [CrossRef] [PubMed]

19. Wargocki, P.; Lagercrantz, L.; Witterseh, T.; Sundell, J.; Wyon, D.P.; Fanger, P.O. Subjective perceptions, symptom intensity and performance: A comparison of two independent studies, both changing similarly the pollution load of an office. *Indoor Air* **2002**, *12*, 74. [CrossRef]

20. Brauer, C.; Mikkelsen, S. The influence of individual and contextual psychosocial work factors on the perception of the indoor environment at work: A multilevel analysis. *Int. Arch. Occup. Environ. Health* **2010**, *83*, 639–651. [CrossRef]

21. Magnavita, N. Work-related symptoms in indoor environments: A puzzling problem for the occupational physician. *Int. Arch. Occup. Environ. Health* **2015**, *88*, 185–196. [CrossRef]

22. Finell, E.; Haverinen-Shaughnessy, U.; Tolvanen, A.; Laaksonen, S.; Karvonen, S.; Sund, R.; Saaristo, V.; Luopa, P.; Ståhl, T.; Putus, T.; et al. The associations of indoor environment and psychosocial factors on the subjective evaluation of indoor air quality among lower secondary school students: A multilevel analysis. *Indoor Air* **2017**, *27*, 329–337. [CrossRef]

23. Lahtinen, M.; Lappalainen, S.; Reijula, K. Multiprofessional teams resolving indoor-air problems-emphasis on the psychosocial perspective. *Scand. J. Work. Environ. Health* **2008**, *34*, 30–34.

24. Carrer, P.; Wolkoff, P. Assessment of indoor air quality problems in office-like environments: Role of occupational health services. *Int. J. Environ. Res. Public Health* **2018**, *15*, 741. [CrossRef] [PubMed]

25. Tähtinen, K.; Lappalainen, S.; Karvala, K.; Remes, J.; Salonen, H. Association between four-level categorisation of indoor exposure and perceived indoor air quality. *Int. J. Environ. Res. Public Health* **2018**, *15*, 679. [CrossRef] [PubMed]

26. Finnish Ministry of the Environment, Department of Built Environment. *D2 the National Building Code of Finland, Health. Indoor Climate and Ventilation of Buildings, Regulations and Guidelines*; Department of Built Environment: Helsinki, Finland, 2012.

27. Andersson, K. Epidemiological approach to indoor air problems. *Indoor Air* **1998**, *8*, 32–39. [CrossRef]

28. Hellgren, U.-M.; Palomaki, E.; Lahtinen, M.; Riuttala, H.; Reijula, K. Complaints and symptoms among hospital staff in relation to indoor air and the condition and need for repairs in hospital buildings. *Scand. J. Work Environ. Health* **2008**, *34*, 58–63.

29. Reijula, K.; Sundman-Digert, C. Assessment of indoor air problems at work with a questionnaire. *Occup. Environ. Med.* **2004**, *61*, 33–38. [PubMed]

30. Elo, A.-L.; Leppänen, A.; Jahkola, A. Validity of a single-item measure of stress symptoms. *Scand. J. Work Environ. Health* **2003**, *29*, 444–451. [CrossRef]

31. Tähtinen, K.; Lappalainen, S.; Karvala, K.; Salonen, H. A comprehensive approach to evaluating the urgency of iaq measures. In Proceedings of the 15th Conference of the International Society of Indoor Air Quality & Climate (ISIAQ), Philadelphia, PA, USA, 22–27 July 2018.

32. Lahtinen, M.; Sundman-Digert, C.; Reijula, K. Psychosocial work environment and indoor air problems: A questionnaire as a means of problem diagnosis. *Occup. Environ. Med.* **2004**, *61*, 143. [CrossRef]

33.   Bakke, J.; Moen, B.; Wieslander, G.; Norbäck, D. Gender and the physical and psychosocial work environments are related to indoor air symptoms. *J. Occup. Environ. Med.* **2007**, *49*, 641–650. [CrossRef]

34.   Finell, E.; Seppälä, T. Indoor air problems and experiences of injustice in the workplace: A quantitative and a qualitative study. *Indoor Air* **2018**, *28*, 125–134. [CrossRef]

Permissions

All chapters in this book were first published in MDPI; hereby published with permission under the Creative Commons Attribution License or equivalent. Every chapter published in this book has been scrutinized by our experts. Their significance has been extensively debated. The topics covered herein carry significant findings which will fuel the growth of the discipline. They may even be implemented as practical applications or may be referred to as a beginning point for another development.

The contributors of this book come from diverse backgrounds, making this book a truly international effort. This book will bring forth new frontiers with its revolutionizing research information and detailed analysis of the nascent developments around the world.

We would like to thank all the contributing authors for lending their expertise to make the book truly unique. They have played a crucial role in the development of this book. Without their invaluable contributions this book wouldn't have been possible. They have made vital efforts to compile up to date information on the varied aspects of this subject to make this book a valuable addition to the collection of many professionals and students.

This book was conceptualized with the vision of imparting up-to-date information and advanced data in this field. To ensure the same, a matchless editorial board was set up. Every individual on the board went through rigorous rounds of assessment to prove their worth. After which they invested a large part of their time researching and compiling the most relevant data for our readers.

The editorial board has been involved in producing this book since its inception. They have spent rigorous hours researching and exploring the diverse topics which have resulted in the successful publishing of this book. They have passed on their knowledge of decades through this book. To expedite this challenging task, the publisher supported the team at every step. A small team of assistant editors was also appointed to further simplify the editing procedure and attain best results for the readers.

Apart from the editorial board, the designing team has also invested a significant amount of their time in understanding the subject and creating the most relevant covers. They scrutinized every image to scout for the most suitable representation of the subject and create an appropriate cover for the book.

The publishing team has been an ardent support to the editorial, designing and production team. Their endless efforts to recruit the best for this project, has resulted in the accomplishment of this book. They are a veteran in the field of academics and their pool of knowledge is as vast as their experience in printing. Their expertise and guidance has proved useful at every step. Their uncompromising quality standards have made this book an exceptional effort. Their encouragement from time to time has been an inspiration for everyone.

The publisher and the editorial board hope that this book will prove to be a valuable piece of knowledge for researchers, students, practitioners and scholars across the globe.

List of Contributors

Sungroul Kim, Sujung Park and Jeongeun Lee
Department of Environmental Health Sciences, Soonchunhyang University, Asan 31538, Korea

Gaetano Settimo
Environment and Health Department, Istituto Superiore di Sanità, 00161 Rome, Italy

Marco Gola and Stefano Capolongo
Architecture, Built Environment and Construction Engineering Department, Politecnico di Milano, 20133 Milan, Italy

Pietro Grisoli and Marco Albertoni
Department of Drug Sciences, Laboratory of Microbiology, University of Pavia, 27100 Pavia, Italy

Marinella Rodolfi
Department of Earth and Environmental Sciences, Mycology Section, University of Pavia, 27100 Pavia, Italy

Wioletta Rogula-Kozłowska and Karolina Bralewska
The Main School of Fire Service, Safety Engineering Institute, 01629 Warsaw, Poland

Izabela Jureczko
Power Research & Testing Company ENERGOPOMIAR Ltd., 44100 Gliwice, Poland

Ioannis Sakellaris
Department of Mechanical Engineering, University of Western Macedonia, Sialvera & Bakola Str., 50100 Kozani, Greece

Dikaia Saraga
Department of Mechanical Engineering, University of Western Macedonia, Sialvera & Bakola Str., 50100 Kozani, Greece
Environmental Research Laboratory, INRASTES, National Center for Scientific Research "DEMOKRITOS", Aghia Paraskevi Attikis, 15310 Athens, Greece

Corinne Mandin
Université Paris Est, CSTB-Centre Scientifique et Technique du Bâtiment, 84 avenue Jean Jaurès, 77447 Marne-la-Vallée CEDEX 2, France

Yvonne de Kluizenaar
The Netherlands Organization for Applied Scientific Research (TNO), 2509 JE The Hague, The Netherlands

Serena Fossati
ISGlobal, Institute for Global Health, 08036 Barcelona, Spain

Andrea Spinazzè and Andrea Cattaneo
Department of Science and High Technology, University of Insubria, Via Valleggio 11, 22100 Como, Italy

Tamas Szigeti and Victor Mihucz
Cooperative Research Centre for Environmental Sciences, Eötvös Loránd University, Pázmány Péter sétány 1/A, H-1117 Budapest, Hungary

Eduardo de Oliveira Fernandes
Institute of Science and Innovation in Mechanical Engineering and Industrial Management, Rua Dr. Roberto Frias s/n, 4200-465 Porto, Portugal

Krystallia Kalimeri and John Bartzis
Department of Mechanical Engineering, University of Western Macedonia, Sialvera & Bakola Str., 50100 Kozani, Greece

Paolo Carrer
Department of Biomedical and Clinical Sciences-Hospital "L. Sacco", University of Milan, via G.B. Grassi 74, 20157 Milano, Italy

Miguel Ángel Campano-Laborda, Samuel Domínguez-Amarillo, Jesica Fernández-Agüera and Ignacio Acosta
Instituto Universitario de Arquitectura y Ciencias de la Construcción, Escuela Técnica Superior de Arquitectura, Universidad de Sevilla, 41012 Seville, Spain

Anna M. Marcellon, Alessandra Chiominto, Simona Di Renzi, Paola Melis, Renata Sisto, Maria C. D'Ovidio and Emilia Paba
Department of Occupational and Environmental Medicine, Epidemiology and Hygiene, Italian Workers' Compensation Authority (INAIL), Via Fontana Candida 1, Monte Porzio Catone, 00078 Rome, Italy

Annarita Wirz
Santa Lucia Foundation IRCCS, Via Ardeatina 306, 00142 Rome, Italy

Maria C. Riviello
Santa Lucia Foundation IRCCS, Via Ardeatina 306, 00142 Rome, Italy
Institute of Cell Biology and Neurobiology, National Research Council (CNR), Via E. Ramarini 32, Monterotondo, 00015 Rome, Italy

Stefania Massari
Department of Occupational and Environmental Medicine, Epidemiology and Hygiene, ItalianWorkers' Compensation Authority (INAIL), Via Stefano Gradi 55, 00143 Rome, Italy

Antonio Pacitto and Luca Stabile
Department of Civil and Mechanical Engineering, University of Cassino and Southern Lazio, 03043 Cassino, Italy

Stefania Russo
FIMP-Federazione Italiana Medici Pediatri, 00185 Roma, Italy

Giorgio Buonanno
Department of Civil and Mechanical Engineering, University of Cassino and Southern Lazio, 03043 Cassino, Italy
International Laboratory for Air Quality and Health, Queensland University of Technology, Brisbane, QLD 4000, Australia

Zhengguo Yang, Yuto Lim and Yasuo Tan
School of Information Science, Japan Advanced Institute of Science and Technology 1-1 Asahidai, Nomi, Ishikawa 923-1292, Japan

Pamela Dalton
Monell Chemical Senses Center, Philadelphia, PA 19104, USA

Anna-Sara Claeson
Department of Psychology, Umeå University, 90187 Umeå, Sweden

Steve Horenziak
The Procter & Gamble Company, Cincinnati, OH 45202, USA

Michał Piasecki and Krystyna Barbara Kostyrko
Department of Thermal Physic, Acoustic and Environment, Building Research Institute, Filtrowa 1, 00-611 Warsaw, Poland

Kanittha Pamonpol
Environmental Science and Technology Program, Faculty of Science and Technology, Valaya Alongkorn Rajabhat University under the Royal Patronage, 1 Moo 20 Paholyothin Road, Klong Nueng, Khlong Luang, Pathum Thani 13180, Thailand

Thanita Areerob
Faculty of Technology and Environment, Prince of Songkla University, Phuket Campus 83120, Thailand

Kritana Prueksakorn
Faculty of Technology and Environment, Prince of Songkla University, Phuket Campus 83120, Thailand
Andaman Environment and Natural Disaster Research Center, Faculty of Technology and Environment, Prince of Songkla University, Phuket Campus 83120, Thailand

Anna Mainka and Elwira Zajusz-Zubek
Department of Air Protection, Silesian University of Technology, 22B Konarskiego St., 44-100 Gliwice, Poland

Katja Tähtinen
Department of Civil Engineering, Aalto University, 02150 Espoo, Finland
Finnish Institute of Occupational Health, Healthy Workspaces, 00032 Työterveyslaitos, Finland

Sanna Lappalainen, Kirsi Karvala and Marjaana Lahtinen
Finnish Institute of Occupational Health, Healthy Workspaces, 00032 Työterveyslaitos, Finland

Heidi Salonen
Department of Civil Engineering, Aalto University, 02150 Espoo, Finland

Index

Printed in the USA
CPSIA information can be obtained
at www.ICGtesting.com
JSHW061340150424
61201JS00005B/81